C0-DAN-155

LINCOLN LABORATORY PUBLICATIONS

An Introduction to the Theory of
RANDOM SIGNALS AND NOISE

LINCOLN LABORATORY PUBLICATIONS

An Introduction to the Theory of
RANDOM SIGNALS AND NOISE

Wilbur B. Davenport, Jr., and **William L. Root**

Lincoln Laboratory, Massachusetts Institute of Technology

McGRAW-HILL BOOK COMPANY, INC.

New York   Toronto   London

1958

AN INTRODUCTION TO THE THEORY OF RANDOM SIGNALS
AND NOISE

Copyright © 1958 by the McGraw-Hill Book Company, Inc. Printed in the
United States of America. All rights reserved. This book, or parts thereof,
may not be reproduced in any form without permission of the publishers.

Library of Congress Catalog Card Number 57-10020

II

THE MAPLE PRESS COMPANY, YORK, PA.

PREFACE

During the past decade there has been a rapidly growing realization amongst engineers of the power of the methods and concepts of mathematical statistics when applied to problems arising in the transmission and processing of information. The use of the statistical approach has resulted not only in a better understanding of the theory of communications but also in the practical development of new equipment. We have tried here to provide an *introduction* to the statistical *theory* underlying a study of signals and noises in communications systems; in particular, we have tried to emphasize techniques as well as results.

This book is an outgrowth of a set of lecture notes prepared by the first-named author in 1952 for a first-year graduate course on the statistical theory of noise and modulation given in the Department of Electrical Engineering at M.I.T. The material has recently been used in substantially its present form as a text for an informal course at the M.I.T. Lincoln Laboratory. With some expansion, it could form the basis of a two-semester, first-year graduate course. Alternatively, by cutting out some parts, say Arts. 6-4 and 9-5, all of Chap. 11, and parts of Chap. 14, the remaining material could probably be covered in a single semester. Prerequisites would include some electrical network theory and an advanced engineering calculus course containing Fourier series and integrals.

It would be impracticable to acknowledge all those who have contributed to this book. However, we do wish to mention David Middleton, with whom we have had many stimulating discussions, and our colleagues at M.I.T., who have made many valuable comments and criticisms. In particular, we wish to thank Morton Loewenthal for the time and effort he spent in making a critical review of the entire manuscript.

WILBUR B. DAVENPORT, JR.
WILLIAM L. ROOT

v

CONTENTS

CHAPTER 1

INTRODUCTION

1-1. Communications Systems and Statistics

A simple communications system is the cascade of an information source, a communication link, and an information user where, as shown in Fig. 1-1, the communication link consists of a transmitter, a channel,

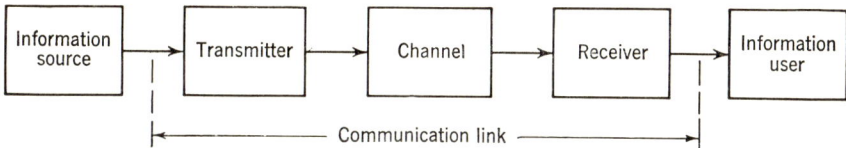

FIG. 1-1. A communications system.

and a receiver. The transmitter modulates or encodes information supplied by the source into a signal form suitable to the channel, the channel conveys the signal over the space intervening between the transmitter and receiver, and the receiver demodulates or decodes the signal into a form suitable to the user. Typical communication links are (1) the frequency-modulated, very-high-frequency radio link that brings entertainment to the listener, (2) the time-division multiplex radio link that transmits various data to some controlled device, e.g., a ship or an airplane, (3) the Teletype machine, connecting wire or cable, and associated terminal apparatus that convey telegrams from one point to another, and (4) the eye and pertinent portions of the nervous system that transmit a visual image to the brain.

Randomness or unpredictability can enter a communications system in three ways: the information generated by the source may not be completely predictable, the communication link may be randomly disturbed, and the user may misinterpret the information presented to it. It is in fact a fundamental tenet of information theory that the output of a source *must* be unpredictable to some degree in order to impart information at all; if the source output were completely predictable, the user could, without communication from the source, state at any instant the entire future output of the source. Disturbances in the communication link can occur in many ways. Both the transmitter and the receiver can add

1

noise.   If the channel is a radio channel, it can, for example, add atmospheric noise, galactic noise, or man-made interference.   In addition, it might be subject to a randomly variable multipath propagation mode which causes a single transmitted signal to appear as a multiplicity of interfering signals at the receiver.

Even though the signals and noises in a communications system are random, the behavior of that system can be determined to a considerable degree if various average properties of the signals and noises are known. Such properties might include the average intensity or power, the distribution of power with frequency, and the distribution of the instantaneous amplitude.   The determination of the relations among the various average properties falls into the domain of probability theory and statistics.   The purpose of this book is to introduce to the reader the application of statistical techniques to the study of communications systems.†
In general we shall assume the statistical properties of the signals and noises to be given; the study of optimum statistical properties for signals may be found in works on information theory‡ and will not be discussed here.

1-2. The Book

Approximately the first half of this book is devoted to a development of those elements of probability theory and statistics which are particularly pertinent to a study of random signals and noises in communications systems.   The remainder is devoted to applications.

Survey of the Text. In Chap. 2, the *probability* of an *event*, i.e., the outcome of an experiment, is introduced in terms of the relative frequency of occurrence of that event and a set of axioms of probability theory is presented.   The probabilities of multiple events are then discussed and the concept of *statistical independence* is mentioned.   The representation of an event by a point in a *sample space* is made in Chap. 3, and a *random variable* is defined there as a function over a sample space.   Probabilities are then introduced on the sample space, and probability distribution functions and probability density functions of random variables are defined.   The probability concepts are next extended to random functions of time by the definition of a *random process* as a family of random variables indexed by the parameter t.

The notion of *statistical average* is introduced in Chap. 4 in terms of the common arithmetic average, and certain statistical averages are considered.   In particular, the *characteristic function* of the probability distribution function of the random variable x is defined to be the statistical

† An excellent short introduction to the subject matter of this book may be found in Bennett (II).   (See Bibliography.)
‡ For example Shannon (I) and (II).

average of $\exp(jvx)$ and is shown to be the Fourier transform of the probability density function of x. The *correlation coefficient* of the two random variables x and y is next stated to be the statistical average of a normalized product of x and y and is related to the problem of obtaining the best mean-square prediction of y given x. The random variables x and y are then said to be *linearly independent* if their correlation coefficient is zero. Finally, in Chap. 4 the relation between time averages and statistical averages is investigated.

In Chap. 5, sampling is introduced and the *sample mean* in particular is discussed at some length. A simple form of the central limit theorem is derived, and the relation between the relative frequency of occurrence of an event and the probability of that event is further studied.

The *spectral density*, i.e., the distribution of power with frequency, of a function of time is considered in Chap. 6 and is shown to be the Fourier transform of the autocorrelation function of the time function. The concept of spectral density is then extended to random processes, and the problem of representing a random process by a series of orthogonal time functions with linearly independent random coefficients is discussed.

The determination of the statistical properties of a physical process is illustrated in Chap. 7 by a study of the shot noise generated in thermionic vacuum tubes. First, the properties of shot noise in temperature-limited diodes are found in a manner similar to that used by Rice (I), and then the results so obtained are extended to space-charge-limited tubes.

One of the most commonly occurring and thoroughly studied classes of random processes is that of the gaussian processes. The statistical properties of these processes are reviewed in Chap. 8. In particular, the properties of a narrow-band gaussian random process are considered in some detail, as are the joint statistics of a sine wave and a narrow-band gaussian process.

The techniques developed in Chaps. 2 through 8 are applied to a study of the passage of random signals and noises through linear systems in Chaps. 9, 10, and 11. The analysis of the response of a linear system to an ordinary function of time is reviewed in Chap. 9 and extended there to random time functions. The correlation functions and spectral densities at the output of a linear system in response to random inputs are then determined; and the problem of obtaining the probability density function of the output of a linear system is treated. These results are applied in Chap. 10 to a study of noise in amplifiers. *Noise figure* is defined, and some of its properties are discussed. The synthesis of *optimum* linear systems is discussed in Chap. 11. In particular, the theory of least-mean-square error smoothing and predicting, using either the infinite past of the input or only a finite portion of the past, is investigated.

The passage of random processes through a class of nonlinear devices which have no memory is considered in Chaps. 12 and 13. In Chap. 12, this problem is treated directly as a transformation of variables using the nonlinear transfer characteristic of the device in question, and specific results are obtained for the full-wave square-law detector and the half-wave linear detector. In Chap. 13, the *transfer function* of a nonlinear device is defined to be the Fourier transform of the transfer characteristic of the device. The transfer function is then used to determine the auto-correlation function and spectral density of the output of a nonlinear device in response to an input consisting of the sum of a sine wave and a gaussian random process. Particular results are then obtained for the class of νth-law nonlinear devices.

Chap. 14 presents an introduction to the application of statistical-hypothesis testing and parameter estimation to problems of signal detection and extraction. The statistical principles needed are developed, including the Neyman-Pearson hypothesis test and other tests involving the likelihood ratio and the maximum-likelihood method of parameter estimation. Applications are made to radar and to radio-communications systems using a binary alphabet.

The Bibliography. The various sources referred to in this book have been collected in the Bibliography at the end of the book. Only those sources were included which seemed to us to be particularly pertinent either to the text itself or to the problems at the ends of the chapters; no attempt was made to be all-inclusive. Extensive lists of references may be found in the bibliographies of Chessin, Green, and Stumpers and in the books of Blanc-Lapierre and Fortet, Bunimovich, Cramér, Doob, Gnedenko and Kolmogorov, and Solodovnikov (see Bibliography).

CHAPTER 2

PROBABILITY

The fields of mathematics pertinent to a study of random signals and noise are probability theory and statistics. The purpose of Chaps. 2, 3, and 4 is to treat the relevant portions of the probability calculus in sufficient detail so that a reader previously unfamiliar with the subject may become acquainted with the tools he needs for the main part of the book. This will be done without any pretense of mathematical rigor. For anyone wanting to devote time to a careful study of mathematical probability, there are several excellent texts, particularly Cramér (I), Feller (I), and Loève (I).

2-1. Introduction

One way to approach the notion of probability is through the phenomenon of *statistical regularity*. There are many repeating situations in nature for which we can predict in advance from previous experience roughly what will happen, or what will happen on the average, but not exactly what will happen. We say in such cases that the occurrences are *random;* in fact, the reason for our inability to predict exactly may be that (1) we do not know all the causal forces at work, (2) we do not have enough data about the conditions of the problem, (3) the forces are so complicated that calculation of their combined effect is unfeasible, or possibly (4) there is some basic indeterminacy in the physical world. Whatever the reason for the randomness, in very many situations leading to random occurrences, a definite average pattern of results may be observed when the situation is re-created a great number of times. For example, it is a common observation that if a good coin is flipped many times it will turn up heads on about half the flips. The tendency of repeated similar experiments to result in the convergence of over-all averages as more and more trials are made is called *statistical regularity*. It should be realized, however, that our belief in statistical regularity is an induction and is not subject to mathematical proof.

A conventional method of introducing mathematics into the study of random events is, first, to suppose that there are identifiable systems subject to statistical regularity; second, to form a mathematical model (i.e., a set of axioms and the propositions they imply) incorporating the features of statistical regularity; and, last, to apply deductions made with the mathematics to real systems. The mathematical model most widely accepted and used is called *mathematical probability theory* and results from the axioms stated by Kolmogorov (I, Chap. 1).

Let us now use the idea of statistical regularity to explain probability. We first of all *prescribe a basic experiment*, for example, the observation of the result of the throw of a die or the measurement of the amplitude of a noise voltage at a given instant of time. Secondly, we *specify all the possible outcomes* of our basic experiment. For example, the possible outcomes of the throw of a die are that the upper face shows one, two, three, four, five, or six dots, whereas in the case of a noise-voltage measurement, the instantaneous amplitude might assume any value between plus and minus infinity. Next, we *repeat the basic experiment a large number of times* under uniform conditions and observe the results.

Consider now one of the possible outcomes of our basic experiment— say, the rolling of a two with a die. To this event we want to attach a nonnegative real number called the *probability of occurrence*. Suppose that in a large number N of repetitions of our basic experiment, the event of interest (A) occurs $n(A)$ times. The relative frequency of occurrence of that event in those N repetitions is then $n(A)/N$. If there is a practical certainty (i.e., a strong belief based upon practice) that the measured relative frequency will tend to a limit as the number of repetitions of the experiment increases without limit, we would like to say that the event (A) has a definite probability of occurrence $P(A)$ and take $P(A)$ to be that limit, i.e.,

$$\frac{n(A)}{N} \to P(A) \qquad \text{as } N \to \infty$$

Unfortunately, this simple approach involves many difficulties. For example, one obvious difficulty is that, strictly speaking, the limit can never be found (since one would never live long enough), even though in some cases (as in gambling problems) there may be good reason to suppose that the limit exists and is known. Therefore, rather than define a probability as the limit of a relative frequency, we shall define probability abstractly so that probabilities behave like limits of relative frequencies. An important after-the-fact justification of this procedure is that it leads to so-called *laws of large numbers*, according to which, roughly, in certain very general circumstances the mathematical counterpart of an empirical relative frequency does converge to the appropriate probability, and hence an empirical relative frequency may be used to estimate a probability.† We shall discuss one form of the law of large numbers in Art. 5-5.

2-2. Fundamentals

Having introduced the concept of probability and its relation to relative frequency, we shall now define probability and consider some of its properties.

First, however, we need to extend our notion of an *event*. It makes perfectly good sense in speaking of the roll of a die to say that "the face with three dots or the face with four dots turned up." Thus we want to speak of compound events $(A$ or $B)$ where (A) and (B) are events.‡ Also we want to speak of events $(A$ and $B)$, i.e., the simultaneous occurrence of events (A) and (B). For example, in the roll of a die, let (A) be the event "at most four dots turn up" and (B) the event "at least four dots turn up." Then $(A$ and $B)$ is the event "four dots turn up." Finally, it is useful, if (A) is an event, to consider the event (not A). Thus if (A) is the event that one dot turns up in the roll of a die, (not A) is the event that two or more dots turn up.

In a succession of repetitions of an experiment, if after each trial we can determine whether or not (A) happened and whether or not (B) happened, then we can also determine whether or not $(A$ and $B)$, $(A$ or $B)$, (not A), and (not B) happened. One may then calculate an empirical frequency ratio for the occurrence of each of the events (A), (B), $(A$ and $B)$, $(A$ or $B)$, (not A), and (not B). It therefore seems reasonable to require:

AXIOM I. *To each event* (A) *of a class of possible events of a basic experiment, there is assigned a nonnegative real number* $P(A)$ *called the probability of that event. If this class includes the event* (A) *and the event* (B), *it also includes the events* $(A$ and $B)$, $(A$ or $B)$, *and* (not A).

It follows from this axiom that a probability is defined for the *certain event* (i.e., an event which must occur), since for any event (A), $(A$ or not A) is the certain event. Also, a probability is defined for the "null event," since for any event (A), $(A$ and not $A)$ is the null event.

The relative frequency of occurrence of a certain event is unity. Thus, it seems reasonable to require:

AXIOM II. *The probability of the certain event is unity.*

We say two events (A) and (B) are *disjoint* or *mutually exclusive* if

---

† The above discussion touches on a difficult and controversial subject, the foundations of probability, a detailed discussion of which is beyond the scope of this book. For brief readable treatments of this subject see Carnap (I, Chap. II), Cramér (I, Chap. 13), and Jeffreys (I, Chap. I).

‡ We shall generally take "A or B" to mean "either A or B or both."

they are in such relation to each other that if one occurs, the other cannot possibly occur. In the roll of a die, the events that the die turns up two dots and that it turns up three dots are disjoint. In any case, (A) and (not A) are disjoint. Suppose (A) and (B) are disjoint events that can occur as the result of a given basic experiment. Let the basic experiment be repeated N times with (A) occurring $n(A)$ times and (B) occurring $n(B)$ times. Since (A) cannot occur when (B) does, and vice versa, the number of times $(A$ or $B)$ occurred is $n(A) + n(B)$. Thus,

$$\frac{n(A \text{ or } B)}{N} = \frac{n(A)}{N} + \frac{n(B)}{N}$$

This relation holds as $N \to \infty$; hence we are led to require:

AXIOM III. *If (A) and (B) are mutually exclusive events, then*

$$P(A \text{ or } B) = P(A) + P(B) \tag{2-1}$$

A consequence of this axiom is that if $A_1, A_2, \ldots, A_K$ are K mutually exclusive events, then

$$P(A_1 \text{ or } A_2 \text{ or } \cdots \text{ or } A_K) = \sum_{k=1}^{K} P(A_k) \tag{2-2}$$

This is easily shown by successive applications of Axiom III. A consequence of Axioms II and III is

$$0 \leq P(A) \leq 1 \tag{2-3}$$

for any event (A); for

$$P(A) + P(\text{not } A) = P(\text{certain event}) = 1$$

and $P(\text{not } A)$ is nonnegative. It also follows from Axioms II and III that

$$P(\text{null event}) = 0 \tag{2-4}$$

Note that if it is possible to decompose the certain event into a set of mutually exclusive events $A_1, \ldots, A_K$, then

$$\sum_{k=1}^{K} P(A_k) = 1 \tag{2-5}$$

These axioms are self-consistent and are adequate for a satisfactory theory of probability to cover cases in which the number of events is finite. However, if the number of possible events is infinite, these axioms are inadequate and some additional properties are required. The following axiom suffices:†

† See Kolmogorov (I, Chap. II).

AXIOM IV. *If $P(A_i)$ is defined for each of a class of events $(A_1), (A_2), \ldots$, then $P(A_1$ or A_2 or $\cdots)$ is defined; if $(A_1), (A_2), \ldots$ are mutually exclusive events and the probability of each one is defined, then*

$$P(A_1 \text{ or } A_2 \text{ or } \cdots) = \sum_{i=1}^{\infty} P(A_i) \qquad (2\text{-}6)$$

One point should perhaps be emphasized. Although the axioms imply that the probability of the null event is zero, they do not imply that if the probability of an event is zero, it is the null event. The null event is the mathematical counterpart of an impossible event; thus, in interpretation, the mathematical theory assigns probability zero to anything impossible, but does not say that if an event has probability zero it is impossible. That this is a reasonable state of things can be seen from the frequency-ratio interpretation of probability. It is entirely conceivable that there be an event (A) such that $n(A)/N \to 0$ even though $n(A)$ does not remain zero. A common example of this is the following: Let the basic experiment consist of choosing randomly and without bias a point on a line one unit long. Then the choice of a particular point is an event which we require to be exactly as probable as the choice of any other point. Thus if one point has non-zero probability, all must have non-zero probability; but this cannot be, for then the sum of the probabilities of these disjoint events would add up to more than one, in violation of Axiom II. Thus every choice one makes must be an event of probability zero.

It follows also, of course, that although the certain event must have probability one, an event of probability one need not necessarily happen.

We conclude with a simple example. Let a basic experiment have six possible mutually exclusive outcomes, which we shall call simply events 1, 2, 3, 4, 5, and 6. The class of events to which we shall assign probabilities consists of these six events plus any combination of these events of the form (— and — and . . .). It will be noted that this class satisfies Axiom I. Now if we assign probabilities to each of the events 1, 2, 3, 4, 5, and 6, then by Axiom III a probability will be defined for every event in the class considered. We are free to assign any probabilities we choose to events 1, 2, 3, 4, 5, and 6 as long as they are nonnegative and add to one. Thus, for example, we may take

$$P(1) = \tfrac{1}{2} = \text{probability of event 1}$$
$$P(2) = \tfrac{1}{4}$$
$$P(3) = \tfrac{1}{8}$$
$$P(4) = \tfrac{1}{16}$$
$$P(5) = \tfrac{1}{16}$$
$$P(6) = 0$$

Then $P(1 \text{ or } 3 \text{ or } 6) = \tfrac{5}{8}$, $P(2 \text{ or } 3) = \tfrac{3}{8}$, etc.

Another consistent assignment of probabilities to this same class of events is

$$P(1) = P(2) = P(3) = P(4) = P(5) = P(6) = \tfrac{1}{6}$$

We cannot decide on the basis of mathematical probability theory which of these assignments applies to the experiment of the rolling of a particular die; both choices are valid mathematically.

2-3. Joint Probabilities

So far, we have been concerned primarily with the outcomes of a single basic experiment. In many interesting problems, however, we might instead be concerned with the outcomes of several different basic experiments, for example, the amplitudes of a noise-voltage wave at several different instants of time, or the outcome of the throw of a pair of dice. In the first case, we might wish to know the probability that the noise at one instant of time t_1 exceeds a certain value x_1 and that the noise at another time t_2 exceeds another value x_2. In the second case we might wish to know the probability that one die shows two dots and that the other shows five dots.

Probabilities relating to such combined experiments are known as *joint probabilities*. That these probabilities have the same basic properties as those discussed in the previous section can be seen by realizing that a joint experiment consisting of the combination of one experiment having the possible outcomes (A_k) with another having the possible outcomes (B_m) might just as well be considered as a single experiment having the possible outcomes $(A_k$ and $B_m)$. Therefore, if the probability that the kth outcome of experiment A and the mth outcome of experiment B *both* occur is denoted by $P(A_k,B_m)$, it follows from Eq. (2-3) that

$$0 \leq P(A_k,B_m) \leq 1 \tag{2-7}$$

It further follows from the discussion leading to Eq. (2-5) that if there are K possible outcomes (A_k) and M possible outcomes (B_m), all of which are mutually exclusive, we must then obtain the result that

$$\sum_{m=1}^{M} \sum_{k=1}^{K} P(A_k,B_m) = 1 \tag{2-8}$$

as we are dealing with an event that must occur. Both these results may obviously be extended to cases in which we deal with combinations of more than just two basic experiments.

A new problem now arises, however, when we become interested in the relations between the joint probabilities of the combined experiment and the elementary probabilities of the basic experiments making up the combined experiment. For example, in the above case, we might wish to know the relations between the joint probabilities $P(A_k,B_m)$ and the elementary probabilities $P(A_k)$ and $P(B_m)$. To this end, let us consider the probability that the kth outcome of experiment A occurs and that *any* one of the possible outcomes of experiment B occurs. If all the possible outcomes of experiment B are mutually exclusive, it then follows from Axiom III that

$$P(A_k,B_1 \text{ or } B_2 \text{ or } \cdots \text{ or } B_M) = \sum_{m=1}^{M} P(A_k,B_m)$$

This is simply the probability that the event A_k occurs irrespective of the outcome of experiment B; i.e., it is simply the probability $P(A_k)$. Thus

$$P(A_k) = \sum_{m=1}^{M} P(A_k,B_m) \qquad (2\text{-}9a)$$

when all the possible outcomes of experiment B are mutually exclusive. In a similar manner,

$$P(B_m) = \sum_{k=1}^{K} P(A_k,B_m) \qquad (2\text{-}9b)$$

when all the possible outcomes of experiment A are mutually exclusive. Thus we have shown the important fact that the elementary probabilities of the component basic experiments making up a combined experiment may be derived from the joint probabilities of that combined experiment.

It further follows from Eqs. (2-9a and b) that

$$P(A_k) \geq P(A_k,B_m) \qquad \text{and} \qquad P(B_m) \geq P(A_k,B_m) \qquad (2\text{-}10)$$

for any value of k and m, since the joint probabilities $P(A_k,B_m)$ are nonnegative.

2-4. Conditional Probabilities

In the preceding section, we introduced joint probabilities pertaining to the results of combined experiments and showed the relation between the joint probabilities of the combined events and the elementary probabilities of the basic events. It is also of interest to answer the question: "What is the probability of occurrence of the event (A) if we know that the event (B) has occurred?" We will study probabilities of this type— "conditional probabilities"—in this section.

Consider now a combined experiment consisting of two basic experiments, one giving rise to the basic events (A_k) and the other to the basic events (B_m). Suppose that the combined experiment is repeated N times, that the basic event (A_k) occurs $n(A_k)$ times in that sequence of N experiment repetitions, that the basic event (B_m) occurs $n(B_m)$ times, and that the joint event (A_k,B_m) occurs $n(A_k,B_m)$ times.

For the moment, let us focus our attention on those $n(A_k)$ experiments in each of which the event (A_k) has occurred. In each of these, some one of the events (B_m) has also occurred; in particular the event

(B_m) occurred $n(A_k,B_m)$ times in this subset of experiments. Thus the relative frequency of occurrence of the event (B_m) under the assumption that the event (A_k) also occurred is $n(A_k,B_m)/n(A_k)$. Such a relative frequency is called a *conditional relative frequency*, since it deals with a specified condition or hypothesis. It may also be expressed in the form

$$\frac{n(A_k,B_m)}{n(A_k)} = \frac{n(A_k,B_m)/N}{n(A_k)/N}$$

and hence is equal to the ratio of the relative frequency of occurrence of the joint event (A_k,B_m) to the relative frequency of occurrence of the hypothesis event (A_k). With this fact in mind, we are able to define *conditional probability.*†

DEFINITION. *The conditional probability $P(B|A)$ of occurrence of the event (B) subject to the hypothesis of the occurrence of the event (A) is defined as the ratio of the probability of occurrence of the joint event (A,B) to the probability of occurrence of the hypothesis event (A):*

$$P(B|A) = \frac{P(A,B)}{P(A)} \tag{2-11}$$

The conditional probability is undefined if $P(A)$ is zero.

On rewriting Eq. (2-11) as

$$P(A,B) = P(B|A)P(A) \tag{2-12}$$

we see that the joint probability of two events may be expressed as the product of the conditional probability of one event, given the other, times the elementary probability of the other.

Conditional probability as defined above has essentially the same properties as the various probabilities previously introduced. For example, consider the combined experiment giving rise to the joint events (A_k,B_m) in which the basic events (B_m) are mutually exclusive. From the definition above, the conditional probability of the event $(B_j$ or $B_m)$ subject to the hypothesis of the occurrence of the event (A_k) is

$$P(B_j \text{ or } B_m|A_k) = \frac{P(A_k,B_j \text{ or } B_m)}{P(A_k)}$$

From our previous discussion of the joint probabilities of mutually exclusive events, it follows that

$$\frac{P(A_k,B_j \text{ or } B_m)}{P(A_k)} = \frac{P(A_k,B_j) + P(A_k,B_m)}{P(A_k)}$$

† There are some subtleties connected with the definition of conditional probability which can only be discussed in terms of measure theory and hence are beyond the scope of this book. See Kolmogorov (I, Chap. V).

The right-hand side is simply the sum of $P(B_j|A_k)$ and $P(B_m|A_k)$, hence

$$P(B_j \text{ or } B_m|A_k) = P(B_j|A_k) + P(B_m|A_k) \qquad (2\text{-}13)$$

i.e., conditional probabilities of mutually exclusive events are additive.

Furthermore, if the events (B_m) form a set of M mutually exclusive events, we obtain

$$P(B_1 \text{ or } \cdots \text{ or } B_M|A_k) = \frac{\displaystyle\sum_{m=1}^{M} P(A_k,B_m)}{P(A_k)} = \sum_{m=1}^{M} P(B_m|A_k)$$

and if these M events (B_m) comprise the certain event, it follows from Eq. (2-9a) that the numerator of the middle term is simply $P(A_k)$, hence in this case

$$\sum_{m=1}^{M} P(B_m|A_k) = 1 \qquad (2\text{-}14)$$

It follows from Eq. (2-10) and the defining equation (2-11) that conditional probabilities are also bounded by zero and one:

$$0 \le P(B_m|A_k) \le 1 \qquad (2\text{-}15)$$

and that a conditional probability is at least equal to the corresponding joint probability:

$$P(B|A) \ge P(A,B) \qquad (2\text{-}16)$$

since the hypothesis probability $P(A)$ is bounded by zero and one.

2-5. Statistical Independence

The conditional probability $P(B|A)$ is the probability of occurrence of the event (B) assuming the occurrence of the event (A). Suppose now that this conditional probability is simply equal to the elementary probability of occurrence of the event (B):

$$P(B|A) = P(B)$$

It then follows from Eq. (2-12) that the probability of occurrence of the joint event (A,B) is equal to the product of the elementary probabilities of the events (A) and (B):

$$P(A,B) = P(A)P(B) \qquad (2\text{-}17)$$

and hence that $\qquad\qquad P(A|B) = P(A)$

i.e., the conditional probability of the event (A) assuming the occurrence of the event (B) is simply equal to the elementary probability of the event (A). Thus we see that in this case a knowledge of the occurrence of one event tells us no more about the probability of occurrence of the other event than we knew without that knowledge. Events (A) and (B), which satisfy such relations, are said to be *statistically independent events*.

When more than two events are to be considered, the situation becomes more complicated. For example,† consider an experiment having four mutually exclusive outcomes (A_1), (A_2), (A_3), and (A_4), all with the same probability of occurrence, $\frac{1}{4}$. Let us now define three new events (B_j) by the relations

$$(B_1) = (A_1 \text{ or } A_2)$$
$$(B_2) = (A_1 \text{ or } A_3)$$
$$(B_3) = (A_1 \text{ or } A_4)$$

Since the events (A_m) are mutually exclusive, it follows that

$$P(B_1) = P(A_1) + P(A_2) = \frac{1}{2}$$

and similarly that

$$P(B_2) = \frac{1}{2} = P(B_3)$$

Consider now the joint occurrence of the events (B_1) and (B_2). Since the events (A_m) are mutually exclusive, the event (B_1,B_2) occurs if and only if (A_1) occurs. Hence

$$P(B_1,B_2) = P(A_1) = \frac{1}{4}$$

Similarly

$$P(B_1,B_3) = \frac{1}{4} = P(B_2,B_3)$$

Since the elementary probabilities of the events (B_j) are all $\frac{1}{2}$, we have thus shown that

$$P(B_1,B_2) = P(B_1)P(B_2)$$
$$P(B_1,B_3) = P(B_1)P(B_3)$$
$$P(B_2,B_3) = P(B_2)P(B_3)$$

and hence that the events (B_j) are independent by pairs. However, note that if we know that any two of the events (B_j) have occurred, we also know that the experiment outcome was (A_1) and hence that the remaining (B_j) event must also have occurred. Thus, for example,

$$P(B_3|B_1,B_2) = 1 \neq P(B_3) = \frac{1}{2}$$

Thus we see that the knowledge that the $(N > 2)$ events in a given set are pairwise statistically independent is not sufficient to guarantee that three or more of those events are independent in a sense that satisfies our intuition. We must therefore extend the definition of statistical independence to cover the case of more than two events.

If, in addition to obtaining pairwise independence, the joint proba-

† This example was first pointed out by Serge Bernstein. See Kolmogorov (I, p. 11, footnote 12).

bility of the three events in the preceding example was equal to the product of the elementary probabilities of those events, it would also have turned out that

$$P(B_3|B_1,B_2) = \frac{P(B_1)P(B_2)P(B_3)}{P(B_1)P(B_2)} = P(B_3)$$

and similarly for the other conditional probabilities. In this case, then, the three events could justifiably be called statistically independent. With this fact in mind, statistical independence for the case of N events is defined as follows:

DEFINITION. *N events* (A_n) *are said to be statistically independent events if for all combinations* $1 \leq i < j < k \cdots \leq N$ *the following relations are satisfied:*

$$P(A_i,A_j) = P(A_i)P(A_j)$$
$$P(A_i,A_j,A_k) = P(A_i)P(A_j)P(A_k) \qquad (2\text{-}18)$$
$$\cdots\cdots\cdots\cdots\cdots\cdots\cdots\cdots$$
$$P(A_1, A_2, \ldots, A_N) = P(A_1)P(A_2) \cdots P(A_N)$$

Let us now turn our attention to the experiments giving rise to our events. In particular, let us consider the case of M experiments $A^{(m)}$, the mth of which has N_m mutually exclusive outcomes. If we so desired, we could consider the entire set of outcomes as a set of N events where

$$N = \sum_{m=1}^{M} N_m$$

The conditions of the preceding definition would then apply to the determination of whether or not these events are statistically independent. However, we are also interested in deciding whether or not the experiments themselves are statistically independent. The following definition applies to this case:

DEFINITION. *M experiments* $A^{(m)}$, *the* m*th of which has* N_m *mutually exclusive outcomes* $A_{n_m}^{(m)}$, *are said to be statistically independent experiments if for each set of* M *integers,* $n_1, n_2, \ldots, n_M$, *the following relation is satisfied:*

$$P[A_{n_1}^{(1)},A_{n_2}^{(2)}, \ldots, A_{n_M}^{(M)}] = P[A_{n_1}^{(1)}]P[A_{n_2}^{(2)}] \cdots P[A_{n_M}^{(M)}] \quad (2\text{-}19)$$

The simplicity of this set of relations as compared to the similar relations—Eqs. (2-18)—for events follows from the fact that the joint probabilities for any $K - 1$ of K *experiments* may be derived from the joint probabilities of the outcomes of the K *experiments*. For example, suppose that we have M experiments for which Eqs. (2-19) are satisfied. Let us sum up these equations over n_M. It then follows from Eq. (2-9) that

$$\sum_{n_M=1}^{N_M} P[A_{n_1}{}^{(1)}, A_{n_2}{}^{(2)}, \ldots, A_{n_M}{}^{(M)}] = P[A_{n_1}{}^{(1)}, A_{n_2}{}^{(2)}, \ldots, A_{n_{M-1}}{}^{(M-1)}]$$

On summing the right-hand side of Eq. (2-19), remembering that

$$\sum_{n_M=1}^{N_M} P[A_{n_M}{}^{(M)}] = 1$$

we see that if Eq. (2-19) is satisfied, it then follows that the relation

$$P[A_{n_1}{}^{(1)}, A_{n_2}{}^{(2)}, \ldots, A_{n_{M-1}}{}^{(M-1)}] = P[A_{n_1}{}^{(1)}]P[A_{n_2}{}^{(2)}] \cdots P[A_{n_{M-1}}{}^{(M-1)}]$$

is also satisfied. This result is simply Eq. (2-19) for $M - 1$ experiments. The process may be continued in order to show that, if Eq. (2-19) is satisfied for M experiments, probability relations of the form of Eq. (2-19) are satisfied for any $K < M$ experiments.

2-6. Examples

Let us now consider a few examples involving the concepts introduced in the preceding sections. These examples are of the so-called "combinatorial" type and thus are typical of a large class of probability problems. Feller (I) contains an excellent detailed discussion of problems of this kind.

Example 2-6.1. Card Drawings. Consider the problem of drawing from a deck of cards. The deck has 52 cards and is divided into 4 different suits, with 13 cards in each suit, ranging from the two up through the ace. We will *assume* that the deck has been well shuffled and that all the cards present are equally likely to be drawn.

Suppose that a single card is drawn from a full deck. What is the probability that that card is the king of diamonds? We assumed above that the various events (A_i) representing the drawing of particular cards are all equally likely to occur, hence all $P(A_i)$ equal some number p. The drawings of different cards are mutually exclusive events, hence

$$\sum_{i=1}^{52} P(A_i) = 52p = 1$$

Therefore, $p = \frac{1}{52}$ for any card, and in particular,

$$P(\text{king of diamonds}) = \frac{1}{52}$$

Suppose now that we ask, "What is the probability that the single card drawn is a king of any one of the four suits?" Since there are four kings, and since these events are mutually exclusive,

$P(\text{king}) = P(\text{king of spades}) + P(\text{king of hearts}) + P(\text{king of diamonds})$
$$+ P(\text{king of clubs})$$

Hence: $$P(\text{king}) = \frac{4}{52} = \frac{1}{13}$$

In general we see that when we are dealing with a set of mutually exclusive basic events, all of which are equally likely, the probability of any event (basic or compound)

is equal to the ratio of the number of basic events satisfying the conditions of the event in question to the total number of possible basic events.

Suppose next that we draw two cards from a full deck. What is the probability that we have drawn a king and a queen, not necessarily of the same suit? This event can occur in two ways: either a king may be drawn first and then a queen, or a queen may be drawn first and then a king. In symbols:

$$P(\text{king and queen}) = P(\text{king, queen}) + P(\text{queen, king})$$

From our discussion of conditional probability it follows that

$$P(\text{king, queen}) = P(\text{queen}|\text{king})P(\text{king})$$
and $$\qquad P(\text{queen, king}) = P(\text{king}|\text{queen})P(\text{queen})$$

Assuming that a king (queen) has been drawn, 51 cards remain in which are contained all four queens (kings). Hence

$$P(\text{queen}|\text{king}) = \tfrac{4}{51} = P(\text{king}|\text{queen})$$

and, using our previous results,

$$P(\text{king and queen}) = \tfrac{4}{51}\,\tfrac{1}{13} + \tfrac{4}{51}\,\tfrac{1}{13} = \tfrac{8}{663}$$

This result, of course, could also have been obtained directly by taking the ratio of the number of favorable basic events to the total number of possible basic events.

Example 2-6.2. Coin Tossings. Let us consider next coin tossings, for which we shall *assume* that successive tossings are statistically independent experiments. However, we shall make no assumption as to whether the coin is fair or not and shall write

$$P(H) = p \qquad \text{and} \qquad P(T) = q = 1 - p$$

since a (head) and a (tail) are mutually exclusive events. Such tossings are known as *Bernoulli trials*.

Suppose that the coin is tossed N times. What is the probability $P(nH)$ that n heads will appear? In order to answer this question, let us first consider a *particular* sequence of N tosses in which "head" occurred n times. Since the successive experiments making up this sequence were assumed to be statistically independent, the probability of occurrence of our particular sequence $P_i(nH)$ is simply the product of the probabilities of occurrence of n heads and $(N - n)$ tails:

$$P_i(nH) = p^n q^{(N-n)}$$

The particular sequence above is not the only possible sequence having n heads in N tosses. The probability of obtaining any one of the various possible sequences of this kind is equal to the total number possible of such sequences times the probability of obtaining a particular one—since we are dealing with mutually exclusive (compound) events all having the same probability of occurrence.

Let us now determine the number of different possible sequences of N tosses, each of which results in n heads. If the results of the N tosses were all different, there would have been

$$N(N - 1) \cdots 3 \cdot 2 \cdot 1 = N!$$

different possible sequences. However, not all the results of the N tosses are different; n are heads and $(N - n)$ are tails. Thus in the $N!$ sequences, there are $n!$ duplications since it is not possible to tell one head from another, and there are $(N - n)!$ duplications since it is not possible to tell one tail from another. The total possible number of different sequences in which there are n heads in N tosses is therefore given by the *binomial coefficient*

$$\binom{N}{n} = \frac{N!}{n!(N-n)!} \tag{2-20}$$

The total probability of obtaining any one of the various possible sequences of n heads in N tosses is therefore

$$P(nH) = \binom{N}{n} p^n q^{(N-n)} \tag{2-21}$$

The set of probabilities corresponding to the various possible values of n(i.e., $n = 0$, $1, \ldots ,N$) is known as the *binomial distribution*.

2-7. Problems

An extensive collection of problems covering the topics of this chapter may be found in Feller (I, Chaps. 1–5).

1. Experiment A has three mutually exclusive possible outcomes (A_m), with the probabilities of occurrence $P(A_m)$. Let the compound events (B) and (C) be defined by

$$(B) = (A_1 \text{ or } A_2)$$
$$(C) = (A_1 \text{ or } A_3)$$

Determine the relation among $P(B)$, $P(C)$, and $P(B \text{ or } C)$.

2. Experiment A has three mutually exclusive possible outcomes (A_m), and experiment B has two mutually exclusive possible outcomes (B_n). The joint probabilities $P(A_m,B_n)$ are:

$$P(A_1,B_1) = 0.2 \quad P(A_1,B_2) = 0.1$$
$$P(A_2,B_1) = 0.1 \quad P(A_2,B_2) = 0.2$$
$$P(A_3,B_1) = 0.1 \quad P(A_3,B_2) = 0.3$$

Determine the probabilities $P(A_m)$ and $P(B_n)$ for all values of m and n.

3. For the experiments of Prob. 2, determine the conditional probabilities $P(A_m|B_n)$ and $P(B_n|A_m)$ for all values of m and n.

4. Show whether or not the experiments of Prob. 2 are statistically independent.

5. Let K be the total number of dots showing up when a pair of unbiased dice are thrown. Determine $P(K)$ for each possible value of K.

6. Evaluate the probability of occurrence of n heads in 10 independent tosses of a coin for each possible value of n when the probability of occurrence p of a head in a single toss is $\frac{1}{10}$.

7. Repeat Prob. 6 for the case of an unbiased coin ($p = \frac{1}{2}$).

8. Determine the probability that *at most* $n < N$ heads occur in N independent tosses of a coin. Evaluate for $N = 10$, $n = 5$, and $p = \frac{1}{2}$.

9. Determine the probability that *at least* $n < N$ heads occur in N independent tosses of a coin. Evaluate for $N = 10$, $n = 5$, and $p = \frac{1}{2}$.

10. Show the relation between the probabilities of Probs. 8 and 9.

11. The experiment A has M mutually exclusive possible outcomes A_m, and the experiment B has N mutually exclusive possible outcomes B_n. Show that $P[B_n|A_m]$ may be expressed in terms of $P[A_m|B_n]$ and $P[B_n]$ by the relation

$$P[B_n|A_m] = \frac{P[A_m|B_n]P[B_n]}{\sum_{i=1}^{N} P[A_m|B_i]P[B_i]} \tag{2-22}$$

This relation is known as *Bayes' rule*.

CHAPTER 3

RANDOM VARIABLES AND PROBABILITY DISTRIBUTIONS

3-1. Definitions

Sample Points and Sample Spaces. In the preceding chapter, we discussed experiments, events (i.e., possible outcomes of experiments), and probabilities of events. In such discussions, it is often convenient to think of an experiment and its possible outcomes as defining a space and its points. With each *basic* possible outcome we may associate a point called the *sample point*. The totality of sample points, corresponding to the aggregate of all possible outcomes of the experiment, is called the *sample space* corresponding to the experiment in question. In general, an event may correspond either to a single sample point or to a set of sample points. For example, there are six possible outcomes of the throw of a die: the showing of one, two, three, four, five, or six dots on the upper face. To each of these we assign a sample point; our sample space then consists of six sample points. The basic event "a six shows" corresponds to a single sample point, whereas the compound event "an even number of dots shows" corresponds to the set of three sample points representing the showing of two, four, and six dots respectively.

In many problems of interest it is possible to represent numerically each possible outcome of an experiment. In some cases a single number will suffice, as in the measurement of a noise voltage or the throw of a die. In others a set of numbers may be required. For example, three numbers are needed for the specification of the instantaneous position of a moving gas particle (the three spatial coordinate values). When such a numerical representation of outcomes is possible, the set of numbers specifying a given outcome may be thought of as the coordinate values of a vector which specifies the position of the corresponding sample point in the sample space. Thus if K numbers are required to specify each possible outcome, each sample point will have K coordinate values, and the sample space will be a *K-dimensional vector space*. We will generally confine ourselves to sample spaces of this type.

The probability of a given event may now be thought of as assigning a weight or mass to the corresponding sample point (or set of sample points). In our sample-point and sample-space terminology, the probability $P(A)$ that the outcome of a given experiment will be the event (A) may be expressed as the probability $P(S_A)$ that the sample point s, corresponding to the outcome of the experiment, falls in the subset of sample points S_A corresponding to the event (A):†

$$P(A) = P(s \, \varepsilon \, S_A) = P(S_A)$$

Random Variable. A real-valued function $x(s)$ defined on a sample space of points s will be called a *random variable*‡ if for every real number a the set of points s for which $x(s) \leq a$ is one of the class of admissable sets for which a probability is defined. This condition is called *measurability* and is almost always satisfied in practice. A complex-valued function $z(s) = x(s) + jy(s)$ defined on a sample space will be called a *complex random variable* if $x(s)$ and $y(s)$ are both measurable. Similarly, a function which assigns a vector to each point of a sample space will be called a *vector random variable* or a *random vector*.

It was pointed out above that a sample space representing the outcomes of the throw of a die is a set of six points which may be taken to be the integers 1, . . . , 6. If now we identify the point k with the event that k dots show when the die is thrown, the function $x(k) = k$ is a random variable such that $x(k)$ equals the number of dots which show when the die is thrown. The functions $g(k) = k^2$ and $h(k) = \exp(k^2)$ are also random variables on this space.

Another example of a random variable is the real-valued function of a real variable whose value represents a noise voltage as measured at a given instant of time. Here one takes the real line as a sample space. A related variable is that defined to be unity when the noise voltage being measured lies between V and $V + \Delta V$ volts and defined to be zero otherwise. It should be realized that a function of a random variable is a random variable.

3-2. Probability Distribution Functions

Consider now the real random variable $x(s)$ such that the range of x is the real line (i.e., $-\infty \leq x \leq +\infty$). Consider a point X on the real line. The function of X whose value is the probability $P(x \leq X)$ that the random variable x is less than or equal to X is called the *probability distribution function* of the random variable x. Since probabilities are always bounded by zero and one, the extremal values of the probability distribution function must also be zero and one:

---

† The notation $s \, \varepsilon \, S_A$ means that the point s is an element of the point set S_A.
‡ The use of the term "random variable" for a function is dictated by tradition.

$$P(x \leq -\infty) = 0 \qquad \text{and} \qquad P(x \leq +\infty) = 1 \qquad (3\text{-}1)$$

It further follows that the probability that the random variable x falls in the interval $a < x \leq b$ is simply the difference between the values of the probability distribution function obtained at the endpoints of the interval:

$$P(x \leq b) - P(x \leq a) = P(a < x \leq b) \geq 0 \qquad (3\text{-}2)$$

The right-hand inequality follows from the nonnegativeness of probabilities. Equation (3-2) shows that $P(x \leq b) \geq P(x \leq a)$ when $b \geq a$ and hence that the probability distribution function is a nondecreasing function of X.

Paralleling the single variable case, we may now define a joint probability distribution function by $P(x \leq X, y \leq Y)$, the probability that the random variable x is less than or equal to a specified value X and that the random variable y is less than or equal to a specified value Y. Such a distribution function may be defined whether or not x and y are random variables on the same sample space. Furthermore, whether x and y are considered to be two separate one-dimensional random variables or the components of a two-dimensional random variable is immaterial. In either case, the joint sample space is a two-dimensional sample space (the xy plane), and the joint probability distribution function is the probability that a result of an experiment will correspond to a sample point falling in the quadrant $(-\infty \leq x \leq X, -\infty \leq y \leq Y)$ of that sample space.

The extremal values of the joint probability distribution function are obviously

$$P(x \leq -\infty, y \leq Y) = 0 = P(x \leq X, y \leq -\infty) \qquad (3\text{-}3a)$$

and
$$P(x \leq +\infty, y \leq +\infty) = 1 \qquad (3\text{-}3b)$$

as each of the first extremes represents an impossibility, whereas the second represents a certainty.

The probability that the random variable x is less than or equal to a specified value X whereas the random variable y assumes any possible value whatever is simply the probability that $x \leq X$ irrespective of the value of y. The latter probability is, by definition, the value at X of the probability distribution function of the random variable x. Thus

$$P(x \leq X, y \leq +\infty) = P(x \leq X) \qquad (3\text{-}4a)$$

Geometrically, this is the probability that a sample point falls in the half-plane $(-\infty \leq x \leq X)$. Similarly,

$$P(x \leq +\infty, y \leq Y) = P(y \leq Y) \qquad (3\text{-}4b)$$

Thus we see that, as with the joint probabilities, the joint probability distribution function determines the probability distribution functions of the component variables. The distribution functions of the component variables are often designated as the *marginal distribution functions*.

These definitions and results may be extended in a more or less obvious way to the case of K-dimensional random variables.

3-3. Discrete Random Variables

We will call the random variable x a *discrete random variable* if x can take on only a finite number of values in any finite interval. Thus, for

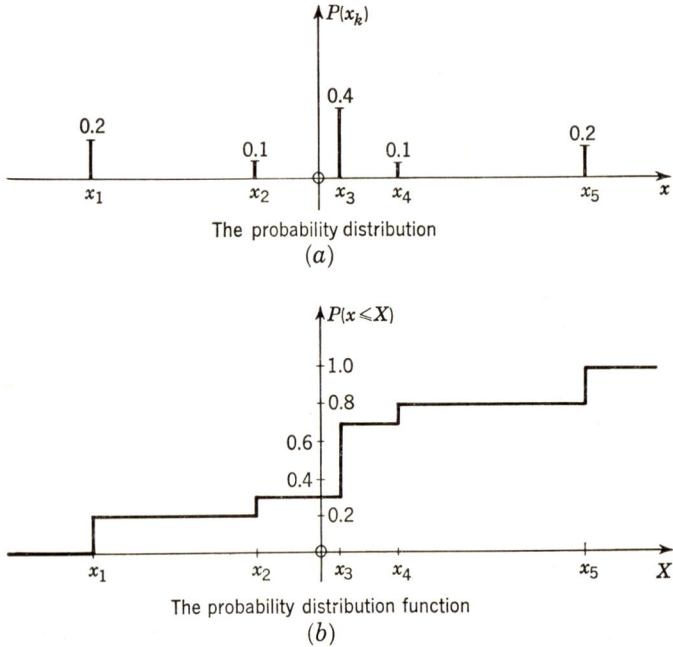

The probability distribution

(a)

The probability distribution function

(b)

FIG. 3-1. A discrete probability distribution and probability distribution function.

example, the random variable defined as the number of heads appearing in N tosses of a coin is a discrete random variable. The complete set of probabilities $P(x_k)$ associated with the possible values x_k of x is called the *probability distribution* of the discrete random variable x. It follows from the definition of the probability distribution function that in the discrete case

$$P(x \leq X) = \sum_{x_k \leq X} P(x_k) \tag{3-5}$$

and hence that

$$P(x \leq +\infty) = \sum_{\text{all } k} P(x_k) = 1 \tag{3-6}$$

The probability distribution and associated probability distribution function for one particular discrete random variable are shown in Fig. 3-1.

The joint probability distribution

(a)

The joint probability distribution function

(b)

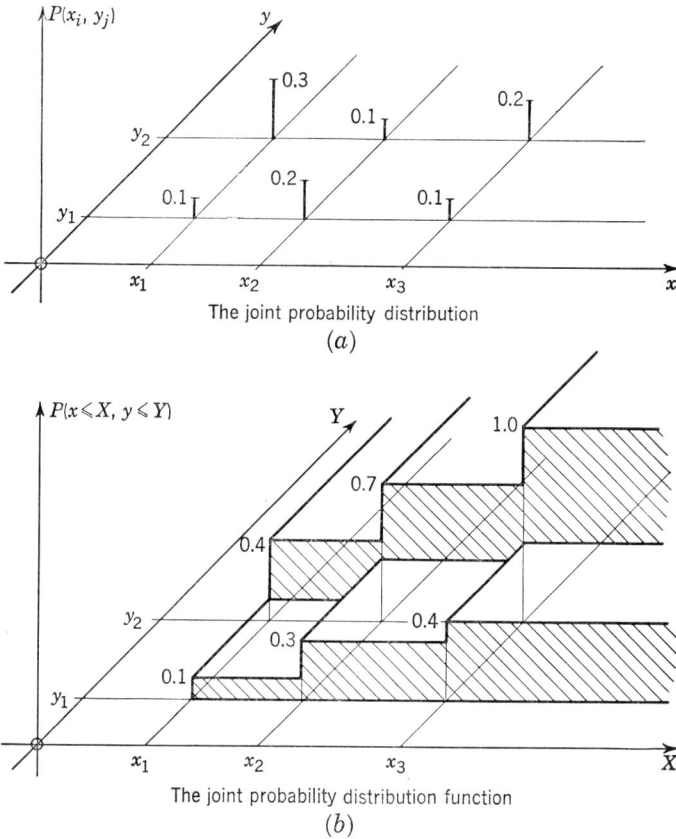

Fig. 3-2. A two-dimensional discrete probability distribution and corresponding probability distribution function.

If the probability distribution of a two-dimensional discrete random variable is given by the set of probabilities $P(x_k, y_m)$, the corresponding joint probability distribution function is given by the expression

$$P(x \leq X, y \leq Y) = \sum_{x_k \leq X} \sum_{y_m \leq Y} P(x_k, y_m) \tag{3-7}$$

Therefore

$$P(x \leq +\infty, y \leq +\infty) = \sum_{\text{all } k} \sum_{\text{all } m} P(x_k, y_m) = 1 \tag{3-8}$$

An example of the joint probability distribution and the corresponding joint probability distribution function of a two-dimensional discrete random variable is shown in Fig. 3-2.

It follows from our definition of a sample point that distinct sample points correspond to mutually exclusive events. The various results obtained in Chap. 2 for such events may therefore be applied directly to the study of discrete random variables. Thus Eq. (3-6) follows from Eq. (2-5), and Eq. (3-8) follows from Eq. (2-8). The relations between the joint probability distribution and the marginal distributions for discrete random variables follow directly from Eq. (2-9), and are

$$P(x_k) = \sum_{\text{all } m} P(x_k, y_m) \quad \text{and} \quad P(y_m) = \sum_{\text{all } k} P(x_k, y_m) \quad (3\text{-}9)$$

Similarly we see from Eq. (2-12) that

$$P(x_k, y_m) = P(y_m | x_k) P(x_k) = P(x_k | y_m) P(y_m) \quad (3\text{-}10)$$

and from Eq. (2-14) that

$$\sum_{\text{all } k} P(x_k | y_m) = 1 = \sum_{\text{all } m} P(y_m | x_k) \quad (3\text{-}11)$$

3-4. Continuous Random Variables

Not all random variables are discrete. For example, the random variable which represents the value of a thermal noise voltage at a specified instant of time may assume *any* value between plus and minus infinity. Nevertheless it can be shown that, since a probability distribution function is a bounded, nondecreasing function, any probability distribution function may be decomposed into two parts:† a staircase function having jump discontinuities at those points X for which $P(x = X) > 0$ (i.e., a discrete probability distribution function as shown in Fig. 3-1), and a part which is everywhere continuous. A random variable for which the probability distribution function is everywhere continuous will be called a *continuous random variable*. Thus we see that any random variable may be thought of as having a discrete part and a continuous part.

Probability Density Functions. Any continuous distribution function can be approximated as closely as we like by a nondecreasing staircase function, which can then be regarded as the probability distribution function of a discrete random variable. Thus a continuous random variable can always be approximated by a discrete one. However, a more direct method of analysis is made possible when the probability distribution function is not only continuous but also differentiable with a continuous derivative everywhere except possibly at a discrete set of points. In such a case we define a *probability density function $p(x)$* as the derivative of the probability distribution function

† See Cramér (I, Arts. 6.2 and 6.6).

$$p(X) = \frac{dP\ (x \le X)}{dX} \tag{3-12}$$

such that

$$P(x \le X) = \int_{-\infty}^{X} p(x)\ dx \tag{3-13}$$

Although it is necessary to realize that there exist continuous random variables which do not have a probability density function,† we may generally safely ignore such pathological cases.

From the definition of the derivative as a limit, it follows that

$$p(X) = \lim_{\Delta X \to 0} \frac{P(x \le X) - P(x \le X - \Delta X)}{\Delta X} \tag{3-14a}$$

or, using Eq. (3-2),

$$p(X) = \lim_{\Delta X \to 0} \frac{P(X - \Delta X < x \le X)}{\Delta X} \tag{3-14b}$$

We may therefore write in differential notation

$$p(X)\ dX = P(X - dX < x \le X) \tag{3-15}$$

and interpret $p(X)\ dX$ as the probability that the random variable x has a value falling in the interval $(X - dX < x \le X)$. Since the probability distribution function is a nondecreasing function, the probability density function is always nonnegative:

$$p(x) \ge 0 \tag{3-16}$$

An example of the probability distribution function and the corresponding probability density function of a continuous random variable is shown in Fig. 3-3.

From Eq. (3-2) and Eq. (3-13) it follows that the probability that a continuous random variable has a value falling in the interval $(a < x \le b)$ is given by the integral of the probability density function over that interval:

$$P(a < x \le b) = \int_{a}^{b} p(x)\ dx \tag{3-17}$$

In particular, when the interval in question is the infinite interval $(-\infty, +\infty)$, we obtain, using Eq. (3-1),

$$\int_{-\infty}^{+\infty} p(x)\ dx = 1 \tag{3-18}$$

This result is the continuous-random-variable expression of the fact that the probability of a certain event is unity. Since by definition the probability distribution function of a continuous random variable has no jumps, it is apparent that, for any x_0

† See, for example, Titchmarsh (I, Art. 11.72).

$$P(x = x_o) = 0 \qquad\qquad (3\text{-}19)$$

That is, the probability that a *continuous* random variable takes on any *specific* value is zero. Obviously, however, this event is not impossible.

Impulse Functions. It would be convenient to use a single system of notation to cover both the case of the discrete random variable and that of the continuous random variable and thus to handle in a simple way the situation in which we must deal with mixed random variables having

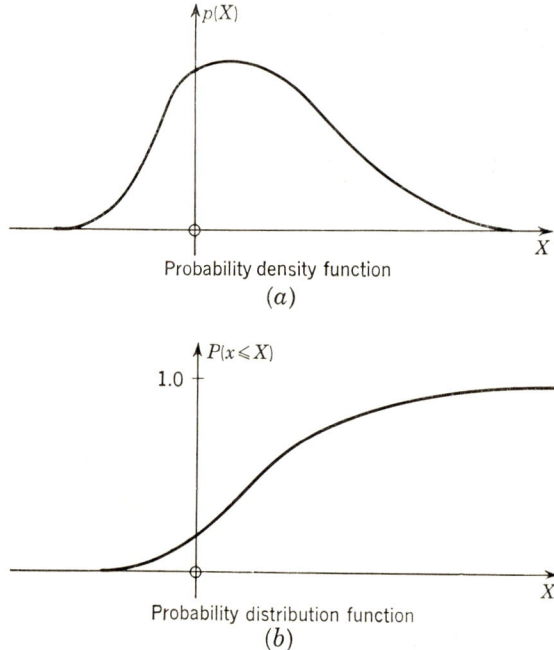

Probability density function

(a)

Probability distribution function

(b)

FIG. 3-3. An example of the probability density function and probability distribution function of a continuous random variable.

both discrete and continuous components. These aims may be accomplished either by introducing a more general form of integration† or by admitting the use of impulse functions and ignoring the cases in which the continuous part of the distribution does not have a derivative. We shall use the second of these methods.

Let us now apply the limiting expression for the probability-density function to a discrete random variable. Suppose, for convenience, that

---

† Specifically the Stieltjes integral. An averaging operation with respect to any probability distribution can be written in terms of Stieltjes integrals even when the continuous part of the probability distribution function does not possess a derivative. See Cramér (I, Chap. 7).

the discrete random variable in question has M possible values x_m with probabilities $P(x_m)$. Then

$$\lim_{\Delta X \to 0} P(X - \Delta X < x \leq X) = \begin{cases} P(x_m) & \text{if } X = x_m \\ 0 & \text{otherwise} \end{cases}$$

Using this result in Eq. (3-14b) we see that

$$p(X) = \begin{cases} \infty & \text{if } X = x_m \\ 0 & \text{otherwise} \end{cases}$$

Further, if ϵ is an arbitrarily small positive number (and smaller than the smallest distance between adjacent values of x_m), we also have the result

$$\int_{X-\epsilon}^{X+\epsilon} p(x)\,dx = \begin{cases} P(x_m) & \text{if } X = x_m \\ 0 & \text{otherwise} \end{cases}$$

Appendix 1 contains an account of some of the properties of impulse and step functions. Reference to this material shows that the results presented above are just the defining equations for impulse functions. Thus the probability density function of a discrete random variable may be thought of as consisting of an impulse of weight (i.e., area) $P(x_m)$ at each of the possible values x_m of that variable. We may therefore write

$$p(x) = \sum_{m=1}^{M} P(x_m)\delta(x - x_m) \tag{3-20}$$

as the probability density function for a discrete random variable which has M possible values x_m with probabilities $P(x_m)$. The function $\delta(x - x_m)$ in Eq. (3-20) is the unit impulse function centered upon $x = x_m$.

The probability distribution function for a discrete random variable may now be obtained by substituting Eq. (3-20) into Eq. (3-13). Interchanging order of integration and finite summation, we find that

$$P(x \leq X) = \sum_{m=1}^{M} P(x_m) \int_{-\infty}^{X} \delta(x - x_m)\,dx$$

As shown in Appendix 1, the integral in this equation is the unit step function $U(X - x_m)$ defined by the relations

$$U(X - x_m) = \begin{cases} 1 & \text{when } X \geq x_m \\ 0 & \text{otherwise} \end{cases}$$

Thus we may write

$$P(x \leq X) = \sum_{m=1}^{M} P(x_m) U(X - x_m) \tag{3-21}$$

as the probability distribution function of our discrete random variable. The graph of such a probability distribution function has the form of a staircase, as shown in Fig. 3-1*b*.

We have shown above that we may extend the concept of the probability density function to cover the case of the discrete random variable by admitting the use of impulse functions. Henceforth we will use, therefore, the probability density function, if convenient, whether we are concerned with continuous, discrete, or mixed random variables.

Joint Probability Density Functions. In the case of a single random variable, the probability density function was defined as the derivative of the probability distribution function. Similarly, in the two-dimensional case, if the joint probability distribution function is everywhere continuous and possesses a continuous mixed second partial derivative everywhere except possibly on a finite set of curves, we may define the *joint probability density function* as this second derivative

$$p(X,Y) = \frac{\partial^2}{\partial X \, \partial Y} P(x \leq X, y \leq Y) \tag{3-22}$$

Then
$$P(x \leq X, y \leq Y) = \int_{-\infty}^{Y} \int_{-\infty}^{X} p(x,y) \, dx \, dy \tag{3-23}$$

As before, we will ignore the pathological cases and restrict ourselves to those cases in which the partial derivatives exist either in the usual sense or in the form of impulse functions (e.g., in the case of discrete random variables).

From the limiting definition of the partial derivative, we may write

$$p(X,Y) = \lim_{\substack{\Delta X \to 0 \\ \Delta Y \to 0}} \frac{1}{\Delta X \, \Delta Y} \begin{bmatrix} P(x \leq X, y \leq Y) - P(x \leq X - \Delta X, y \leq Y) \\ -P(x \leq X, y \leq Y - \Delta Y) \\ + P(x \leq X - \Delta X, y \leq Y - \Delta Y) \end{bmatrix}$$

and hence
$$p(X,Y) = \lim_{\substack{\Delta X \to 0 \\ \Delta Y \to 0}} \frac{P(X - \Delta X < x \leq X, Y - \Delta Y < y \leq Y)}{\Delta X \, \Delta Y}$$

or, in differential notation,

$$p(X,Y) \, dX \, dY = P(X - dX < x \leq X, Y - dY < y \leq Y) \tag{3-24}$$

Thus $p(X,Y) \, dX \, dY$ may be interpreted as the probability that a sample point falls in an incremental area $dX \, dY$ about the point (X,Y) in our sample space. It follows that the joint probability density function is always nonnegative,

$$p(x,y) \geq 0 \tag{3-25}$$

since the joint probability distribution function is a nondecreasing function of its arguments.

From our definition of the joint probability density function, it follows that the probability $P(s \in R)$ that a sample point s falls in a region R of the sample space is given by the integral of the joint probability density function over that region:

$$P(s \in R) = \iint_R p(x,y) \, dx \, dy \tag{3-26}$$

In particular, when the region in question is the entire sample space (i.e., the entire xy plane), we obtain

$$\int_{-\infty}^{+\infty} \int_{-\infty}^{+\infty} p(x,y) \, dx \, dy = 1 \tag{3-27}$$

as we are concerned here with a certainty. On the other hand, if we allow only one of the upper limits to recede to infinity, we obtain, on application of Eqs. (3-4),

$$\int_{-\infty}^{+\infty} \int_{-\infty}^{X} p(x,y) \, dx \, dy = P(x \leq X) \tag{3-28a}$$

and

$$\int_{-\infty}^{Y} \int_{-\infty}^{+\infty} p(x,y) \, dx \, dy = P(y \leq Y) \tag{3-28b}$$

The derivatives of the right-hand sides of these equations are simply the probability density functions $p(X)$ and $p(Y)$ respectively. Therefore, by differentiating both sides of Eq. (3-28), we obtain

$$\int_{-\infty}^{+\infty} p(X,y) \, dy = p(X) \tag{3-29a}$$

and

$$\int_{-\infty}^{+\infty} p(x,Y) \, dx = p(Y) \tag{3-29b}$$

since here the derivative of an integral with respect to the upper limit is the integrand evaluated at that limit. Eqs. (3-29) are the continuous-random-variable equivalents of the discrete-random-variable Eqs. (3-9).

As with probability distribution functions, the various definitions and results above may be extended to the case of k-dimensional random variables.†

Conditional Probability Density Functions. Let us consider now the probability that the random variable y is less than or equal to a particular value Y, subject to the hypothesis that a second random variable x has a value falling in the interval $(X - \Delta X < x \leq X)$. It follows from the definition of conditional probability, Eq. (2-11), that

$$P(y \leq Y | X - \Delta X < x \leq X) = \frac{P(X - \Delta X < x \leq X, y \leq Y)}{P(X - \Delta X < x \leq X)}$$

† See for example Cramér (I, Arts. 8.4 and 22.1).

The probabilities on the right-hand side may be expressed in terms of the probability densities defined above.   Thus

$$P(y \leq Y | X - \Delta X < x \leq X) = \frac{\int_{-\infty}^{Y} \int_{X-\Delta X}^{X} p(x,y)\, dx\, dy}{\int_{X-\Delta X}^{X} p(x)\, dx}$$

Assuming that the probability densities are continuous functions of x in the stated interval and that $p(X) > 0$, the integrals over x can be replaced by the integrands evaluated at $x = X$ and multiplied by ΔX as $\Delta X \to 0$. The incremental width ΔX is a factor of both the numerator and the denominator and therefore may be canceled.   The limit of the left-hand side as $\Delta X \to 0$ becomes the probability that $y \leq Y$ subject to the hypothesis that the random variable x assumes the value X.   Thus we get

$$P(y \leq Y | X) = \frac{\int_{-\infty}^{Y} p(X,y)\, dy}{p(X)} \tag{3-30}$$

The probability $P(y \leq Y | X)$ defined here will be called the *conditional probability distribution function* of the random variable y *subject to the hypothesis that* $x = X$.

If now the usual continuity requirements for the joint probability density are satisfied, we may define a *conditional probability density function* $p(Y|X)$ to be the derivative of the conditional probability distribution function:

$$p(Y|X) = \frac{dP\,(y \leq Y|X)}{dY} \tag{3-31}$$

Then $$P(y \leq Y|X) = \int_{-\infty}^{Y} p(y|X)\, dy \tag{3-32}$$

Differentiating Eq. (3-30) with respect to Y, we get

$$p(Y|X) = \frac{p(X,Y)}{p(X)} \tag{3-33a}$$

or, in product form,

$$p(X,Y) = p(Y|X)p(X) \tag{3-33b}$$

This result is the continuous-random-variable version of the factoring of a joint probability into the product of a conditional probability and an elementary probability.

As with all distribution functions, the conditional probability distribution function $P(y \leq Y|X)$ is a nondecreasing function (of Y).   Therefore the conditional probability density function is nonnegative:

$$p(Y|X) \geq 0 \tag{3-34}$$

From Eq. (3-32) it follows that the probability that y has a value falling in the interval $(a < y \leq b)$, subject to the hypothesis that $x = X$, is given by the integral of the conditional probability density function over that interval:

$$P(a < y \leq b|X) = \int_a^b p(y|X)\, dy \tag{3-35}$$

If now the limits recede to infinity, we obtain

$$\int_{-\infty}^{+\infty} p(y|X)\, dy = 1 \tag{3-36}$$

as here we have a certainty.

The conditional probability density function $p(X|Y)$ may, of course, be defined in the same way as was $p(Y|X)$, and corresponding results may be obtained for it.

3-5. Statistically Independent Random Variables

In Art. 2-5 we defined experiments to be statistically independent when their joint probabilities were expressible as products of their respective elementary probabilities. Thus if the experiments A and B have K and M mutually exclusive outcomes A_k and B_m respectively, we called them statistically independent if the relations

$$P(A_k,B_m) = P(A_k)P(B_m)$$

were satisfied for all values of k and m. Suppose now that we characterize the outcomes of the experiments A and B by the discrete random variables x and y so that x assumes the value x_k whenever the event A_k occurs and y assumes the value y_m whenever the event B_m occurs. Then

$$P(x_k) = P(A_k), \qquad P(y_m) = P(B_m), \qquad \text{and} \qquad P(x_k,y_m) = P(A_k,B_m)$$

We now say that x and y are *statistically independent random variables*, if and only if the experiments they characterize are statistically independent. Thus, the discrete random variables x and y, having K values x_k and M values y_m respectively, are said to be statistically independent if and only if the relation

$$P(x_k,y_m) = P(x_k)P(y_m) \tag{3-37}$$

is satisfied for all values of k and m.

From Eqs. (3-7) and (3-37) we find that

$$P(x \leq X, y \leq Y) = \sum_{x_k \leq X} \sum_{y_m \leq Y} P(x_k)P(y_m)$$

when x and y are statistically independent. Then, using Eq. (3-5), it follows that the equation

$$P(x \leq X, y \leq Y) = P(x \leq X)P(y \leq Y) \tag{3-38}$$

is satisfied for all values of X and Y so long as x and y are statistically independent random variables. Conversely, it is easy to show that if Eq. (3-38) is satisfied for all X and Y, then Eq. (3-37) is satisfied for all k and m. Thus the discrete random variables x and y are statistically independent if and only if Eq. (3-38) is satisfied for all values of X and Y.

Although Eq. (3-37) applies only to discrete random variables, Eq. (3-38) may be satisfied whether the random variables x and y are discrete, continuous, or mixed. We will therefore base our general definition of statistically independent random variables on Eq. (3-38). Thus:

DEFINITION. *The random variables x, y, . . . , and z are said to be statistically independent random variables if and only if the equation*

$$P(x \leq X, y \leq Y, \ldots , z \leq Z)$$
$$= P(x \leq X)P(y \leq Y) \cdots P(z \leq Z) \quad (3\text{-}39)$$

is satisfied for all values of X, Y, . . . , and Z.

Suppose now that the random variables x and y are continuous random variables and that the various probability densities exist. Eq. (3-38) may then be written

$$\int_{-\infty}^{Y} \int_{-\infty}^{X} p(x,y)\, dx\, dy = \int_{-\infty}^{X} p(x)\, dx \int_{-\infty}^{Y} p(y)\, dy$$

If now we take the mixed second partial derivative of both sides (with respect to X and Y), we get

$$p(X,Y) = p(X)p(Y)$$

Conversely, if we integrate this result, we can obtain Eq. (3-38). The factoring of the probability densities is therefore a necessary and sufficient condition for the statistical independence of x and y. Similarly it may be shown that the random variables x, y, . . . , and z are statistically independent if and only if the equation

$$p(x,y, \ldots , z) = p(x)p(y) \cdots p(z) \quad (3\text{-}40)$$

is satisfied for all values of x, y, . . . , and z.

3-6. Functions of Random Variables

One of the basic questions arising in applications of probability theory is, "Given a random variable x and its probability functions, what are the probability functions of a new random variable y defined as some function of x, say $y = g(x)$?" This question arises, for example, in the determination of the probability density function of the output of the detector in a radio receiver. In this section we shall attempt to show how such a question may be answered.

Single Variable Case. Suppose that x is a real random variable. We can take it to be the coordinate variable of the sample space S_x, which consists of the real line with some assignment of probabilities. Let the points in S_x be denoted by s. Then x is the function $x(s) = s$. Let y be a single-valued real function of a real variable, and consider the function of s given by $y[x(s)]$. Since for each s, $y[x(s)]$ is a real number, $y[x(s)]$ is (subject, of course, to the condition that it be measurable) a new random variable defined on the sample space S_x. On the other hand, $y(x)$ could also be considered as providing a mapping between S_x and a new sample space S_y, where S_y is the set of real numbers which constitutes the range of values taken on by $y[x(s)]$ as s ranges over S_x. Then if A is a set of points in S_x, there corresponds to A a set of points B in S_y given by the condition: $s_o \in A$ if and only if $y[x(s_o)] \in B$. Hence

$$P(s \in A) = P[x(s) \in A] = P\{y[x(s)] \in B\}$$

Since we usually suppress the sample space variable s in writing expressions with random variables, we may write $P(x \in A) = P[y(x) \in B]$. This result establishes the fundamental relation between the probability functions of x and those of $y(x)$. It is important to note that it doesn't matter whether we regard $y(x)$ as a random variable on S_x or y as the coordinate random variable on S_y.

The probability distribution function of y may now be obtained by choosing B to be the semi-infinite interval $(-\infty < y \leq Y)$. If then $A(Y)$ is the set of points in S_x which corresponds to the set of points $(-\infty < y \leq Y)$ in S_y, we obtain

$$P(y \leq Y) = P[x \in A(Y)] \tag{3-41}$$

as the relation between the probability distribution function of y and a probability function of x. The probability density function of y may then be obtained by differentiation:

$$p(Y) = \frac{dP[x \in A(Y)]}{dY} \tag{3-42}$$

Suppose now that the probability density function of x, $p_1(x)$, exists and is continuous almost everywhere, that y is a differentiable monotonic function of x, and that its derivative dy/dx vanishes only at isolated points. In this case we can obtain a direct relation between $p_1(x)$ and the probability density function of y, $p_2(y)$, for it follows from our assumptions that x is a single-valued function of y. Thus to the point Y in S_y there corresponds a unique point X in S_x, and the mapping is said to be *one-to-one*. Assuming, for the moment, that y is a monotonically increasing function of x, it then follows that the interval $(-\infty < x \leq X(Y))$ in

S_x corresponds to the interval $(-\infty < y \leq Y)$ in S_y. Using this fact in Eq. (3-42) we get

$$p_2(Y) = \frac{d}{dY} \int_{-\infty}^{X(Y)} p_1(x)\, dx = p_1(X) \frac{dX}{dY}$$

Since a similar result (except for a minus sign) is obtained when y is a monotonically decreasing function of x, we can write

$$p_2(Y) = p_1(X) \left| \frac{dX}{dY} \right| \tag{3-43}$$

as the desired relation between the probability density functions of y and x.

If the various conditions specified above are not satisfied, incorrect results may well be obtained through the use of Eq. (3-43). For example, suppose that y is constant over a certain interval in S_x (as would be the case for a half-wave rectifier); then x is not a single-valued function of y and Eq. (3-43) is meaningless as it stands.

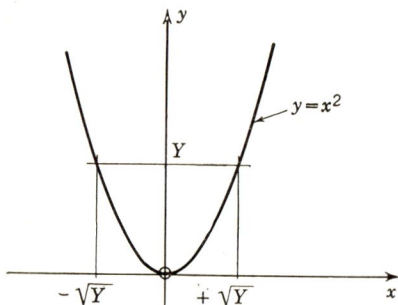

FIG. 3-4. The square-law transformation function.

Example 3-6.1. The technique of determining the probability density function of a function of a random variable can perhaps best be understood by studying a specific example. Consider the case of the square-law transformation

$$y = x^2 \tag{3-44}$$

which is plotted in Fig. 3-4. This transformation function would be obtained, for example, in the study of a full-wave square-law detector. An examination of Fig. 3-4 shows, first of all, that y can never be negative. Therefore

$$P(y \leq Y) = 0 \qquad \text{for } Y < 0$$

and hence

$$p_2(Y) = 0 \qquad \text{for } Y < 0 \tag{3-45}$$

Further reference to Fig. 3-4 shows that $A(Y)$ is the set of points $(-\sqrt{Y} \leq x \leq +\sqrt{Y})$ when $Y \geq 0$. Therefore

$$P(y \leq Y) = P[-\sqrt{Y} \leq x \leq \sqrt{Y}]$$
$$= P[x \leq +\sqrt{Y}] - P[x < -\sqrt{Y}]$$

and hence

$$P(y \leq Y) = \int_{-\infty}^{+\sqrt{Y}} p_1(x)\, dx - \int_{-\infty}^{-\sqrt{Y}} p_1(x)\, dx$$

If now we take the derivative of both sides with respect to Y, we obtain

$$p_2(Y) = \frac{p_1(x = +\sqrt{Y}) + p_1(x = -\sqrt{Y})}{2\sqrt{Y}} \qquad \text{for } Y \geq 0 \tag{3-46}$$

The combination of Eq. (3-45) and Eq. (3-46) then gives the required probability density function for y in terms of that of x.

If next we specify the form of $p(x)$, say as the gaussian probability density

$$p_1(x) = \frac{\exp(-x^2/2)}{(2\pi)^{\frac{1}{2}}}$$

we may evaluate Eqs. (3-45) and (3-46) and get an explicit result for $p_2(y)$:

$$p_2(y) = \begin{cases} \dfrac{\exp(-y/2)}{(2\pi y)^{\frac{1}{2}}} & \text{for } y \geq 0 \\ 0 & \text{for } y < 0 \end{cases} \tag{3-47}$$

The probability density function given by Eq. (3-47) is called a *chi-squared* density function when written as a function of $y = \chi^2$. The densities $p_1(x)$ and $p_2(y)$ are shown in Fig. 3-5.

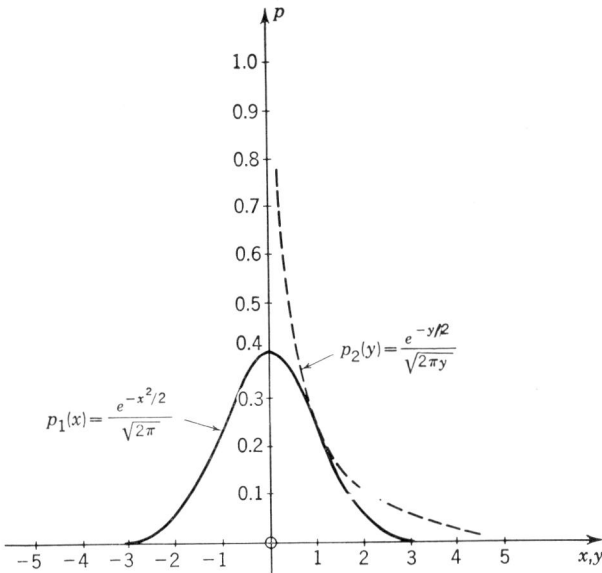

$$p_1(x) = \frac{e^{-x^2/2}}{\sqrt{2\pi}}$$

$$p_2(y) = \frac{e^{-y/2}}{\sqrt{2\pi y}}$$

FIG. 3-5. Gaussian and chi-squared probability density functions.

Multiple Variable Case. The ideas involved in determining the probability functions for functions of multiple random variables are the same as those in the single variable case. Thus if $(x_1, x_2, \ldots, x_N)$ are a set of real random variables whose joint probability functions are known, the equations

$$y_m = g_m(x_1, x_2, \ldots, x_N) \qquad m = 1, 2, \ldots, M$$

define a mapping from the sample space of the random variables x_n to the sample space of the random variables y_m. The probability functions of the random variables y_m may then be found by determining the point sets in the sample space of the y_m which are the mappings of given point sets in the sample space of the x_n, and equating probabilities. Note that the number of random variables y_m need not equal the number of random variables x_n.

Example 3-6.2. Suppose that a random variable z is defined as the sum of the real random variables x and y:

$$z = g(x,y) = x + y \qquad (3\text{-}48)$$

and that the joint probability functions for x and y are known. The problem is to determine the probability density function $p_3(z)$ of the random variable z.

Since x and y are real random variables, their joint sample space may be taken as the (x,y) plane consisting of the points $(-\infty \leq x \leq +\infty, -\infty \leq y \leq +\infty)$. Also, since x and y are real, z must be real, too, and its sample space may be taken as the real line $(-\infty \leq z \leq +\infty)$. As usual, we shall determine the probability density function by taking the derivative of the probability distribution function. Hence we

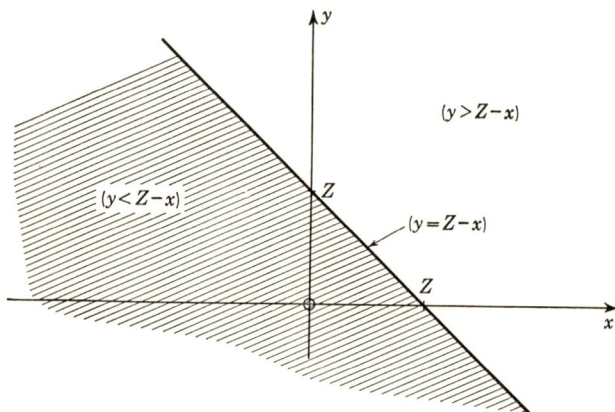

Fig. 3-6.

need to know the set of points in the (x,y) plane which corresponds to the set of points $(z \leq Z)$ on the real z line. It follows from the definition of z that the points in question are those located in the half-plane $(y \leq Z - x)$ which is shown in Fig. 3-6. Thus

$$P(z \leq Z) = P(y \leq Z - x)$$

If the joint probability density function for x and y exists and is continuous, we may write

$$P(z \leq Z) = \int_{-\infty}^{+\infty} \int_{-\infty}^{Z-x} p(x,y) \, dy \, dx$$

and, on differentiating with respect to Z, we get

$$p_3(Z) = \int_{-\infty}^{+\infty} p(x, y = Z - x) \, dx \qquad (3\text{-}49)$$

If now x and y are statistically independent random variables, their joint probability density function is given by the product of their individual probability density functions. Hence in this case,

$$p_3(Z) = \int_{-\infty}^{+\infty} p_1(x) p_2(y = Z - x) \, dx \qquad (3\text{-}50)$$

The probability density function of the sum of two statistically independent random variables is therefore given by the convolution of their respective probability density functions.

The Jacobian. As in the corresponding single-variable case, a direct relationship may be obtained between the probability density functions of the old and new random variables when they are related by a one-to-one mapping. For example, suppose that not only are the new random variables y_m defined as single-valued continuous functions of the old random variables x_n (with continuous partial derivatives everywhere):

$$y_m = g_m(x_1, x_2, \ldots, x_N) \qquad m = 1, 2, \ldots, N \qquad (3\text{-}51)$$

but also the old random variables may be expressed as single-valued continuous functions of the new:

$$x_n = f_n(y_1, y_2, \ldots, y_N) \qquad n = 1, 2, \ldots, N \qquad (3\text{-}52)$$

(where the numbers of old and new variables are now equal). In this case, to each point in the sample space of the x_n there corresponds one and only one point in the sample space of the y_m, and the functional relations between the old and new random variables provide a one-to-one mapping.

Further, suppose that A is an arbitrary closed region in the sample space of the x_n and that B is its mapping in the sample space of the y_m. The probability that a sample point falls in A then equals the probability that its mapping falls in B. Assuming that the joint probability density function of the x_n exists, we may write

$$\int_A \cdots \int p_1(x_1, \ldots, x_N)\, dx_1 \cdots dx_N$$
$$= \int_B \cdots \int p_2(y_1, \ldots, y_N)\, dy_1 \cdots dy_N$$

The problem before us now is simply that of evaluation of a multiple integral by means of a change of variables. It therefore follows from our assumed properties of the mapping that†

$$\int_A \cdots \int p_1(x_1, \ldots, x_N)\, dx_1 \cdots dx_N$$
$$= \int_B \cdots \int p_1(x_1 = f_1, \ldots, x_N = f_N)|J|\, dy_1 \cdots dy_N$$

† E.g., see Courant (I, Vol. 2, pp. 247–254).

in which the f_m are given by Eq. (3-52) and in which J is the *Jacobian* of the transformation

$$J = \frac{\partial(x_1, \ldots, x_N)}{\partial(y_1, \ldots, y_N)} = \begin{vmatrix} \dfrac{\partial f_1}{\partial y_1}, & \cdots & , \dfrac{\partial f_N}{\partial y_1} \\ & \cdot & \\ & \cdot & \\ & \cdot & \\ \dfrac{\partial f_1}{\partial y_N}, & \cdots & , \dfrac{\partial f_N}{\partial y_N} \end{vmatrix} \qquad (3\text{-}53)$$

The old and new joint probability density functions are therefore related by the equation

$$p_2(y_1, \ldots, y_N) = p_1(x_1 = f_1, \ldots, x_N = f_N)|J| \qquad (3\text{-}54)$$

When $N = 1$, Eq. (3-54) reduces to the result previously derived in the single-variable case, Eq. (3-43).

3-7. Random Processes†

Up to this point we have not mentioned the possible time dependence of any of our probability functions. This omission was deliberate, for in many cases time enters the picture only implicitly, if at all, and has no particular bearing on the problem at hand. However, there are many cases of interest in which the time dependence of the probability functions is of importance, as, for example, in problems involving random signals and noises. We will here attempt to show how our previous notions of probability and random variable may be extended to cover such situations.

One of the basic ideas involved in our concept of probability is the determination of the relative frequency of occurrence of a given outcome in a large number ($\to \infty$) of repetitions of the basic experiment. The idea of "repetition" of an experiment commonly implies performing the experiment, repeating it at a later time, repeating it at a still later time, etc. However, the factor of importance in a relative-frequency determination is not performance of experiments in time succession but rather performance of a large number of experiments. We could satisfy our requirements by *simultaneously* performing a large number of identically prepared experiments as well as by performing them in time sequence.

Let us now consider a sequence of N throws of a die. Suppose that on the nth throw $k(n)$ dots show on the upper face of the die; $k(n)$ can assume any integer value between one and six. An appropriate sample

† A detailed discussion of the various possible definitions of a random process is beyond the scope of this book. For a thorough treatment see Doob (II, Chap. I, Arts. 5 and 6; Chap. II, Arts. 1 and 2). See also Grenander (I, Arts. 1.2–1.4).

space for the nth throw therefore consists of the six points 1, . . . , 6, and an appropriate random variable is $k(n)$ itself. Similarly, an appropriate joint sample space for a sequence of N throws is the N-dimensional vector space in which the sample point coordinates are $k(1)$, $k(2)$, . . . , $k(N)$, and the appropriate random variable is the position vector of the sample point. As the number of throws in a sequence increases, so does the dimensionality of the joint sample space. The probability that $k(1)$ dots will appear on the first throw can be estimated from the results of simultaneously throwing a large number of dice by determining the corresponding relative frequency. Similarly, by repeating the process, we can estimate the probability that $k(n)$ dots will appear on the nth throw and also the joint probability that $k(1)$ dots will appear on the first throw, that $k(2)$ dots will appear on the second throw, . . . , and that $k(N)$ dots will appear on the Nth throw. The specification of such a set of experiments with the corresponding probability functions and random variables (for all N) is said to define a *random (stochastic) process* and, in particular, a *discrete-parameter random process.*†

Consider next a particular sequence of throws of a die. The numbers $k(n)$ obtained in this sequence of throws are called *sample values* and may be thought of as defining a particular function of the index n. Such a function, several examples of which are shown in Fig. 3-7a, is called a *sample function* of the random process in question. The set of all possible such sample functions, together with a probability law, is called the *ensemble* of the sample functions. It will often be convenient to consider a random process to be defined by its ensemble of sample functions as well as by the random variables and probability functions discussed above.

The concepts introduced in our discussion of a discrete-parameter random process may be extended to apply to a continuous-parameter random process in a more or less obvious manner. For example, consider the thermal-agitation noise voltage $x(t)$ generated in a resistor. In particular, consider the measurement $x(t_1)$ of this voltage at an instant of time t_1. This voltage could take on any value between plus and minus infinity. The sample space corresponding to our measurement would then be the real line $(-\infty \leq x \leq +\infty)$, and an appropriate random variable x_1 would be the measured voltage. If we had available a large number of identical resistors, we could simultaneously measure a set of values $x(t_1)$ and from these data determine the relative frequency of occurrence of events of the type $[X - \Delta X < x(t_1) \leq X]$. The relative frequencies so obtained would provide measures of the probabilities $P[X - \Delta X < x_1 \leq X]$ and hence of the probability density function $p(x_1)$. Similarly, we could make measurements at N instants of time, say t_1 through t_N, giving sample values of the random variables x_1 through x_{N} and obtain

† The discrete parameter being the "throw" index n.

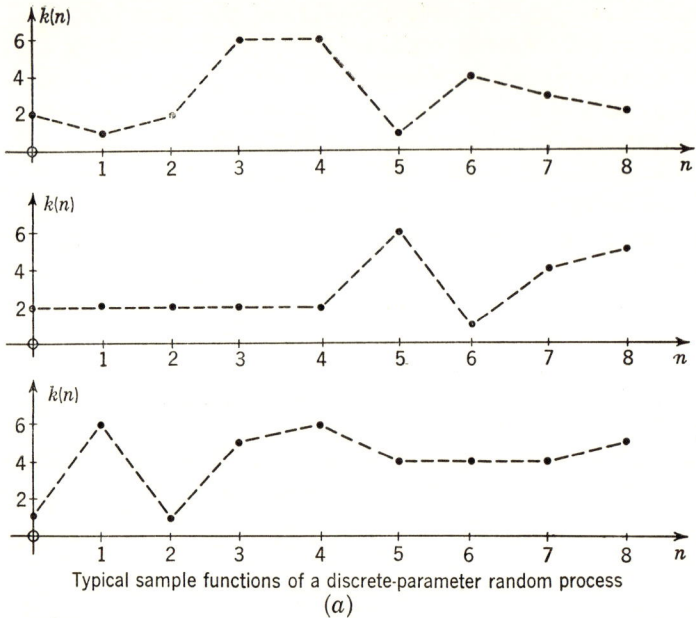

Typical sample functions of a discrete-parameter random process

(a)

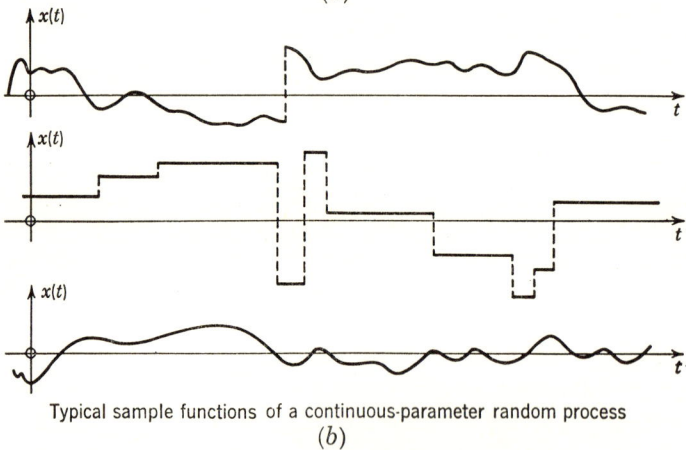

Typical sample functions of a continuous-parameter random process

(b)

Fig. 3-7.

a measure of the joint probability density function $p(x_1, \ldots, x_N)$. The specification of all such sets of random variables and of their joint probability distribution functions for all values of N defines a *continuous-parameter*† *random process*. The output voltage, as a function of time, of a particular resistor would then be a sample function of our continuous-

---

† "Continuous parameter" refers to the fact that the parameter t may assume any real value and has nothing to do with the fact that the random variables in question may be continuous random variables.

parameter random process. Typical examples of such sample functions are shown in Fig. 3-7b. The ensemble of all possible such sample functions could, as in the discrete-parameter case, be thought of as defining the random process.

Probability Relations. Once we have defined a random process and its various probability functions and random variables as outlined above, we may apply directly all the probability relations we have previously defined and derived. Thus, if x_1 through x_N are the random variables pertaining to the random process in question, the probability density functions $p(x_1, x_2, \ldots, x_N)$ have the following properties. First of all,

$$0 \le p(x_1, x_2, \ldots, x_N) \tag{3-55}$$

for all values of N, since all probability densities must be nonnegative. Further, we must have the relation

$$\int_{-\infty}^{+\infty} \cdots \int_{-\infty}^{+\infty} p(x_1, x_2, \ldots, x_N)\, dx_1 \cdots dx_N = 1 \tag{3-56}$$

corresponding to the fact that the probability that *any* possible event occurs is unity. Also, we must have the relation, following from Eq. (3-29),

$$\int_{-\infty}^{+\infty} \cdots \int_{-\infty}^{+\infty} p(x_1, \ldots, x_k, x_{k+1}, \ldots, x_N)\,dx_{k+1} \cdots dx_N$$
$$= p(x_1, \ldots, x_k) \qquad k < N \tag{3-57}$$

The conditional probability density functions are given, as usual, by

$$p(x_1, \ldots, x_k \mid x_{k+1}, \ldots, x_N) = \frac{p(x_1, \ldots, x_N)}{p(x_{k+1}, \ldots, x_N)} \tag{3-58}$$

and must satisfy the relation

$$\int_{-\infty}^{+\infty} p(x_N \mid x_1, \ldots, x_{N-1})\, dx_N = 1 \tag{3-59}$$

as again we are dealing with a certainty.

Suppose that x_1 through x_N are the random variables pertaining to one random process and y_1 through y_M are the random variables pertaining to another random process. The two processes are said to be *statistically independent random processes* if

$$p(x_1, \ldots, x_N, y_1, \ldots, y_M) = p(x_1, \ldots, x_N)p(y_1, \ldots, y_M) \tag{3-60}$$

for all sets of random variables x_n and y_m and for all values of N and M.

Stationary Random Processes. Let us now reconsider the various outputs of the large number of identically prepared thermal-agitation noise-voltage sources discussed above. Just as we could previously estimate the probability density function $p(x_{t_i})$ from the data obtained at the time instant t_1, so now we can estimate the probability density function $p(x_{t_1+t})$ from data obtained at a different time instant $t_1 + t$. Similarly, just as we estimate the joint probability density function $p(x_{t_1}, \ldots, x_{t_N})$ from the data obtained at the N instants of time t_1 through t_N, so may we now estimate the joint probability density function $p(x_{t_1+t}, \ldots, x_{t_N+t})$ from the data obtained at the N instants of time $(t_1 + t)$ through $(t_N + t)$. Obviously one of two situations exists: either these two sets of probability density functions are identical or they are not. If the probability functions corresponding to the time instants t_n *are* identical to the probability functions corresponding to the time instants $(t_n + t)$ for all values of N and t, then those probability functions are invariant under a shift of the time origin. A random process characterized by such a set of probability functions is said to be a *stationary random process;* others are said to be *nonstationary.*

As an example of a stationary random process, consider the thermal-agitation noise voltage generated in a resistor of resistance R at a temperature T, which is connected in parallel with a condensor of capacitance C. In our later studies, we shall see that the probability functions of the noise voltage developed across the parallel RC combination are completely determined by the values of R, T, and C. If the values of R, T, and C are fixed for all time, the noise voltage corresponds to a stationary random process. On the other hand, if, say, T varies with time, then the process is nonstationary.

3-8. Problems

1. The random variable x has an *exponential probability density function*

$$p(x) = a \exp(-b|x|)$$

where a and b are constants.

  a. Determine the required relation between the constants a and b.
  b. Determine the probability distribution function of x, $P(x \leq X)$.
  c. Plot $p(x)$ and $P(x \leq X)$ for the case $a = 1$.

2. The random variable y has the *Cauchy probability density function*

$$p(y) = \frac{c}{d^2 + y^2} \tag{3-61}$$

where c and d are constants.

  a. Determine the relation necessary between the constants c and d.
  b. Determine the probability distribution function of y, $P(y \leq Y)$.
  c. Plot $p(y)$ and $P(y \leq Y)$ for the case $d = 1$.

3. Show that the joint probability density function of the random variables $(x_1, \ldots, x_N)$ may be expressed in terms of the conditional probability density functions $p(x_n|x_1, \ldots, x_n - 1)$ by the relation

$$p(x_1, \ldots, x_N) = p(x_N|x_1, \ldots, x_{N-1}) \cdots p(x_3|x_1, x_2)p(x_2|x_1)p(x_1) \tag{3-62}$$

4. Consider the discrete random variables x and y, having K possible values x_k and

M possible values y_m respectively. Show that if the equation

$$P(x \leq X, y \leq Y) = P(x \leq X)P(y \leq Y)$$

is satisfied for all values of X and Y, then the equations

$$P(x_k, y_m) = P(x_k)P(y_m)$$

are satisfied for all values of k and m, and hence the random variables x and y are statistically independent.

5. Let x and y be statistically independent discrete random variables; x has three possible values $(0,1,3)$ with the probabilities $(\frac{1}{2}, \frac{3}{8}, \frac{1}{8})$, and y has the two possible values $(0,1)$ with the probabilities $(\frac{1}{3}, \frac{2}{3})$. Determine, and plot, the joint probability distribution function of the random variables x and y.

6. Let the random variables x and y be those defined in Prob. 5 above. A new random variable v is defined as the sum of x and y:

$$v = x + y$$

Determine the probability distribution function of v, $P(v \leq V)$, and plot.

7. Let the probability density function of the random variable x be $p(x)$. The random variable y is defined in terms of x as

$$y = ax + b$$

where a and b are constants. Determine the relation between $p(y)$ and $p(x)$.

8. Let x and y be statistically independent random variables. The new random variables u and v are defined by the equations

$$u = ax + b \text{ and } v = cy + d$$

where a, b, c, and d are constants. Show that the random variables u and v are also statistically independent random variables.

9. Let x be a *gaussian* random variable having the probability density function

$$p(x) = \frac{\exp(-x^2/2)}{(2\pi)^{1/2}}$$

The random variable y is defined as $y = x^3$. Determine and plot the probability density function of the random variable y.

10. Let x be the gaussian random variable of Prob. 9. The random variable y is defined by the equation

$$y = \begin{cases} x^{1/n} & \text{when } x \geq 0 \\ -(-x)^{1/n} & \text{when } x < 0 \end{cases}$$

where n is a positive constant. Determine and plot the probability density function of the random variable y for the cases $n = 1$, 2, and infinity.

11. The random process $x(t)$ is defined by

$$x(t) = \sin(wt + \theta)$$

where w is constant and where θ is a random variable having the probability density function

$$p(\theta) = \begin{cases} \dfrac{1}{2\pi} & \text{for } 0 \leq \theta \leq 2\pi \\ 0 & \text{otherwise} \end{cases}$$

a. Determine the probability distribution function $P(x_t \leq X)$.
b. Determine the probability density function $p(x_t)$.
c. Plot the results of **a** and **b**.

12. Let x and y be statistically independent random variables. A new random variable z is defined as the product of x and y:

$$z = xy$$

Derive an expression for the probability density function $p(z)$ of the new random variable z in terms of the probability density functions $p(x)$ and $p(y)$ of the original random variables x and y.

13. Let x and y be statistically independent random variables with the probability density functions

$$p(x) = \frac{1}{\pi \sqrt{1 - x^2}} \quad \text{and} \quad p(y) = \begin{cases} y \exp\left(\frac{-y^2}{2}\right) & \text{for } y \geq 0 \\ 0 & \text{for } y < 0 \end{cases}$$

Show that their product has a gaussian probability density function.

14. Let x and y be statistically independent random variables with the uniform probability density functions

$$p(x) = \begin{cases} \dfrac{1}{A} & \text{for } 0 \leq x \leq A \\ 0 & \text{otherwise} \end{cases} \quad \text{and} \quad p(y) = \begin{cases} \dfrac{1}{B} & \text{for } 0 \leq y \leq B \\ 0 & \text{otherwise} \end{cases}$$

Determine and plot for $A = 1$ and $B = 2$ the probability density function of their product.

15. Let x and y be statistically independent random variables with the probability density functions

$$p(x) = \frac{\exp(-x^2/2)}{(2\pi)^{1/2}} \quad \text{and} \quad p(y) = \begin{cases} y \exp\left(\frac{-y^2}{2}\right) & \text{for } y \geq 0 \\ 0 & \text{for } y < 0 \end{cases}$$

Show that their product has an exponential probability density function (as in Prob. 1).

CHAPTER 4

AVERAGES

In the two preceding chapters we introduced the concept of probability, defined various probability functions and random variables, and studied some of the properties of those functions. We pointed out that probability functions describe the long-run behavior of random variables (i.e., behavior on the average). In fact it can be shown that the various averages of random variables (e.g., the mean, the mean square, etc.) can be determined through the use of the probability functions. It is this aspect of probability theory which we shall investigate in this chapter. In particular, we shall define the process of statistical averaging, study a few averages of interest, and finally investigate the relation between time and statistical averages.

4-1. Statistical Averages

Let us first consider a discrete random variable $y = g(x)$, where x is a discrete random variable taking on any one of M possible values x_m, and $g(x)$ is a single-valued function of x. We want now to determine the average of $g(x)$. To this end, let us return for the moment to the relative-frequency interpretation of probability. Suppose that we repeat N times a basic experiment which defines the random variable x and that the event corresponding to x_m occurs $n(x_m)$ times in these N repetitions. The *arithmetic average* of $g(x)$ is, in this case,

$$g(x)_{av} = \frac{g(x_1)n(x_1) + \cdots + g(x_M)n(x_M)}{N}$$

$$= \sum_{m=1}^{M} g(x_m) \frac{n(x_m)}{N}$$

If we believe that repetitions of this experiment will exhibit statistical regularity, we would expect (as pointed out in Chap. 2) this average to converge to a limit as $N \to \infty$. Since the probability $P(x_m)$ represents the limit of the relative frequency $n(x_m)/N$, we are led to define the *statistical average*† (or *expectation*) $E[g(x)]$ of the discrete random variable $g(x)$ by the equation

$$E[g(x)] = \sum_{m=1}^{M} g(x_m)P(x_m) \qquad (4\text{-}1)$$

The statistical average then represents the limit of the arithmetic average.

If $x_{k_1}, \ldots, x_{k_j}$ are all the mutually exclusive values of x for which $g(x)$ takes on the value y_k, then the probability of y_k is $P(x_{k_1}) + \cdots + P(x_{k_j})$, and Eq. (4-1) may be written

$$E[g(x)] = E(y) = \sum_{k=1}^{K} y_k P(y_k) \qquad (4\text{-}2)$$

Eqs. (4-1) and (4-2) give the same value for the statistical average of $g(x)$; conceptually they are different because Eq. (4-1) refers to the sample space of x and Eq. (4-2) refers to the sample space of y. The simplest application of Eq. (4-1) is that in which $g(x) = x$. $E(x)$ is usually called the mean value of x and denoted by m_x.

Now suppose x is a continuous random variable with the probability density $p_1(x)$ and $g(x)$ is a single-valued function of x. Again we want to determine the average of the random variable $g(x)$. Let x be approximated by a discrete random variable x' which takes on the values x_m with probability $p_1(x_m) \Delta x_m$, where the M intervals Δx_m partition the sample space of x. Then by Eq. (4-1)

$$E[g(x')] = \sum_{m=1}^{M} g(x_m)p_1(x_m) \, \Delta x_m$$

If we let all the $\Delta x_m \to 0$, thus forcing $M \to \infty$, the limiting value of this sum is the integral given below in Eq. (4-3), at least for piecewise continuous $g(x)$ and $p_1(x)$. This procedure suggests that we define the statistical average of the continuous random variable $g(x)$ by the equation

$$E[g(x)] = \int_{-\infty}^{+\infty} g(x)p_1(x) \, dx \qquad (4\text{-}3)$$

Equation (4-3) may be extended to apply to the case where x is a mixed discrete and continuous random variable by allowing $p_1(x)$ to contain impulse functions. The statistical average of $y = g(x)$ can also be expressed in terms of probabilities on the sample space of y; if the probability density $p_2(y)$ is defined, then

$$E[g(x)] = E(y) = \int_{-\infty}^{+\infty} y p_2(y) \, dy \qquad (4\text{-}4)$$

† Also known as the *mean, mathematical expectation, stochastic average,* and *ensemble average.* For convenience, we will sometimes denote the statistical average by a bar over the quantity being averaged, thus: $\overline{g(x)}$.

Multiple Random Variables. So far we have studied only statistical averages of functions of a single random variable. Let us next consider those cases where two or more random variables are involved. First suppose that the random variables x and y are discrete random variables having the M possible values x_m and K possible values y_k respectively. The problem before us is to determine the average of a single-valued function, say $g(x,y)$, of the two random variables x and y. Functions of interest might be the sum $x + y$, the product xy, and so forth.

Even though we are here talking about two random variables x and y, the statistical average of $g(x,y)$ is really already defined by Eq. (4-1). Since x can take on M possible values and y can take on K, the pair (x,y) can take on MK possible values and the double indices (m,k) can be replaced by a single index which runs from 1 to MK. Equation (4-1) then applies. However, there is no point in actually changing the indices; evidently

$$E[g(x,y)] = \sum_{k=1}^{K} \sum_{m=1}^{M} g(x_m,y_k)P(x_m,y_k) \qquad (4\text{-}5)$$

The corresponding expression when x and y are continuous (or mixed) random variables is

$$E[g(x,y)] = \int_{-\infty}^{+\infty} \int_{-\infty}^{+\infty} g(x,y)p(x,y)\,dx\,dy \qquad (4\text{-}6)$$

Equations (4-5) and (4-6) may be extended to more than two random variables in a fairly obvious manner.

Example 4-1.1. Let us now apply the above expressions to several functions $g(x,y)$ of interest. First, let $g(x,y) = ax + by$ where a and b are constants. Direct application of Eq. (4-6) gives the result that

$$E(ax + by) = a \int_{-\infty}^{+\infty} \int_{-\infty}^{+\infty} xp(x,y)\,dx\,dy + b \int_{-\infty}^{+\infty} \int_{-\infty}^{+\infty} yp(x,y)\,dx\,dy$$

The first term on the right-hand side is a times the statistical average of the function $g_1(x,y) = x$, and the second is b times the statistical average of the function $g_2(x,y) = y$. Thus

$$E(ax + by) = aE(x) + bE(y) \qquad (4\text{-}7)$$

That this result is consistent with our original definition, Eq. (4-3), follows from the properties of the joint probability density function, specifically from application of Eq. (3-29a). We may easily extend the result in Eq. (4-7) to the case of N random variables x_n to show that

$$E\left(\sum_{n=1}^{N} a_n x_n\right) = \sum_{n=1}^{N} a_n E(x_n) \qquad (4\text{-}8)$$

where the a_n are constants. Thus, *the mean of a weighted sum of random variables is equal to the weighted sum of their means.*

Example 4-1.2. Next let us determine the statistical average of the product of a function of the random variable x times a function of the random variable y. From Eq. (4-6),

$$E[h(x)f(y)] = \int_{-\infty}^{+\infty} \int_{-\infty}^{+\infty} h(x)f(y)p(x,y)\,dx\,dy$$

Generally, this is as far as one may proceed until the form of the joint probability density is specified. However, a simplification does occur when x and y are statistically independent. In this case, the joint probability density factors into the product of the probability densities of x and y alone, and we get

$$E[h(x)f(y)] = \int_{-\infty}^{+\infty} h(x)p_1(x)\,dx \int_{-\infty}^{+\infty} f(y)p_2(y)\,dy$$

Therefore
$$E[h(x)f(y)] = E[h(x)]E[f(y)] \tag{4-9}$$

when x and y are statistically independent random variables. This result may easily be extended to the case of more than two random variables, hence *the mean of a product of statistically independent random variables is equal to the product of the means of those random variables.*

Random Processes. The preceding comments about statistical averages of random variables apply equally well to statistical averages of random processes. For example, let x_t be the random variable that refers to the possible values which can be assumed at the time instant t by the sample functions $x(t)$ of a given random process. We then define the statistical average of a function of the given random process to be the statistical average of that function of x_t:

$$E[g(x_t)] = \int_{-\infty}^{+\infty} g(x_t)p(x_t)\,dx_t$$

It should be noted that the statistical average of a random process may well be a variable function of time, since, in general, the probability density of x_t need not be the same as that of $x_{t+\tau}$. For example, suppose the sample functions $x(t)$ of one random process are related to those $y(t)$ of another by the equation

$$x(t) = y(t)\cos t$$

Since $\cos t$ is simply a number for fixed t,

$$E(x_t) = E(y_t)\cos t$$

Suppose now that the process with sample functions $y(t)$ is stationary. The probability density $p(y_t)$ is then the same for all values of t, and $E(y_t)$ has the same value m_y for all t (i.e., the statistical average of a stationary random process is a *constant* function of time). Then

$$E(x_t) = m_y\cos t \neq E(x_{t+\tau}) = m_y\cos(t+\tau)$$

unless τ is a multiple of 2π and $E(x_t)$ is not a constant function of t, as it would be if the corresponding process were stationary.

4-2. Moments

One set of averages of functions of a random variable x which is of particular interest is that consisting of the moments of x in which the nth *moment* of the probability distribution of the random variable x is defined to be the statistical average of the nth power of x:

$$E(x^n) = \int_{-\infty}^{+\infty} x^n p(x)\ dx \tag{4-10}$$

and the nth moment of a random process is similarly defined. In a stationary random process, the first moment $m = E(x)$ (i.e., the mean) gives the steady (or d-c) component; the second moment $E(x^2)$ (i.e., the mean square) gives the *intensity* and provides a measure of the average power carried by a sample function.

The nth central moment μ_n of the probability distribution of the random variable x is defined as

$$\mu_n = E[(x - m)^n] = \int_{-\infty}^{+\infty} (x - m)^n p(x)\ dx \tag{4-11}$$

It then follows from the binomial expansion of $(x - m)^n$ that the central moments and the moments are related by the equation

$$\mu_n = \sum_{r=0}^{n} (-1)^r \binom{n}{r} m^r E(x^{n-r}) \tag{4-12}$$

The second central moment which is often of particular interest is commonly denoted by a special symbol σ^2:

$$\sigma_x^2 = \mu_2 = E(x - m)^2 \tag{4-13}$$

and given a special name *variance* or *dispersion*. The variance of a stationary random process is the intensity of the varying component (i.e., the a-c component). The positive square root σ of the variance is known as the *standard deviation* of the random variable x. The standard deviation of a random process is then the root-mean-square (rms) value of the a-c component of that process.

Multiple Random Variables. The concept of moments of a probability distribution may, of course, be extended to joint probability distributions. For example, an $(n + k)$th order *joint moment* (or *cross moment*) of the joint probability distribution of the random variables x and y is

$$E(x^n y^k) = \int_{-\infty}^{+\infty} \int_{-\infty}^{+\infty} x^n y^k p(x,y)\ dx\ dy \tag{4-14}$$

The corresponding *joint central moments* are

$$\mu_{nk} = E[(x - m_x)^n (y - m_y)^k] \tag{4-15}$$

where m_x is the mean of x and m_y is the mean of y. The joint moment

$$\mu_{11} = E[(x - m_x)(y - m_y)]$$

is called the *covariance* of the random variables x and y. It is of particular interest to us and will be discussed in some detail in Art. 4-4.

4-3. Characteristic Functions†

Another statistical average of considerable importance is the *characteristic function* $M_x(jv)$ of the probability distribution of the real random variable x, which is defined to be the statistical average of $\exp(jvx)$:

$$M_x(jv) = E[\exp(jvx)] = \int_{-\infty}^{+\infty} \exp(jvx)p(x)\,dx \tag{4-16}$$

where v is real. Since $p(x)$ is nonnegative and $\exp(jvx)$ has unit magnitude,

$$\left| \int_{-\infty}^{+\infty} \exp(jvx)p(x)\,dx \right| \leq \int_{-\infty}^{+\infty} p(x)\,dx = 1$$

Hence the characteristic function always exists, and

$$|M_x(jv)| \leq M_x(0) = 1 \tag{4-17}$$

It follows from the definition of the Fourier integral that the characteristic function of the probability distribution of a random variable x is the Fourier transform of the probability density function of that random variable. Therefore, under suitable conditions,‡ we may use the inverse Fourier transformation

$$p(x) = \frac{1}{2\pi} \int_{-\infty}^{+\infty} M_x(jv) \exp(-jvx)\,dv \tag{4-18}$$

to obtain the probability density function of a random variable when we know its characteristic function. It is often easier to determine the characteristic function first and then transform to obtain the probability density than it is to determine the probability density directly.

When x is a discrete random variable, Eq. (4-16) becomes

$$M_x(jv) = \sum_m P(x_m) \exp(jvx_m) \tag{4-19}$$

Moment Generation. Suppose now that we take the derivative of the characteristic function with respect to v,

$$\frac{dM_x(jv)}{dv} = j \int_{-\infty}^{+\infty} x \exp(jvx)p(x)\,dx$$

† For a more detailed discussion of the properties of the characteristic function, see Cramér (I, Chap. 10) and Doob (II, Chap. I, Art. 11).

‡ See, for example, Courant (I, Vol. II, Chap. 4, Appendix Art. 5) or Guillemin (II, Chap. VII, Art. 19).

If both sides of this equation are evaluated at $v = 0$, the integral becomes the first moment of the random variable x. Hence, on solving for that moment, we get

$$m_x = -j \left. \frac{dM_x(jv)}{dv} \right]_{v=0} \tag{4-20}$$

and we see that the first moment may be obtained by differentiation of the characteristic function.

Similarly, by taking the nth derivative of the characteristic function with respect to v, we obtain

$$\frac{d^n M_x(jv)}{dv^n} = j^n \int_{-\infty}^{+\infty} x^n \exp(jvx) p(x) \, dx$$

Evaluating at $v = 0$, the integral becomes the nth moment of the random variable x. Therefore

$$E(x^n) = (-j)^n \left. \frac{d^n M_x(jv)}{dv^n} \right]_{v=0} \tag{4-21}$$

The differentiation required by Eq. (4-21) is sometimes easier to carry out than the integration required by Eq. (4-10).

Suppose that the characteristic function has a Taylor-series expansion:

$$M_x(jv) = \sum_{n=0}^{\infty} \left. \frac{d^n M_x(jv)}{dv^n} \right]_{v=0} \frac{v^n}{n!}$$

It then follows, using Eq. (4-21), that

$$M_x(jv) = \sum_{n=0}^{\infty} E(x^n) \frac{(jv)^n}{n!} \tag{4-22}$$

Therefore, if the characteristic function of a random variable has a Taylor-series expansion valid in some interval about the origin, it is uniquely determined in this interval by the moments of the random variable. If the moments of all orders of the distribution of a random variable do not exist but the moment of order n does exist, then all moments of order less than n also exist and the characteristic function can be expanded in a Taylor series with remainder of order n.† If the moments do uniquely determine the characteristic function, they also uniquely determine the probability distribution function.‡ A simple condition for this uniqueness is the following: if a probability distribution has moments of all orders, and if the series in Eq. (4-22) converges

---

† See Courant (I, Vol. I, Chap. VI, Art. 2).
‡ See Cramér (I, p. 93), or Loève (I, p. 186).

absolutely for some $v > 0$, then this is the only probability distribution with these moments.†

Another average closely related to the characteristic function is often discussed in statistics literature.‡ This function, called the *moment-generating function* $M_x(u)$ is defined to be the statistical average of $\exp(ux)$:

$$M_x(u) = E[\exp(ux)] \tag{4-23}$$

where u is taken to be real. This function has essentially the same moment-generating properties as the characteristic function. In particular, it is easy to show that

$$E(x^n) = \frac{d^n M_x(u)}{du^n} \bigg]_{u=0} \tag{4-24}$$

An essential difference between the characteristic function and the moment-generating function is that the characteristic function always exists, whereas the moment-generating function exists only if all of the moments exist.

Joint Characteristic Functions. The statistical average of $\exp(jv_1x + jv_2y)$

$$M(jv_1, jv_2) = E[\exp(jv_1x + jv_2y)]$$
$$= \int_{-\infty}^{+\infty} \int_{-\infty}^{+\infty} \exp(jv_1x + jv_2y) p(x,y) \, dx \, dy \tag{4-25}$$

is the *joint characteristic function* of the joint probability distribution of the random variables x and y. It therefore follows that the joint characteristic function is the two-dimensional Fourier transform of the joint probability density function of x and y, and we may use the inverse Fourier transformation

$$p(x,y) = \frac{1}{(2\pi)^2} \int_{-\infty}^{+\infty} \int_{-\infty}^{+\infty} M(jv_1, jv_2) \exp(-jv_1x - jv_2y) \, dv_1 \, dv_2 \tag{4-26}$$

to obtain the joint probability density function of a pair of random variables when we know their joint characteristic function. Similarly, the Nth-order joint characteristic function and the Nth-order joint probability density function form an N-dimensional Fourier transform pair.§

Having defined the joint characteristic function, let us determine some of its properties. First we find that

$$M(0,0) = 1$$

It further follows from Eq. (4-25) that

$$|M(jv_1, jv_2)| \leq M(0,0) = 1 \tag{4-27}$$

---

† See Cramér (I, p. 176).
‡ See, for example, Mood (I, Art. 5.3).
§ See Cramér (I, Art. 10.6).

Therefore, the joint characteristic function always exists, and it assumes its greatest magnitude at the origin of the (v_1,v_2) plane.

Suppose now that we set only v_2 equal to zero. We may then write

$$M(jv_1,0) = \int_{-\infty}^{+\infty} \exp(jv_1x)\, dx \int_{-\infty}^{+\infty} p(x,y)\, dy$$

From the properties of the joint probability density function, it follows that the integral over y is simply the probability density function $p_1(x)$. Thus we obtain the relation

$$M(jv_1,0) = M_x(jv_1) \tag{4-28}$$

between the joint characteristic function for x and y and the characteristic function for x alone. Similarly, it can be shown that

$$M(0,jv_2) = M_y(jv_2) \tag{4-29}$$

Suppose next that we take the nth partial derivative of the joint characteristic function with respect to v_1 and the kth partial derivative with respect to v_2; thus

$$\frac{\partial^{n+k} M(jv_1,jv_2)}{\partial v_1{}^n\, \partial v_2{}^k} = j^{n+k} \int_{-\infty}^{+\infty} \int_{-\infty}^{+\infty} x^n y^k \exp(jv_1x + jv_2y) p(x,y)\, dx\, dy$$

Setting both v_1 and v_2 equal to zero, we note that the double integral becomes the joint moment $E(x^n y^k)$. Therefore

$$E(x^n y^k) = (-j)^{n+k} \frac{\partial^{n+k} M(jv_1,jv_2)}{\partial v_1{}^n\, \partial v_2{}^k}\bigg]_{v_1,v_2=0} \tag{4-30}$$

Thus the various joint moments of the random variables x and y may be obtained from their joint characteristic function through successive differentiations.

A series expansion of the joint characteristic function in terms of the various joint moments may be obtained through the use of power-series expansions of the exponentials in the integral expression for $M(jv_1,jv_2)$. Thus

$$M(jv_1,jv_2) = \int_{-\infty}^{+\infty} \int_{-\infty}^{+\infty} \left[\sum_{n=0}^{\infty} \frac{(jv_1x)^n}{n!}\right]\left[\sum_{k=0}^{\infty} \frac{(jv_2y)^k}{k!}\right] p(x,y)\, dx\, dy$$

or, on interchanging orders of integration and summation,

$$M(jv_1,jv_2) = \sum_{n=0}^{\infty} \sum_{k=0}^{\infty} \frac{(jv_1)^n (jv_2)^k}{n!k!} \int_{-\infty}^{+\infty} \int_{-\infty}^{+\infty} x^n y^k p(x,y)\, dx\, dy$$

The double integral in this equation is simply the joint moment $E(x^n y^k)$; hence

$$M(jv_1, jv_2) = \sum_{n=0}^{\infty} \sum_{k=0}^{\infty} E(x^n y^k) \frac{(jv_1)^n (jv_2)^k}{n! k!} \qquad (4\text{-}31)$$

Statistically Independent Random Variables. The N-dimensional joint characteristic function for the N random variables x_n is by definition

$$M(jv_1, \ldots, jv_N) = E\left[\exp\left(j \sum_{n=1}^{N} v_n x_n\right)\right] = E\left[\prod_{n=1}^{N} \exp(jv_n x_n)\right]$$

If the N random variables x_n are statistically independent, the average of the product is equal to the product of the respective averages. Since the statistical average of $\exp(jv_n x_n)$ is the characteristic function for x_n, we obtain the result that

$$M(jv_1, \ldots, jv_N) = \prod_{n=1}^{N} M_{x_n}(jv_n) \qquad (4\text{-}32)$$

Thus the N-dimensional joint characteristic function for N independent random variables is equal to the N-fold product of their individual characteristic functions. Conversely, it may easily be shown that the N random variables x_n are necessarily statistically independent if their joint characteristic function is equal to the product of their respective characteristic functions.

Example 4-3.1. Let us next consider the random variable y, defined to be the sum of N statistically independent random variables x_n:

$$y = \sum_{n=1}^{N} x_n$$

The characteristic function for y is then

$$M_y(jv) = E\left[\exp\left(jv \sum_{n=1}^{N} x_n\right)\right] = E\left[\prod_{n=1}^{N} \exp(jv x_n)\right]$$

Since the random variables x_n are statistically independent,

$$M_y(jv) = \prod_{n=1}^{N} M_{x_n}(jv) \qquad (4\text{-}33)$$

and we see that the characteristic function for a sum of statistically independent random variables is equal to the product of their respective characteristic functions. Do not, however, confuse the results contained in Eq. (4-32) and Eq. (4-33) because of their similarity of form. Equation (4-32) expresses the joint characteristic function of the N random variables x_n and is a function of the N variables v_n. On the other hand, Eq. (4-33) expresses the characteristic function of a single random variable y and is a function of the single variable v.

Example 4-3.2. Let a random variable z be defined as the sum of two statistically independent random variables x and y: $z = x + y$. It follows from Eq. (4-33) that

$$M_z(jv) = M_x(jv)M_y(jv) \tag{4-34}$$

The probability density function for z may now be obtained by taking the inverse Fourier transform of both sides of Eq. (4-34). Thus

$$p_3(z) = \frac{1}{2\pi} \int_{-\infty}^{+\infty} M_x(jv)M_y(jv) \exp(-jvz)\, dv$$

On substituting for $M_x(jv)$ from Eq. (4-16) and rearranging the order of integration, we get

$$p_3(z) = \int_{-\infty}^{+\infty} p_1(x) \left\{ \frac{1}{2\pi} \int_{-\infty}^{+\infty} M_y(jv) \exp[-jv(z-x)]\, dv \right\} dx$$

Hence, using Eq. (4-18),

$$p_3(z) = \int_{-\infty}^{+\infty} p_1(x)p_2(y = z - x)\, dx \tag{4-35}$$

This result is precisely the same as that previously derived, Eq. (3-50), directly from probability distribution functions.

4-4. Correlation

As we observed earlier in our discussion of conditional probabilities, it is often desirable to know something about the dependence of one random variable upon another. One way of bringing to light a possible

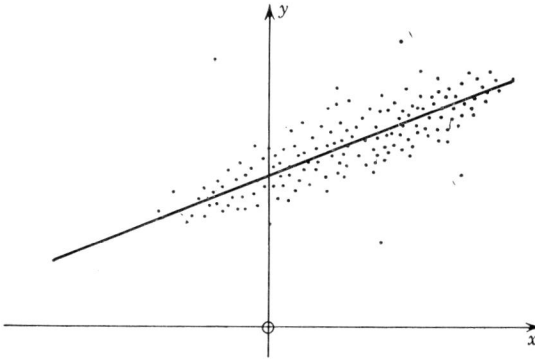

FIG. 4-1. A scatter diagram.

dependence relationship between two real random variables x and y would be to plot the outcomes of various performances of the basic experiment defining these random variables as points in the (x,y) plane (i.e., plot the various sample points measured) and study the resultant figure. Such a plot, which might have the form shown in Fig. 4-1, is known as a *scatter diagram*. If the two random variables x and y were not dependent upon each other, it might be expected that the sample points would be more or less scattered throughout the plane. On the other hand, if the variables were strongly dependent upon each other, we might then expect the sample points to be clustered in the immediate vicinity of a curve describing their functional dependence. The simplest form of dependence is linear dependence, which is of considerable practical importance. In this case, we would expect the sample points to be concentrated along some straight line, as shown in Fig. 4-1, for example.

Suppose that a scatter diagram indicates that the random variables x and y have a strong linear dependence. It is then of interest to determine which straight line

$$y_p = a + bx \tag{4-36}$$

gives the best predicted value y_p of the random variable y based upon a sample value of the random variable x. In order to make such a determination, we must specify just what we mean by "best." One *convenient* criterion of goodness which is useful in many applications is the mean-square difference (error) ϵ_{ms} between the true sample value of the random variable y and its predicted value:

$$\epsilon_{ms} = E[(y - y_p)^2] = E\{[y - (a + bx)]^2\} \tag{4-37}$$

Under this criterion of goodness, the best line of prediction is that line which minimizes the mean-square error. Such a line is often known as the linear mean-square *regression line*.†

Let us now determine those values of the intercept a and the slope b which give the least mean-square error. Differentiating the expression for the mean-square error with respect to a and b and equating the results to zero gives

$$\frac{\partial \epsilon_{ms}}{\partial a} = -2E(y) + 2a + 2bE(x) = 0$$

$$\frac{\partial \epsilon_{ms}}{\partial b} = -2E(xy) + 2aE(x) + 2bE(x^2) = 0$$

Solving for a and b, we obtain

$$b = \frac{E[(x - m_x)(y - m_y)]}{\sigma_x{}^2} = \frac{\mu_{11}}{\sigma_x{}^2}$$

$$a = m_y - m_x \frac{E[(x - m_x)(y - m_y)]}{\sigma_x{}^2} = m_y - \frac{\mu_{11}}{\sigma_x{}^2} m_x$$

If next we substitute these values in Eq. (4-36) and verify the achievement of a true minimum, we find

$$y_p = m_y + \frac{\mu_{11}}{\sigma_x{}^2} (x - m_x) \tag{4-38}$$

† See Cramér (I, Arts. 21.5–7).

as the equation of the best prediction line. From this equation, it follows that the best line of prediction passes through the point (m_x, m_y).

It is convenient to introduce here the *standardized variable* ξ corresponding to the random variable x and defined by the equation

$$\xi = \frac{x - m_x}{\sigma_x} \tag{4-39}$$

Now

$$E(\xi) = 0 \quad \text{and} \quad \sigma_\xi^2 = 1 \tag{4-40}$$

i.e., the standardized variable has a zero mean and a unit standard deviation. In terms of the standardized variable ξ, on defining a standardized prediction $\eta_p = (y_p - m_y)/\sigma_y$, the previously derived expression for the best prediction line assumes the particularly simple form

$$\eta_p = \rho \xi \tag{4-41}$$

where ρ is the *correlation coefficient* defined by the equation

$$\rho = E(\xi\eta) = \frac{\mu_{11}}{\sigma_x \sigma_y} \tag{4-42}$$

in which η is the standardized variable corresponding to y. The correlation coefficient is often known as the *normalized covariance* of the random variables x and y. Equation (4-41) shows that the correlation coefficient is the slope of the line which gives the best predicted value of η based upon a sample value of ξ.

Consider for the moment the mean square of $(\xi \pm \eta)$. Since ξ and η are both real, the square (and hence the mean square) of this function must be nonnegative. Hence

$$E[(\xi \pm \eta)^2] = E(\xi^2) \pm 2E(\xi\eta) + E(\eta^2)$$
$$= 2(1 \pm \rho) \geq 0 \tag{4-43}$$

The greatest possible positive and negative values of the correlation coefficient, therefore, have unit magnitude:

$$-1 \leq \rho \leq 1 \tag{4-44}$$

Suppose that $\rho = +1$. It then follows that

$$E[(\xi - \eta)^2] = \int_{-\infty}^{+\infty} \int_{-\infty}^{+\infty} (\xi - \eta)^2 p(x,y) \, dx \, dy = 0$$

Since both factors of the integrand are nonnegative, this result can hold if $p(x,y)$ is continuous only if $\eta = \xi$ for all values of x and y for which $p(x,y) > 0$; the set of all pairs of values of ξ and η for which $p(x,y) = 0$ occurs with probability zero. Similarly, if $\rho = -1$, then $\eta = -\xi$ for all values of x and y for which $p(x,y) > 0$. The extremal values of $\rho = \pm 1$

therefore correspond to those cases in which $\eta = \pm \xi$ with probability one.

Linear Independence and Statistical Independence. If the correlation coefficient of the real random variables x and y is zero,

$$\rho = 0$$

then those random variables are said to be *uncorrelated* or *linearly independent*. From the defining equation of the correlation coefficient (4-42), it follows that if the joint moment $E(xy)$ of the random variables x and y factors into the product of their respective means

$$E(xy) = E(x)E(y)$$

then the correlation coefficient is zero, and hence x and y are linearly independent. Therefore, if two random variables are statistically independent they are also linearly independent.

On the other hand, however, two random variables that are linearly independent may or may not be statistically independent. This can be shown as follows: It was pointed out in Art. 4-3 that a necessary and sufficient condition for the statistical independence of two random variables is that their joint characteristic function factor into the product of their respective individual characteristic functions:

$$M_{x,y}(jv_1, jv_2) = M_x(jv_1) M_y(jv_2)$$

Suppose now that the joint characteristic function has a Taylor series valid in some region about the origin of the (v_1, v_2) plane. Then $M_x(jv_1)$ and $M_y(jv_2)$ will also have Taylor's expansions, and, substituting these expansions into the above equation, we get

$$\sum_{n=0}^{\infty} \sum_{k=0}^{\infty} E(x^n y^k) \frac{(jv_1)^n (jv_2)^k}{n!\,k!} = \left[\sum_{n=0}^{\infty} E(x^n) \frac{(jv_1)^n}{n!} \right] \left[\sum_{k=0}^{\infty} E(y^k) \frac{(jv_2)^k}{k!} \right]$$

$$= \sum_{n=0}^{\infty} \sum_{k=0}^{\infty} E(x^n) E(y^k) \frac{(jv_1)^n (jv_2)^k}{n!\,k!}$$

Since these series must be equal term by term, it follows that, if a Taylor-series expansion exists for the joint characteristic function, then a necessary and sufficient condition for the statistical independence of two random variables x and y is that their joint moment factors

$$E(x^n y^k) = E(x^n) E(y^k) \tag{4-45}$$

for all positive integral values of n and k. Since linear independence guarantees only factoring of $E(xy)$, it appears that linearly independent random variables are not necessarily also statistically independent, as shown by Prob. 12 of this chapter. However, in the special case where x and y are jointly gaussian, linear independence does imply statistical independence, as we shall see later.

4-5. Correlation Functions

Let x_1 and x_2 be the random variables that refer to the possible values which can be assumed at the time instants t_1 and t_2, respectively, by the sample functions $x(t)$ of a given random process. Our remarks about correlation in the preceding article apply just as well to these two random variables as to any other pair. However, since the joint probability distribution of x_1 and x_2 may change as t_1 and t_2 change, the average $E(x_1x_2)$ relating x_1 and x_2 may well be a function of both time instants. In order to point up this time dependence, we call that average the *autocorrelation function* of the random process and denote it by the symbol $R_x(t_1,t_2)$. Therefore, if the random process in question is real, we have

$$R_x(t_1,t_2) = E(x_1x_2) \qquad (4\text{-}46a)$$

If the random process is complex, we define

$$R_x(t_1,t_2) = E(x_1x_2^*) \qquad (4\text{-}46b)$$

where the asterisk denotes complex conjugate. Eq. (4-46b), of course, reduces to Eq. (4-46a) for a real random process.

Just as $E(x_1x_2)$ is a function of the time instants t_1 and t_2, so is the corresponding correlation coefficient. We shall call the correlation coefficient of x_1 and x_2 the *normalized autocorrelation function* of their random process and denote it by the symbol $\rho_x(t_1,t_2)$. If the random process in question is real,

$$\rho_x(t_1,t_2) = \frac{E[(x_1 - m_1)(x_2 - m_2)]}{\sigma_1\sigma_2} \qquad (4\text{-}47)$$

and hence

$$\rho_x(t_1,t_2) = \frac{R_x(t_1,t_2) - m_1m_2}{\sigma_1\sigma_2} \qquad (4\text{-}48)$$

where $m_1 = E(x_1)$, $m_2 = E(x_2)$, $\sigma_1 = \sigma(x_1)$, and $\sigma_2 = \sigma(x_2)$.

If the given random process is stationary, the joint probability distribution for x_1 and x_2 depends only on the time difference $\tau = t_1 - t_2$ and not on the particular values of t_1 and t_2. The autocorrelation function is then a function only of the time difference τ, and we write

$$R_x(t,t - \tau) = R_x(\tau) = E(x_t x_{t-\tau}^*) \qquad (4\text{-}49)$$

for any t. The corresponding normalized autocorrelation function is

$$\rho_x(\tau) = \frac{R_x(\tau) - m_x^2}{\sigma_x^2} \qquad (4\text{-}50)$$

where $m_x = m_1 = m_2$ and $\sigma_x = \sigma_1 = \sigma_2$, since we are dealing with a stationary random process.

Some random processes are not stationary *in the strict sense* (i.e., their probability distributions are not invariant under shifts of the time origin) but nevertheless have autocorrelation functions which satisfy Eq. (4-49) and means which are constant functions of time. Such random processes are said to be stationary *in the wide sense*.† Obviously, any random process which is stationary in the strict sense is also stationary in the wide sense.

Consider next the two random processes, not necessarily real, whose sample functions are $x(t)$ and $y(t)$ respectively. For each process there exists an autocorrelation function $R_x(t_1,t_2)$ and $R_y(t_1,t_2)$ respectively. In addition, we may define two *cross-correlation functions*

$$R_{xy}(t_1,t_2) = E(x_1 y_2^*) \tag{4-51}$$

and

$$R_{yx}(t_1,t_2) = E(y_1 x_2^*)$$

The correlations between the values of the sample functions $x(t)$ and $y(t)$ at two different instants of time are then specified by the *correlation matrix*

$$\mathbf{R} = \begin{bmatrix} R_x(t_1,t_2) & R_{xy}(t_1,t_2) \\ R_{yx}(t_1,t_2) & R_y(t_1,t_2) \end{bmatrix}$$

In general, an N-by-N correlation matrix is required to specify the correlations for two instants of time among N random processes, or for N instants of time between two random processes.

Some General Properties. Suppose that the random processes with the sample functions $x(t)$ and $y(t)$ are individually and jointly stationary (at least in the wide sense). Then, defining $t' = t - \tau$,

$$E(x_t y_{t-\tau}^*) = E(x_{t'+\tau} y_{t'}^*) = E^*(y_{t'} x_{t'+\tau}^*)$$

Since both processes are stationary, the indicated averages are invariant under a translation of the time origin. Therefore

$$R_{xy}(\tau) = R_{yx}^*(-\tau) \tag{4-52}$$

and

$$R_x(\tau) = R_x^*(-\tau) \tag{4-53}$$

Hence the autocorrelation function of a stationary real random process is an even function of its argument. The cross-correlation functions of two such random processes may or may not be.

Suppose, for the moment, that the real random processes with the sample functions $x(t)$ and $y(t)$ are not necessarily stationary. Since $x(t)$ and $y(t)$ are real functions of time, and since the square of a real func-

† Cf. Doob (II, Chap. II, Art. 8).

tion is nonnegative, it follows that

$$0 \le E\left[\left(\frac{x_1}{E^{\frac12}(x_1{}^2)} \pm \frac{y_2}{E^{\frac12}(y_2{}^2)}\right)^2\right]$$

On expanding, we can then obtain the result

$$|R_{xy}(t_1,t_2)| \le E^{\frac12}(x_1{}^2)E^{\frac12}(y_2{}^2) \tag{4-54a}$$

Similarly,

$$|R_x(t_1,t_2)| \le E^{\frac12}(x_1{}^2)E^{\frac12}(x_2{}^2) \tag{4-54b}$$

For stationary random processes these inequalities become

$$|R_{xy}(\tau)| \le R_x{}^{\frac12}(0)R_y{}^{\frac12}(0) \tag{4-55a}$$

and
$$|R_x(\tau)| \le R_x(0) \tag{4-55b}$$

Example 4-5.1. To get a better understanding of the autocorrelation function, let us now consider a specific example: that of the "random telegraph" wave pictured in Fig. 4-2. This wave may with equal probability assume at any instant of time either

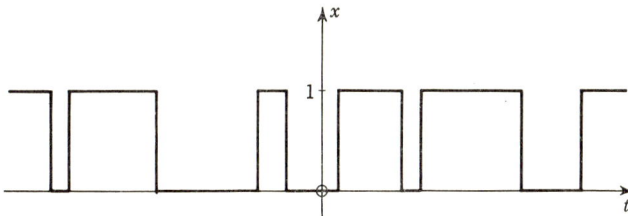

FIG. 4-2. A random telegraph wave.

of the values zero or one, and it makes independent random traversals from one value to the other. Therefore

$$m_x = 0 \cdot P(x = 0) + 1 \cdot P(x = 1) = P(x = 1) = \frac12 \tag{4-56}$$

We shall assume that the probability that k traversals occur in a time interval of length T is given by the Poisson distribution:†

$$P(k,T) = \frac{(aT)^k}{k!} \exp(-aT) \tag{4-57}$$

where a is the average number of traversals per unit time.

Now $x_1 = x_t$ and $x_2 = x_{t-\tau}$ are both discrete random variables having the two possible values zero and one. Therefore

$$R_x(\tau) = (0 \cdot 0)P(x_1 = 0,x_2 = 0) + (0 \cdot 1)P(x_1 = 0,x_2 = 1) + (1 \cdot 0)P(x_1 = 1,$$
$$x_2 = 0) + (1 \cdot 1)P(x_1 = 1,x_2 = 1) = P(x_1 = 1,x_2 = 1)$$

The probability that $x_1 = 1$ and that $x_2 = 1$ is the same as the probability that $x_1 = 1$ and that an even number of traversals occur between t and $t - \tau$. Hence

$$P(x_1 = 1,x_2 = 1) = P(x_1 = 1,k \text{ even}) = P(x_1 = 1)P(k \text{ even})$$

† A derivation of the Poisson distribution is given in Chap. 7, Art. **7-2**.

since the probability that k assumes any particular value is independent of the value of x_1. Since x_1 assumes the values zero and one with equal probability,

$$R_x(\tau) = \tfrac{1}{2}P(k \text{ even})$$

Then, using the Poisson distribution for k,

$$R_x(\tau) = \frac{\exp(-a|\tau|)}{2} \sum_{\substack{k=0 \\ (k \text{ even})}}^{\infty} \frac{(a|\tau|)^k}{k!}$$

We may evaluate the series as follows:

$$\sum_{\substack{k=0 \\ (k \text{ even})}}^{\infty} \frac{(a|\tau|)^k}{k!} = \frac{1}{2} \left[\sum_{k=0}^{\infty} \frac{(a|\tau|)^k}{k!} + \sum_{k=0}^{\infty} \frac{(-a|\tau|)^k}{k!} \right]$$

$$= \tfrac{1}{2} \left[\exp(a|\tau|) + \exp(-a|\tau|) \right]$$

Using this result in the last expression for the autocorrelation function, we get

$$R_x(\tau) = \frac{1}{4} \left[1 + \exp(-2a|\tau|) \right] \tag{4-58}$$

as an expression for the autocorrelation function of the random telegraph wave. A plot is given in Fig. 4-3.

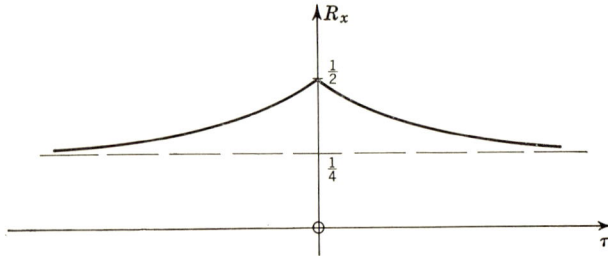

FIG. 4-3. Autocorrelation function of the random telegraph wave.

The limiting values of this autocorrelation function are

$$R_x(0) = E(x^2) = \tfrac{1}{2} \tag{4-59}$$

for $\tau = 0$, and

$$R_x(\infty) = \tfrac{1}{4} = m_x^2 \tag{4-60}$$

for $\tau = \infty$. The right-hand equality in Eq. (4-60) follows from Eq. (4-56).

4-6. Convergence

Let us digress from our discussion of averages per se long enough to derive a few results needed in the remainder of the book. In this article we shall derive an important inequality and then consider the problem of convergence of a sequence of random variables. In the following article we shall discuss integrals of random processes. In Art. 4-8 we shall return to the topic of averages and consider time averages and their relation to statistical averages.

The Chebyshev Inequality. Suppose that y is an arbitrary random variable with a probability density function $p(y)$ such that

$$E(|y|^2) = \int_{-\infty}^{+\infty} |y|^2 p(y) \, dy < \infty$$

Since both $|y|^2$ and $p(y)$ are nonnegative,

$$E(|y|^2) \geq \int_{(|y| \geq \epsilon)} |y|^2 p(y) \, dy$$

where ϵ is an arbitrary positive number. Further, since $|y|^2 \geq \epsilon^2$ at every point in the region of integration,

$$E(|y|^2) \geq \epsilon^2 \int_{(|y| \geq \epsilon)} p(y) \, dy$$

The integral here is the probability that $|y| \geq \epsilon$. Therefore, on solving for this probability, we obtain

$$P[|y| \geq \epsilon] \leq \frac{E(|y|^2)}{\epsilon^2} \tag{4-61}$$

In particular, if y is taken to be the difference between a random variable x and its mean m_x we get the *Chebyshev inequality*

$$P(|x - m_x| \geq \epsilon) \leq \frac{\sigma_x^2}{\epsilon^2} \tag{4-62}$$

Convergence in Mean and in Probability. On a number of subsequent occasions it will be of importance to know whether or not a sequence of random variables x_n converges to a random variable x, and if so, in what sense. We shall define here several kinds of convergence and show some of their interrelations.

Suppose that $E(|x_n|^2) < \infty$ for all n and that $E(|x|^2) < \infty$. The sequence of random variables x_n is said to *converge in the mean* to the random variable x if

$$\lim_{n \to \infty} E(|x_n - x|^2) = 0 \tag{4-63}$$

in which case we write

$$\text{l.i.m.}_{n \to \infty} x_n = x \tag{4-64}$$

where "l.i.m." denotes *limit in the mean*.

If the sequence of random variables x_n is such that

$$\lim_{n \to \infty} P[|x_n - x| \geq \epsilon] = 0 \tag{4-65}$$

for arbitrary $\epsilon > 0$, that sequence of random variables x_n is said to *converge in probability* to the random variable x, and we write

$$p \lim_{n \to \infty} x_n = x \tag{4-66}$$

It then follows from Eqs. (4-61), (4-63), and (4-65), on setting $y = x - x_n$ in Eq. (4-61), that if a sequence of random variables x_n converges in the mean to a random variable x, that sequence also converges in probability to x.

Convergence in Distribution. It can next be shown that, if a sequence of random variables x_n converges in probability to a random variable x, then the probability distribution functions of the x_n converge to the probability distribution function of x at every point of continuity of the latter. The random variables x_n are then said to *converge in distribution* to the random variable x.

Consider first the probability distribution function $P(x_n \leq X)$. The event $(x_n \leq X)$ may occur either when $(x \leq X + \epsilon)$ or when $(x > X + \epsilon)$. Since the latter events are mutually exclusive,

$$P(x_n \leq X) = P(x_n \leq X, x \leq X + \epsilon) + P(x_n \leq X, x > X + \epsilon)$$

Similarly,

$$P(x \leq X + \epsilon) = P(x \leq X + \epsilon, x_n \leq X) + P(x \leq X + \epsilon, x_n > X)$$

Subtraction then gives

$$\begin{aligned} P(x_n \leq X) - P(x \leq X + \epsilon) &= P(x_n \leq X, x > X + \epsilon) \\ &- P(x_n > X, x \leq X + \epsilon) \leq P(x_n \leq X, x > X + \epsilon) \end{aligned}$$

If $(x_n \leq X)$ and $(x > X + \epsilon)$, then $(|x_n - x| > \epsilon)$. This is one way in which the event $(|x_n - x| > \epsilon)$ can occur; hence

$$P(x_n \leq X, x > X + \epsilon) \leq P(|x_n - x| > \epsilon)$$

Therefore, on defining $\delta_n = P(|x_n - x| > \epsilon)$,

$$P(x_n \leq X) \leq P(x \leq X + \epsilon) + \delta_n$$

In a similar manner it can be shown that

$$P(x \leq X - \epsilon) - \delta_n \leq P(x_n \leq X)$$

and hence that

$$P(x \leq X - \epsilon) - \delta_n \leq P(x_n \leq X) \leq P(x \leq X + \epsilon) + \delta_n$$

Now $\delta_n \to 0$ as $n \to \infty$ since x_n converges in probability to x. Hence

$$P(x \leq X - \epsilon) \leq \lim_{n \to \infty} P(x_n \leq X) \leq P(x \leq X + \epsilon)$$

for every $\epsilon > 0$. It therefore follows that at every point of continuity of $P(x \leq X)$

$$\lim_{n \to \infty} P(x_n \leq X) = P(x \leq X) \tag{4-67}$$

which was to be shown.

4-7. Integrals of Random Processes

We shall use integrals of random processes frequently. They arise naturally in many places; for example, as we shall see in Chap. 9, if a system operates on an input function in a way which involves integration and we want to consider what happens when the input is a noise wave, we must deal with integrals of random processes. It is clear what the integration of a random process should mean in such an example; a particular noise input is a sample function of its random process, so the integration of the process should amount to ordinary integration of functions when a particular sample function is considered. For each sample function of the random process in the integrand, the value of the integral is a number; but this number is generally different for each sample function, and over the ensemble of sample functions which make up the random process, the integral takes on a whole set of values. The probability that these values lie in a certain range is equal to the probability of having a sample function in the integrand which would yield a value in this range. Thus we can assign a probability law in a natural way to the values of the integral; that is, the integral of the random process may be considered to be a random variable. In symbols, if we temporarily denote a random process by $x(s,t)$, where s is the probability variable taking values in a sample space S and t is time, we may write

$$y(s) = \int_a^b x(s,t)\ dt \tag{4-68}$$

For each s this may be regarded as an ordinary integral (of a sample function). Since s ranges over S, $y(s)$ is a function on S, i.e., a random variable.

It can, in fact, be shown that under reasonable conditions it is possible to treat the integral of a random process as the ensemble of integrals of its sample functions, so that it has the interpretation which we have just discussed. In particular,† under an appropriate measurability condition, if

$$\int_a^b E[|x(s,t)|]\ dt < \infty \tag{4-69}$$

then all the sample functions except a set of probability zero are absolutely integrable, and, in addition,

$$E\left[\int_a^b x(s,t)\ dt\right] = \int_a^b E[x(s,t)]\ dt \tag{4-70}$$

The limits a and b may be finite or infinite. The measurability condition is one which we may fairly assume to be satisfied in practice. Thus we are free to consider integrals of the sample functions of a random process

† Cf. Doob (II, Theorem 2.7).

whenever the mean value of the process is integrable. Further, we can actually calculate averages involving these integrals by use of Eq. (4-70).

It is customary to suppress the probability variable s, so Eq. (4-68) is written simply

$$y = \int_a^b x(t)\, dt \qquad (4\text{-}71)$$

and Eq. (4-70) is

$$E(y) = E\left[\int_a^b x(t)\, dt\right] = \int_a^b E(x_t)\, dt \qquad (4\text{-}72)$$

Often we have to deal with the integral of a random process weighted by some function. For example, suppose we have

$$y = \int_a^b h(t)x(t)\, dt \qquad (4\text{-}73)$$

where $h(t)$ is a real or complex-valued function of t. Then, according to the theorem referred to, this integral exists with probability one if

$$\int_a^b E[|h(t)x(t)|]\, dt = \int_a^b |h(t)|E[|x(t)|]\, dt < \infty \qquad (4\text{-}74)$$

If the weighting function is also a function of a parameter, say another real variable τ, then we have

$$y(\tau) = \int_a^b h(t,\tau)x(t)\, dt \qquad (4\text{-}75)$$

which defines $y(\tau)$ as a random process.

4-8. Time Averages

In Art. 4-1 we defined the statistical average of a random process with sample functions $x(t)$ to be the function of time $E(x_t)$. This is an average "across the process"; for each t_o it is the mean value of that random variable which describes the possible values which can be assumed by the sample functions at $t = t_o$. It seems natural to consider also averages "along the process," that is, time averages of individual sample functions, and to inquire what relation these averages have to the statistical average. Since in most cases of interest to us the sample functions extend in time to infinity, we define the *time average* $<x(t)>$ of a sample function $x(t)$ by

$$<x(t)> = \lim_{T\to\infty} \frac{1}{2T} \int_{-T}^{+T} x(t)\, dt \qquad (4\text{-}76)$$

if this limit exists. It should be noted that $<x(t)>$ may not exist (i.e., the finite average may not converge) for all or even for any of the sample functions, and that even if the time averages do exist they may be different for different sample functions. Because of these facts and because the statistical average of a nonstationary random process is not generally a constant function of time, we cannot hope to say in general that "time average equals statistical average." Yet in certain cases it seems as though time averages of almost all the sample functions ought to exist and be equal to a constant statistical average.

For example, suppose that the voltage of a noisy diode operating under fixed conditions in time is observed for a long period of time T and that its time average is approximated by measuring the voltage at times kT/K, $k = 1, \ldots, K$, where K is very large, and by averaging over this set of K values. Suppose also that the voltages of K other diodes, identical to the first and operating simultaneously under identical conditions, are measured at one instant and averaged. If in the first case T/K is sufficiently large that the noise voltages measured at times T/K apart have very little interdependence, it appears there is no physical mechanism causing the K measurements on the single diode to average either higher or lower than the simultaneous measurements on the K diodes, and we would expect all the time averages and statistical averages to be equal (in the limit). Note that in this example there is statistical stationarity.

Satisfactory statements of the connection between time and statistical averages have in fact been obtained for stationary random processes. The most important result is the *ergodic theorem*† of Birkoff, which states, primarily, that for a stationary random process, the limit $<x(t)>$ exists for every sample function except for a set of probability zero. Hence we are on safe ground when we speak of the time averages of the sample functions of a stationary process. The time average of a stationary random process is a random variable, but the ergodic theorem states in addition that under a certain condition called *ergodicity* it is equal with probability one to the constant statistical average $E(x_t)$. A precise description of the ergodicity condition is beyond the scope of this book; however, its essence is that each sample function must eventually take on nearly all the modes of behavior of each other sample function in order for a random process to be ergodic. One simple condition which implies ergodicity for the important class of gaussian random processes with continuous autocorrelation functions is‡

$$\int_{-\infty}^{+\infty} |R(\tau)|\, d\tau < +\infty \qquad (4\text{-}77)$$

where $R(\tau)$ is the autocorrelation function of the random process and the

---

† See Khinchin (I, Chaps. 2 and 3), Doob (II, Chaps. 10 and 11, Arts. 1 and 2), or Loève (I, Chap. 9).

‡ Condition (4-77) may be inferred from Doob (II, Chap. 11, Arts. 1 and 8). See also Grenander (I, Art. 5.10). (See Art. 8-4 of this book for the definition of a gaussian random process.)

process is assumed to be stationary with zero mean. In general, however, no condition on $R(\tau)$ can guarantee ergodicity.

It should be noted that if a random process is ergodic, any function of the process (satisfying certain measurability conditions) is also an ergodic random process and therefore has equal time and statistical averages. In particular, if the random process with sample functions $x(t)$ is ergodic, then

$$E(x_t^2) = \lim_{T \to \infty} \frac{1}{2T} \int_{-T}^{+T} x^2(t) \, dt$$

with probability one.

Convergence in Probability. Although we cannot prove the ergodic theorem here, we can prove a lesser statement relating time and statistical averages. Let $x(t)$ be a sample function of a wide-sense stationary random process with a finite mean m_x and a finite variance σ_x^2. The *finite-interval time average* $A_x(T)$ of a sample function is defined by the equation

$$A_x(T) = \frac{1}{2T} \int_{-T}^{+T} x(t) \, dt \tag{4-78}$$

Then if the limit exists,

$$\lim_{T \to \infty} A_x(T) = \,<x(t)> \tag{4-79}$$

For fixed T, as $x(t)$ is allowed to range over all possible sample functions, $A_x(T)$ defines a random variable which we shall also denote by $A_x(T)$. We shall show below that if the autocorrelation function of the random process is absolutely integrable about the square of the mean of the process, i.e., if

$$\int_{-\infty}^{+\infty} |R_x(\tau) - m_x^2| \, d\tau \,<\, \infty \tag{4-80}$$

then the random variables $A_x(T)$ converge in probability to m_x as $T \to \infty$.

First, by taking the statistical average of Eq. (4-78) and interchanging the order of averaging and integration, we find that

$$E[A_x(T)] = \frac{1}{2T} \int_{-T}^{+T} E(x_t) \, dt = m_x \tag{4-81}$$

since $E(x_t) = m_x$. As this result is independent of T, it follows that

$$E[<x(t)>] = m_x \tag{4-82}$$

i.e., the statistical average of the time averages of the sample functions of a wide-sense stationary random process is equal to the mean of that process.

The variance of $A_x(T)$ may now be obtained as follows: From Eqs.

(4-78) and (4-81), interchanging the order of averaging and integration, we obtain

$$\sigma^2[A_x(T)] = \frac{1}{4T^2} \int_{-T}^{+T} dt' \int_{-T}^{+T} E(x_{t'}x_t)\, dt - m_x^2$$

$$= \frac{1}{4T^2} \int_{-T}^{+T} dt' \int_{-T}^{+T} [R_x(t'-t) - m_x^2]\, dt$$

where $R_x(t'-t)$ is the autocorrelation function of the random process in question. Upon defining $\tau = t' - t$, this becomes

$$\sigma^2[A_x(T)] = \frac{1}{4T^2} \int_{-T}^{+T} dt' \int_{t'-T}^{t'+T} [R_x(\tau) - m_x^2]\, d\tau$$

The integration is now over the area of the parallelogram in Fig. 4-4. Since the autocorrelation function is an even functon of τ (and is not a

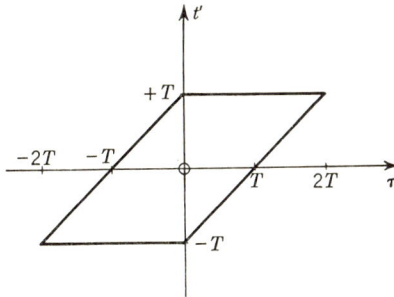

FIG. 4-4. Domain of integration.

function of t' here), the integral over the entire parallelogram is simply twice the integral over that part of the parallelogram which lies to the right of the t' axis. Thus

$$\sigma^2[A_x(T)] = \frac{1}{2T^2} \int_0^{2T} [R_x(\tau) - m_x^2]\, d\tau \int_{-T+\tau}^{+T} dt'$$

and hence $\quad \sigma^2[A_x(T)] = \frac{1}{T} \int_0^{2T} \left(1 - \frac{\tau}{2T}\right) [R_x(\tau) - m_x^2]\, d\tau \qquad (4\text{-}83)$

Since

$$\left| \int_0^{2T} \left(1 - \frac{\tau}{2T}\right) [R_x(\tau) - m_x^2]\, dt \right|$$
$$\leq \int_0^{2T} \left|1 - \frac{\tau}{2T}\right| |R_x(\tau) - m_x^2|\, d\tau \leq \int_0^{2T} |R_x(\tau) - m_x^2|\, d\tau$$

it therefore follows that if

$$\int_{-\infty}^{\infty} |R_x(\tau) - m_x{}^2| \, d\tau < \infty \tag{4-84}$$

then

$$\lim_{N \to \infty} \sigma^2[A_x(T)] = 0 \tag{4-85}$$

and hence, from Art. (4-6), that

$$\underset{T \to \infty}{\text{l.i.m.}} A_x(T) = m_x \tag{4-86}$$

and

$$p \lim_{T \to \infty} A_x(T) = m_x \tag{4-87}$$

It should be noted that this absolute integrability condition is a sufficient condition for the limit in Eq. (4-85), but it may not be a necessary one.

Time Correlation Functions. We shall define the *time autocorrelation function* $\Re_x(\tau)$ of a sample function of a random process by

$$\Re_x(\tau) = \lim_{T \to \infty} \frac{1}{2T} \int_{-T}^{+T} x(t + \tau)x^*(t) \, dt \tag{4-88}$$

If the random process is ergodic, the time autocorrelation function of a sample function of a random process equals the statistical autocorrelation function of that process

$$\Re_x(\tau) = R_x(\tau) \tag{4-89}$$

with probability one. Similarly, we shall define a *time cross-correlation function* $\Re_{xy}(\tau)$ of the sample functions $x(t)$ and $y(t)$ of two random processes by

$$\Re_{xy}(\tau) = \lim_{T \to \infty} \frac{1}{2T} \int_{-T}^{+T} x(t + \tau)y^*(t) \, dt \tag{4-90}$$

If then the given processes are jointly ergodic, it follows that

$$\Re_{xy}(\tau) = R_{xy}(\tau) \tag{4-91}$$

with probability one.

Example 4-8.1. The definition equation (4-88) may be applied to an arbitrary function of time as well as to a sample function of a random process so long as the indicated limit exists. For example, suppose that $x(t)$ is a complex periodic function of time such that its Fourier series converges. We may then write

$$x(t) = \sum_{n=-\infty}^{+\infty} \alpha(jn\omega_o) \exp(jn\omega_o t) \tag{4-92}$$

where n is an integer, ω_o is the fundamental angular frequency of $x(t)$, and the coefficients $\alpha(jn\omega_o)$ are complex constants given by†

$$\alpha(jn\omega_o) = \frac{\omega_o}{2\pi} \int_0^{2\pi/\omega_o} x(t) \exp(-jn\omega_o t) \, dt \tag{4-93}$$

† See for example, Guillemin (II, Chap. VII, Art. 7) or Churchill (I, Sec. 30).

It then follows from Eq. (4-88) that

$$\Re_x(\tau) = \lim_{T \to \infty} \frac{1}{2T} \int_{-T}^{+T} \left[\sum_{n=-\infty}^{+\infty} \alpha^*(jn\omega_o) \exp(-jn\omega_o t) \right]$$
$$\left\{ \sum_{m=-\infty}^{+\infty} \alpha(+jm\omega_o) \exp[+jm\omega_o(t+\tau)] \right\} dt$$

On interchanging orders of integration and summation and using the fact that

$$\lim_{T \to \infty} \frac{1}{2T} \int_{-T}^{+T} \exp[-j(n-m)\omega_o t] \, dt = \lim_{T \to \infty} \frac{\sin[(n-m)\omega_o T]}{(n-m)\omega_o T}$$
$$= \begin{cases} 1 & \text{when } m = n \\ 0 & \text{when } m \neq n \end{cases}$$

we get

$$\Re_x(\tau) = \sum_{n=-\infty}^{+\infty} \left| \alpha(jn\omega_o) \right|^2 \exp(+jn\omega_o\tau) \tag{4-94}$$

When $x(t)$ is a *real* function of time, Eq. (4-94) becomes

$$\Re_x(\tau) = \alpha^2(0) + 2 \sum_{n=1}^{+\infty} \left| \alpha(jn\omega_o) \right|^2 \cos(n\omega_o\tau) \tag{4-95}$$

Thus we see that the autocorrelation function of a periodic function of time is itself periodic, and that the periodicities of the autocorrelation function are identical with those of the original time function. Equation (4-95) further shows that all information as to the relative phases of the various components of $x(t)$ is missing (since only the magnitudes of the coefficients $\alpha(jn\omega_o)$ appear). It therefore follows that all periodic time functions which have the same Fourier coefficient magnitudes and periodicities also have the same autocorrelation function even though their Fourier-coefficient phases (and hence their actual time structures) may be different. This result points up the fact that the correspondence between time functions and auto-correlation functions is a "many-to-one" correspondence.

4-9. Problems

1. The random variable x has an exponential probability density function

$$p(x) = a \exp(-2a|x|)$$

a. Determine the mean and variance of x.
b. Determine the nth moment of x.

2. Prove that if x is a bounded random variable, then the nth moment of x exists for all n.
3. Let x be a random variable and c an arbitrary constant. Determine the value of c which makes $E[(x-c)^2]$ a minimum.

4. The random variable y is defined as the sum of N random variables x_n:

$$y = \sum_{n=1}^{N} x_n$$

Derive an expression for the variance of y when

  a. the random variables x_n are uncorrelated, and
  b. when they are correlated.

Express your results in terms of the variances $\sigma^2(x_n)$.

  5. The random variable x may assume only nonnegative integral values, and the probability that it assumes the specific value m is given by the Poisson probability distribution

$$P(x = m) = \frac{\lambda^m \exp(-\lambda)}{m!} \tag{4-96}$$

  a. Determine the mean and variance of x.
  b. Determine the characteristic function of x.

  6. The discrete random variables x and y each have a Poisson probability distribution (as in Prob. 5). If x and y are statistically independent random variables, determine the probability distribution of their sum $z = x + y$.

  7. Let x be a gaussian random variable with the probability density function

$$p(x) = \frac{\exp[-(x - a)^2/2b^2]}{\sqrt{2\pi}\, b} \tag{4-97}$$

This is also known as the *normal* probability density function.

  a. Derive an expression for the characteristic function of x.
  b. Using a, derive an expression for the mean and standard deviation of x, and express $p(x)$ in terms of these quantities and x.
  c. Using a and b, derive an expression for the nth moment of x for the particular case in which x has a zero mean.

  8. Let x and y be statistically independent gaussian random variables.

  a. Derive an expression for the joint probability density function $p(x,y)$.
  b. Derive an expression for the joint characteristic function for x and y.

  9. Let y be a random variable having the probability density function

$$p(y) = \begin{cases} \dfrac{y^{\frac{n-2}{2}} \exp(-y/2)}{2^{n/2}\Gamma(n/2)} & \text{for } y \geq 0 \\ 0 & \text{for } y < 0 \end{cases} \tag{4-98}$$

where n is a constant and Γ is the gamma function. Normally, this probability density function is written as a function of $y = \chi^2$ and is called the *chi-squared* probability density function of n *degrees* of freedom.

  a. Determine the mean of y.
  b. Determine the characteristic function of y.

10. Let x and y be statistically independent random variables, each having a chi-squared probability density function. Let the number of degrees of freedom be n for x and m for y. Derive an expression for the probability density function of their sum $z = x + y$.

11. Let x be a gaussian random variable with zero mean and standard deviation of unity. Consider the results of making N statistically independent measurements x_n of the random variable x. A new random variable y is then used to characterize this combined experiment and is defined by the equation

$$y = \sum_{n=1}^{N} x_n^2$$

Derive an expression for the probability density function of y.

12. Let z be a random variable such that

$$p(z) = \begin{cases} \dfrac{1}{2\pi} & \text{for } 0 \le z \le 2\pi \\ 0 & \text{otherwise} \end{cases}$$

and let the random variables x and y be defined by the equations

$$x = \sin z \quad \text{and} \quad y = \cos z$$

Show that the random variables x and y are linearly independent but not statistically independent.

13. Let the random variable y be defined as the sum of N statistically independent random variables x_n:

$$y = \sum_{n=1}^{N} x_n$$

where each of the random variables x_n has two possible values: unity, with probability p; and zero, with probability $q = 1 - p$.

a. Determine the characteristic function of x_n.
b. Determine the characteristic function of y.
c. Determine the probability distribution for y.

14. Let x and y be discrete random variables each having the two equally probable values $+1$ and -1.

a. Show that their joint probabilities are symmetric; i.e., show that

$$P(x = 1, y = 1) = P(x = -1, y = -1)$$
$$\text{and} \quad P(x = 1, y = -1) = P(x = -1, y = 1)$$

b. Derive expressions for the joint probabilities in terms of the correlation coefficient ρ_{xy}, relating x and y.

15. Consider the three stationary random processes having the sample functions $x(t)$, $y(t)$, and $z(t)$, respectively. Derive an expression for the autocorrelation function of the *sum* of these three processes under the assumptions

a. that the processes are correlated,
b. that the processes are all uncorrelated,
c. that the processes are uncorrelated and all have zero means.

16. Let $x(t)$ be a periodic square-wave function of time as shown in Fig. 4-5. Derive an expression for the autocorrelation function of $x(t)$.

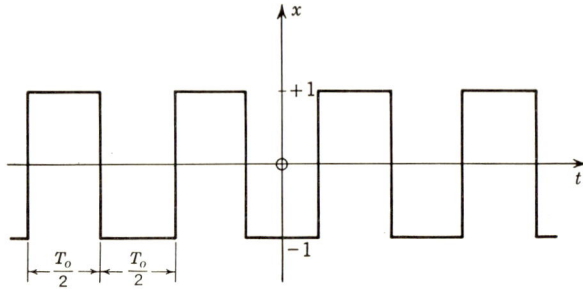

FIG. 4-5. Periodic square wave.

17. Consider an ensemble of pulsed sample functions defined by the equation

$$x(t) = \begin{cases} a_n & \text{for } n \leq t < n + 1 \\ 0 & \text{for } n - 1 \leq t < n \end{cases}$$

where n may assume odd integral values only and where the coefficients a_n are statistically independent random variables each having the two equally likely values zero and one. Determine and plot $<R_x(t,t+\tau)>$.

18. Repeat Prob. 17 for an ensemble in which the coefficients a_n are statistically independent random variables each having the two equally probable values plus one and minus one.

19. Let $x(t)$ be a bounded periodic function of time with a fundamental period of T_0 seconds. Prove that

$$\lim_{T \to \infty} \frac{1}{2T} \int_{-T}^{+T} x(t)\, dt = \frac{1}{T_o} \int_0^{T_0} x(t)\, dt$$

20. Consider the random process defined by the sample functions

$$y(t) = a \cos(t + \phi)$$

where a and ϕ are statistically independent random variables and where

$$p(\phi) = \begin{cases} \dfrac{1}{2\pi} & \text{for } 0 \leq \phi \leq 2\pi \\ 0 & \text{otherwise} \end{cases}$$

a. Derive an expression for the autocorrelation function of this process.
b. Show that $E(y_t) = <y(t)>$

21. Consider the random process defined by the sample functions

$$x(t) = y^2(t)$$

where the sample functions $y(t)$ are those defined in Prob. 20.

a. Show that if a is not constant with probability one,

$$E(x_t) \neq <x(t)>$$

b. Show that the integrability condition stated by Eq. (4-80) is not satisfied by this random process.

22. Consider the constant random process defined by the sample functions

$$x(t) = a$$

where a is a random variable having the two possible values $+1$ and -1 with the probabilities p and q respectively.

a. Determine the autocorrelation function of this process.
b. Show by the direct evaluation that

$$E(x_t) \neq <x(t)>$$

c. Show that the integrability condition stated by Eq. (4-80) is not satisfied by this random process.

23. Consider the stationary (strict-sense) random process whose autocorrelation function is $R_x(\tau)$. Assuming that the various following derivatives exist, show that

$$\frac{dR_x(\tau)}{d\tau} = E\left(x_t^* \frac{dx_{t+\tau}}{d\tau}\right)$$

24. Let $x(t)$ and $y(t)$ be periodic functions of time with incommensurable fundamental periods and zero average values. Show that their time cross-correlation function is zero.

25. Let $x(t)$ and $y(t)$ be different periodic functions of time having the same fundamental period. Show that their cross-correlation function contains only those harmonics which are present in both $x(t)$ and $y(t)$.

CHAPTER 5

SAMPLING

5-1. Introduction

So far we have assumed that the probability distributions of given random variables, or random processes, were known a priori and that the problem was to calculate various functions (e.g., moments, characteristic functions, and so forth) of those probability distributions. In practical situations, however, we often face a quite different problem. After making a set of measurements of a random variable, or random process, having unknown probability distribution functions, we may ask whether or not some of the statistical properties of that random variable or process can be determined from the results of those measurements. For example, we might wish to know if we can determine from measurements the value of the mean or variance.

Consider now the results of N performances of the basic experiment which defines a given random variable; or, alternatively, consider the results of measuring the value of a sample function of a given random process at N different instants of time. The set of N sample values so obtained specifies the result of such a combined experiment and hence a sample point in the N-dimensional sample space which characterizes the combined experiment. Since a given set of results determines only a single point in the joint sample space, we should not expect that it would completely determine the statistical properties of the random variable or random process in question. At best we can hope only to get an *estimate* of those properties. A basic problem, then, is to find a function, or *statistic*, of the results of the measurements which will enable us to obtain a good estimate of the particular statistical property in which we are interested.

In this chapter we shall introduce some of the principles of the theory of sampling by considering the problem of estimating the mean value of a random variable or process and by studying further the relation between relative frequency and probability. For a detailed discussion of the field of sampling theory the reader is referred to statistical literature.†

† See for example Cramér (I, Part III).

5-2. The Sample Mean

Since the problem of estimating the mean of a random variable is essentially identical to that of estimating the mean of a wide-sense stationary random process, we shall consider both problems at the same time. Let x be a real random variable with the finite mean m_x and the finite variance σ_x^2, and let $x(t)$ be the sample functions of a wide-sense stationary real random process which has the same finite mean m_x and finite variance σ_x^2. Suppose now that we make N measurements of either the random variable or a sample function of the random process. Let $x(n)$, where $n = 1, \ldots, N$, be the value of the nth sample of the random variable; let $x(t_n)$ be the value of a sample function of the random process at $t = t_n$; and let x_n be the random variable which describes the possible values that can be taken on by either $x(n)$ or $x(t_n)$. Note that x_n has the same statistical properties as x or as x_t. In particular,

$$E(x_n) = m_x \quad \text{and} \quad \sigma^2(x_n) = \sigma_x^2 \tag{5-1}$$

for all n.

We pointed out in Art. 4-1 that the mean of a random variable was a conceptual extension of the arithmetic average of a sample function. With this fact in mind, let us consider the *sample mean* $\mathfrak{M}$, defined to be the arithmetic average of the N random variables x_n,

$$\mathfrak{M} = \frac{1}{N} \sum_{n=1}^{N} x_n \tag{5-2}$$

as a statistic to be used in the estimation of the desired mean. The statistical average of this sample mean is

$$E(\mathfrak{M}) = \frac{1}{N} \sum_{n=1}^{N} E(x_n) = m_x \tag{5-3}$$

i.e., the statistical average of the sample mean is equal to the mean value of the random variable (or process) under study. A statistic whose expectation equals the quantity being estimated is said to be an *unbiased estimator;* the sample mean is therefore an unbiased estimator for the mean.

5-3. Convergence of the Sample Mean

The sample mean is not the only unbiased estimator of the desired mean; the *single-sample* random variable x_n is also unbiased, as shown by Eq. (5-1), and there are others as well. Whether or not any given statistic is good depends mainly on how likely it is that its sample value is close to the actual value of the statistical property in question. One statistic is better than another if its sample value is more likely to be close to the desired value. Accordingly, we shall now study the convergence properties of the sample mean.

The variance of the sample mean is, from Eqs. (5-2) and (5-3),

$$\sigma^2(\mathfrak{M}) = \left[\frac{1}{N^2} \sum_{n=1}^{N} \sum_{m=1}^{N} E(x_n x_m) \right] - m_x^2$$

where we have interchanged the operations of statistical averaging and summation. A somewhat more convenient form is

$$\sigma^2(\mathfrak{M}) = \frac{1}{N^2} \sum_{n=1}^{N} \sum_{m=1}^{N} [E(x_n x_m) - m_x^2] \qquad (5\text{-}4)$$

In order to proceed further, we must have data as to the correlation between the nth and mth samples, i.e., as to $E(x_n x_m)$. For sampling of a random process, this correlation is simply the autocorrelation function of that process:

$$E(x_n x_m) = R_x(t_n - t_m)$$

For sampling of a random variable, however, no statistical property of the given random variable alone gives us the required information; the correlation must be obtained in one way or another from a knowledge of the method of sampling.

To see the effect of the correlation between samples, suppose first that the samples are so highly correlated that, approximately,

$$E(x_n x_m) = E(x^2) \qquad (5\text{-}5)$$

for all values of n and m. In this case, Eq. (5-4) reduces approximately to

$$\sigma^2(\mathfrak{M}) = \frac{1}{N^2} \sum_{n=1}^{N} \sum_{m=1}^{N} \sigma_x^2 = \sigma_x^2 \qquad (5\text{-}6)$$

Therefore, when the samples are *highly correlated*, the variance of the sample mean is approximately equal to the variance of the random variable (or process) under study, whatever the number of samples. In this case then, a single measurement provides just as good (or bad) an estimate of the desired mean as does any other number of measurements. Whether a good estimate is obtained or not here depends strictly on the properties of the random variable (or process) under study.

Uncorrelated Samples. Next let us suppose that the various samples are uncorrelated. In this case,

$$E(x_n x_m) = E(x_n)E(x_m) = m_x^2 \qquad n \neq m \qquad (5\text{-}7)$$

and all the terms of the double summation in Eq. (5-4) vanish except those for which $n = m$. There are N of these terms, and for each the bracketed expression is simply σ_x^2. Equation (5-4) therefore reduces to

$$\sigma^2(\mathfrak{M}) = \frac{\sigma_x^2}{N} \qquad (5\text{-}8)$$

when the samples are *uncorrelated*. Such samples will be obtained, in particular, when the successive basic experiments are statistically independent.

If now we allow the number of measurements to increase without limit, it follows from Eq. (5-8) that

$$\lim_{N \to \infty} \sigma^2(\mathfrak{M}) = 0 \qquad (5\text{-}9)$$

and hence from Art. 4-6 that the sample mean converges in the mean and hence in probability to the desired mean

$$\underset{N \to \infty}{\text{l.i.m.}} \ \mathfrak{M} = m_x \qquad (5\text{-}10a)$$

and

$$p \lim_{N \to \infty} \mathfrak{M} = m_x \qquad (5\text{-}10b)$$

Such an estimator is said to be a *consistent estimator*.

According to Eq. (5-10), which is one form of the *law of large numbers*, the sample mean becomes a better and better estimator of the desired mean as the number of measurements increases without limit. By this we mean strictly what Eq. (5-10b) says: the probability that the sample mean differs from the desired mean by more than a fixed quantity becomes smaller and smaller as the number of measurements increases. However, it is perfectly possible that the value of the sample mean as obtained from *a particular sequence* of measurements continues to differ from the desired mean by *more* than some fixed amount, as $N \to \infty$.

An estimate of the expected measurement error for finite values of N may be obtained from the Chebyshev inequality, Eq. (4-62):

$$P[|\mathfrak{M} - \overline{\mathfrak{M}}| \geq \epsilon] \leq \frac{\sigma^2(\mathfrak{M})}{\epsilon^2}$$

Using the above expressions for the statistical average and variance of the sample mean, this inequality becomes

$$P[|\mathfrak{M} - m_x| \geq \epsilon] \leq \frac{\sigma_x^2}{N \epsilon^2} \qquad (5\text{-}11)$$

This result holds regardless of the particular probability distribution of the random variable or process being sampled and hence may well give only an excessively high upper bound to the expected error.

Periodic Sampling. For a random process, the case of greatest practical interest is probably that in which the sampling is done periodically in time. Here the N measurements are spaced equally throughout a total measurement interval of length T, with a sampling period $t_o = T/N$. If the measurement interval starts for convenience at $t = 0$, the nth sampling instant occurs at $t_n = nt_o$. Equation (5-4) then becomes

$$\sigma^2(\mathfrak{M}) = \frac{1}{N^2} \sum_{n=1}^{N} \sum_{m=1}^{N} \{R_x[(m-n)t_o] - m_x{}^2\}$$

Consider now the various terms of the double summation. It is perhaps helpful to imagine an N-by-N array of the elements (n,m) making up the double sum. The following can then be observed: First, each of the principal diagonal terms (n,n) is the same and is equal to $\sigma_x{}^2$. Next, each element (n,m) is equal to the element (m,n), since

$$R_x[(m-n)t_o] = R_x[(n-m)t_o]$$

that is, every term above the principal diagonal has an equal counterpart below that diagonal. Further, each element (n,m) is equal to an element $(n+j, m+j)$, as we are sampling a stationary random process, and hence all the $(N-k)$ elements in the kth diagonal above the principal diagonal are equal. It therefore follows that the variance of the sample mean is

$$\sigma^2(\mathfrak{M}) = \frac{\sigma_x{}^2}{N} + \frac{2}{N} \sum_{k=1}^{N-1} \left(1 - \frac{k}{N}\right) [R_x(kt_o) - m_x{}^2] \qquad (5\text{-}12)$$

when a wide-sense stationary random process is periodically sampled.

Let us next determine the effect of increasing the number of measurements without limit, at the same time keeping the total duration of the measurement interval fixed. Here it is convenient to rewrite Eq. (5-12) in the form

$$\sigma^2(\mathfrak{M}) = \frac{\sigma_x{}^2}{N} + \frac{2}{T} \sum_{k=1}^{N-1} \left(1 - \frac{kt_o}{T}\right) [R_x(kt_o) - m_x{}^2]t_o$$

If we now hold T fixed and allow $N \to \infty$ (and hence $t_o \to 0$) in such a way that $kt_o = \tau$, the summation becomes an integral and the term $\sigma_x{}^2/N$ vanishes. Therefore

$$\lim_{N \to \infty} \sigma^2(\mathfrak{M}) = \frac{2}{T} \int_0^T \left(1 - \frac{\tau}{T}\right) [R_x(\tau) - m_x{}^2] \, d\tau \qquad (5\text{-}13)$$

and we see that the variance of the sample mean here will *not* in general become vanishingly small as $N \to \infty$ unless T also increases without limit.

Since when $N \to \infty$ and T remains constant the sample mean becomes a finite time average, i.e., since

$$\lim_{N \to \infty} \mathfrak{M} = \frac{1}{T} \int_0^T x(t) \, dt$$

it is not surprising that Eq. (5-13) gives the same result as that previously obtained in the study of the statistical properties of time averages (Eq. (4-83)).

5-4. The Central Limit Theorem

For statistically independent samples, *the probability distribution of the sample mean tends to become gaussian as the number of statistically independent samples is increased without limit,* regardless of the probability distribution of the random variable or process being sampled as long as it has a finite mean and a finite variance. This result is known as the equal-components case of the *central limit theorem.*

We shall now prove this result by making use of the fact that the characteristic function of a sum of statistically independent random variables is the product of their respective characteristic functions. Let y be the normalized sample mean:

$$y = \frac{\mathfrak{M} - m_x}{\sigma(\mathfrak{M})} \tag{5-14}$$

Since the samples are statistically independent, it follows from Eq. (5-8) and the definition of the sample mean that

$$y = \frac{\left(\dfrac{1}{N} \displaystyle\sum_{n=1}^{N} x_n\right) - m_x}{\sigma_x / N^{1/2}} = \frac{1}{N^{1/2}} \sum_{n=1}^{N} \left(\frac{x_n - m_x}{\sigma_x}\right)$$

If then we define ξ_n to be the normalized random variable $(x_n - m_x)/\sigma_x$, we may write

$$y = \frac{1}{N^{1/2}} \sum_{n=1}^{N} \xi_n \tag{5-15}$$

The characteristic function of y is therefore

$$M_y(jv) = E\left[\exp\left(\frac{jv}{N^{1/2}} \sum_{n=1}^{N} \xi_n\right)\right]$$

$$= E\left[\prod_{n=1}^{N} \exp\left(\frac{jv\xi_n}{N^{1/2}}\right)\right]$$

Since the samples, and hence the normalized samples, are statistically independent random variables, all having the same probability distribution (that of the random variable or process being sampled), it follows from Eq. (4-33) that

$$M_y(jv) = E^N\left[\exp\left(\frac{jv\xi}{N^{1/2}}\right)\right] = M_\xi^N\left(\frac{jv}{N^{1/2}}\right) \tag{5-16}$$

where $\xi = (x - m_x)/\sigma_x$.

We next wish to determine the behavior of the characteristic function of the normalized sample mean as the number of samples increases without limit. To this end, let us expand the characteristic function of ξ in a Taylor series with a remainder.† Thus, since ξ has a zero mean and a unit variance

$$M_\xi\left(\frac{jv}{N^{1/2}}\right) = 1 - \frac{v^2}{2N} + A\left(\frac{v}{N^{1/2}}\right) \tag{5-17}$$

where the remainder term $A(v/N^{1/2})$ is given by‡

$$A\left(\frac{v}{N^{1/2}}\right) = \frac{M_\xi''(\theta v/N^{1/2}) - M_\xi''(0)}{2}\left(\frac{v^2}{N}\right) \tag{5-18}$$

in which M_ξ'' is the second derivative of the characteristic function of ξ and $0 \leq \theta \leq 1$. The existence of the second moment of ξ implies the existence of a continuous second derivative§ of M_ξ, hence

$$\lim_{(v/N^{1/2})\to 0} \left(\frac{N}{v^2}\right) A\left(\frac{v}{N^{1/2}}\right) = 0 \tag{5-19}$$

in particular, as $N \to \infty$ for fixed v. Taking the logarithm of $M_y(jv)$ and using Eq. (5-17), we then get

$$\log M_y(jv) = N \log\left[1 - \frac{v^2}{2N} + A\left(\frac{v}{N^{1/2}}\right)\right]$$

The logarithm on the right side of this equation can be expanded by using the expression

$$\log(1 + z) = z + B(z) \tag{5-20}$$

in which the remainder term is¶

$$B(z) = -\int_0^z \frac{t}{1+t}\,dt \tag{5-21}$$

---

† Cf. Courant (I, Chap. VI, Art. 2).
‡ *Ibid.*, Art. 2, Sec. 3.
§ Cf. Cramér (I, Art. 10.1).
¶ Cf. Courant (I, Chap. VI, Art. 1).

Thus we obtain

$$\log M_y(jv) = -\frac{v^2}{2} + NA\left(\frac{v}{N^{1/2}}\right) + NB\left[\frac{-v^2 + 2NA(v/N^{1/2})}{2N}\right] \quad (5\text{-}22)$$

It follows from Eq. (5-19) that the second term on the right side of Eq. (5-22) vanishes as $N \to \infty$. The third term also vanishes as $N \to \infty$ since, from Eq. (5-21),

$$\left|\frac{B(z)}{z}\right| \leq \frac{1}{z}\int_0^z t\, dt = \frac{z}{2} \to 0 \qquad \text{as } z \to 0$$

Therefore

$$\lim_{N \to \infty} \log M_y(jv) = -\frac{v^2}{2}$$

and hence

$$\lim_{N \to \infty} M_y(jv) = \exp\left(-\frac{v^2}{2}\right) \quad (5\text{-}23)$$

The right side of Eq. (5-23) is the characteristic function of a gaussian random variable which has a zero mean and a variance of unity. Since that limit function is continuous at $v = 0$, it therefore follows that[†]

$$\lim_{N \to \infty} p(y) = \frac{1}{(2\pi)^{1/2}} \exp\left(-\frac{y^2}{2}\right) \quad (5\text{-}24)$$

Thus the limiting form of the probability distribution of the sample mean is gaussian when the individual samples are statistically independent and have a finite mean and a finite variance.

Comments. The tendency of the probability distribution of a sum of random variables to become gaussian as the number of components increases without limit can be shown to hold for much less restrictive assumptions.[‡] For example, if the components x_n having means m_n and variances σ_n^2 are statistically independent and if

$$\lim_{N \to \infty} \frac{\sum_{n=1}^{N} E(|x_n - m_n|^{2+\delta})}{\left(\sum_{n=1}^{N} \sigma_n^2\right)^{1+\delta/2}} = 0 \quad (5\text{-}25)$$

where $\delta > 0$, then it can be shown that the normalized sum random variable has a gaussian probability distribution in the limit as $N \to \infty$. This result is known as *Liapounoff's theorem*.[§] Even the requirement of statistical independence of the components can be waived under certain

[†] See for example Cramér (I, Art. 10.4).
[‡] See for example Gnedenko and Kolmogorov (I, Chap. 4).
[§] See Uspensky (I, Chap. 14, Secs. 2, 3, and 4) or Loève (I, Sec. 20).

conditions, and the limiting distribution of the sum will still be gaussian.† It should be emphasized, however, that the limiting distribution of a sum of random variables is *not always* gaussian and each individual case should be investigated to see whether or not such a theorem applies.

The above statements refer to behavior in the limit as $N \to \infty$. It should be realized that when N is finite, even though apparently large, the gaussian distribution may well give a *poor* approximation to the tails of the probability distribution of the sum, even though the limiting form of the sum distribution is in fact gaussian.‡ This would be true, for example, for the case of a sum of statistically independent Poisson random variables, since the probability distribution of such a sum is also Poisson for any N.

5-5. Relative Frequency

An important special case of sampling is that in which the random variable or process being sampled can assume only the values unity and zero with the probabilities p and $(1 - p)$ respectively. Consider, for example, the determination of the relative frequency of occurrence of some given event (A). Here we define a random variable x_n to have the value unity when the event (A) occurs in the nth sampling and zero when it does not. It therefore follows from the definition of the sample mean, Eq. (5-2), that

$$\mathfrak{M} = \frac{n(A)}{N} \tag{5-26}$$

where $n(A)$ is the random variable which specifies the number of times the event (A) occurs in N samples. In this case, then the sample mean is the random variable corresponding to the relative frequency of occurrence of the event (A) in N samples.

Let p be the probability of occurrence of event (A), then

$$E(x_n) = p \tag{5-27}$$

and hence, from Eq. (5-3),

$$E(\mathfrak{M}) = p \tag{5-28}$$

Thus *the statistical average of the relative frequency of occurrence of an event is equal to the probability of occurrence of that event.*

Uncorrelated Samples. Since

$$\sigma^2(x_n) = E(x_n{}^2) - E^2(x_n)$$
$$= p - p^2 \tag{5-29}$$

---

† See for example, Loève (I, Sec. 28), Uspensky (I, Chap. 14, Sec. 8), or Lévy (I, Chap. 8).

‡ This point is discussed at length in Fry (I, Arts. 82 and 89).

it follows from Eq. (5-8) that the variance of the relative frequency becomes

$$\sigma^2(\mathfrak{M}) = \frac{p(1-p)}{N} \tag{5-30}$$

when the various samples are uncorrelated. An upper bound to this variance may be found by taking the derivative with respect to p, equating to zero, and solving for p. In this way the maximum value of the variance is found to occur when $p = \frac{1}{2}$ and is

$$\sigma^2(\mathfrak{M})_{\max} = \frac{1}{4N} \tag{5-31}$$

If now we substitute these various results into the Chebyshev inequality, Eq. (5-11), we get

$$P\left[\left|\frac{n(A)}{N} - p\right| \geq \epsilon\right] \leq \frac{p(1-p)}{N\epsilon^2} \leq \frac{1}{4N\epsilon^2} \tag{5-32}$$

Since the variance of the relative frequency tends to zero as $N \to \infty$, it follows from Art. 4-6 that *the relative frequency of occurrence of an event converges in probability to the probability of occurrence of that event as the number of uncorrelated samples increases without limit.* This result, known as the Bernoulli theorem, justifies in essence the relative-frequency approach to probability theory introduced in Chap. 2.

Independent Samples. The determination of the probability distribution of the relative frequency becomes simple when the various samples are statistically independent. The situation then becomes the same as that of the coin tossings discussed in Example 2-6.2, and the successive samplings therefore form Bernoulli trials. By paralleling the development of Eq. (2-21), it can be shown that the probability distribution of the relative frequency is the binomial distribution. Thus

$$P\left[\frac{n(A)}{N} = \frac{n}{N}\right] = \binom{N}{n} p^n (1-p)^{N-n} \tag{5-33}$$

when the samples are statistically independent.

5-6. Problems

1. The discrete random variable x has the Poisson probability distribution

$$P(x = m) = \frac{\lambda^m \exp(-\lambda)}{m!} \qquad m = 0, 1, 2, \ldots$$

Let $\mathfrak{M}$ be the sample mean formed from N statistically independent samplings of x.

    a. Determine the mean and variance of the sample mean.
    b. Determine the probability distribution of the sample mean.
    c. Plot the results of **b** for $N = 3$ and $N = 10$.

2. The random variable x has a Cauchy probability density function

$$p(x) = \frac{1}{\pi} \left(\frac{1}{1 + x^2} \right)$$

Let $\mathfrak{M}$ be the sample mean formed from N statistically independent samplings of x.

a. Determine the mean and variance of the sample mean.
b. Determine the probability density function of the sample mean.
c. How does this case fit into the discussion of Art. 5-3 and 5-4?

3. Consider the stationary random process defined by the ensemble of sample functions $x(t)$. Let the autocorrelation function of this process be

$$R_x(\tau) = \begin{cases} m_x{}^2 + \sigma_x{}^2 \left(1 - \dfrac{\tau}{\tau_o} \right) & \text{for } 0 \leq |\tau| \leq \tau_o \\ m_x{}^2 & \text{for } \tau_o \leq |\tau| \end{cases}$$

Suppose that the sample mean of this process is obtained by integrating a sample function over a measurement interval of length $T \geq \tau_o$:

$$\mathfrak{M} = \frac{1}{T} \int_0^T x(t) \, dt$$

Determine the variance of this sample mean.

4. The probability P_N that a measured relative frequency $n(A)/N$ differs by more than ϵ from its corresponding probability $P(A)$ is given by Eq. (5-32).

a. Determine, as a function of $P(A)$, a bound on the required number of measurements N such that

$$P \left[\left| \frac{n(A)}{N} - P(A) \right| \geq \alpha P(A) \right] \leq P_N$$

b. Plot N of part **a** as a function of $P(A)$, assuming $\alpha = 0.05$ and $P_N = 0.1$, for values of $P(A)$ in the interval $(0.01, 0.99)$.

5.† Show that the binomial distribution

$$P(n) = \binom{N}{n} p^n (1 - p)^{N-n}$$

tends to the gaussian distribution

$$P(n) = \frac{1}{\sqrt{2\pi} \, \sigma(n)} \exp \left[-\frac{(n - m_n)^2}{2\sigma^2(n)} \right]$$

where $m_n = E(n)$, when $N \to \infty$ and p remains constant.

6.† Show that the binomial distribution of Prob. 5 tends to the Poisson distribution

$$P(n) = \frac{m_n{}^n}{n!} \exp(-m_n)$$

as $N \to \infty$ and $p \to 0$ such that Np remains constant.

7.† Show that the Poisson distribution of Prob. 6 tends to the gaussian distribution of Prob. 5 as $m_n \to \infty$.

† See Feller (I, Chaps. 6 and 7) and Fry (I, Chap. 8).

CHAPTER 6

SPECTRAL ANALYSIS

6-1. Introduction

The resolution of a known electrical signal into components of different frequencies by means of a Fourier series or Fourier transform is such a useful technique in circuit analysis that one naturally asks if a similar process can be applied to noise waves. To a large extent it can; that is, noise waves are represented by random processes, and at least for stationary (or wide-sense stationary) processes a quite satisfactory theory of Fourier analysis has been developed. In this chapter we shall discuss some parts of this theory.

Fourier Series. It is assumed that the reader has some knowledge of Fourier series and integrals,† so we shall state here only a few basic facts as a reminder and for reference. If $x(t)$ is a real or complex-valued periodic function of a real variable t (e.g., time, if $x(t)$ is to represent a signal) and if $x(t)$ is *absolutely integrable* over a period T, i.e., if

$$\int_0^T |x(t)|\, dt < \infty \tag{6-1}$$

then $x(t)$ has associated with it a Fourier series

$$\hat{x}(t) = \sum_{n=-\infty}^{\infty} a_n e^{jn\omega_o t} \qquad \omega_o = \frac{2\pi}{T} \tag{6-2}$$

where a_n is called the nth *Fourier coefficient* and is given by

$$a_n = \frac{1}{T} \int_0^T x(t) e^{-jn\omega_o t}\, dt \tag{6-3}$$

If any of various conditions are satisfied by $x(t)$, then the sum on the right side of Eq. (6-2) converges in some sense to $x(t)$. In particular, if $x(t)$ is of bounded variation‡ in the interval $0 \le t \le T$, then that sum converges to $x(t)$ everywhere that $x(t)$ is continuous. Since for $x(t)$ to

---

† For example, see Churchill (I, Chaps. III, IV, and V), Guillemin (II, Chap. VII), or Titchmarsh (I, Chap. 13).

‡ Titchmarsh (I, Art. 13.232).

be of bounded variation means roughly that the total rise plus the total fall of the function $x(t)$ is finite in the interval $0 \le t \le T$, this condition on $x(t)$ is usually satisfied in practical problems.

An alternative condition on $x(t)$ is that it be of *integrable square* for $0 \le t \le T$, i.e., that

$$\int_0^T |x(t)|^2 \, dt < \infty \tag{6-4}$$

The sum on the right side of Eq. (6-2) then converges to $x(t)$ in the sense that

$$\lim_{N \to \infty} \int_0^T \left| x(t) - \sum_{n=-N}^{N} a_n e^{jn\omega_0 t} \right|^2 dt = 0 \tag{6-5}$$

This kind of convergence, which means that the mean-square error is going to zero, is called *convergence-in-the-mean*† and is written

$$x(t) = \text{l.i.m.} \sum_{n=-N}^{N} a_n e^{jn\omega_0 t} \tag{6-6}$$

Since the condition of Eq. (6-4) is the requirement that the energy in one period of $x(t)$ is finite, we almost always are dealing with the case where it is satisfied. In fact, we shall assume henceforth that every function we introduce satisfies Eq. (6-4) for every finite number T, unless we state otherwise, whether T is the period of a periodic function or not. For convenience we shall write simply $x(t) = \hat{x}(t)$ rather than $x(t)$ equals a limit-in-the-mean, as in Eq. (6-6).

When the condition of Eq. (6-4) is satisfied, the result is Parseval's theorem,‡ which states that the average energy in the signal is equal to the sum of the average energies in each frequency component; i.e.,

$$\sum_{-\infty}^{\infty} |a_n|^2 = \frac{1}{T} \int_0^T |x^2(t)| \, dt \tag{6-7}$$

Fourier Transforms. The *Fourier transform* $x(t)$ of a function $X(f)$ is defined by

$$x(t) = \int_{-\infty}^{\infty} X(f) e^{j\omega t} \, df \qquad \omega = 2\pi f \tag{6-8}$$

† We have already used in Art. 4-6 the term "convergence-in-the-mean" in connection with convergence of random variables considered as functions on a sample space. It will be observed that the definition here is the same as the previous definition except that the averaging, or integration, is here with respect to the real variable t, whereas there it was with respect to the sample-space variable. The use of the term in both these ways is standard.

‡ Titchmarsh (I, Art. 13.63).

if the integral exists in some sense. The *inverse Fourier transform* of $x(t)$ is defined by

$$\hat{X}(f) = \int_{-\infty}^{\infty} x(t)e^{-j\omega t}\,dt \qquad \omega = 2\pi f \tag{6-9}$$

if it exists. One of the most basic results of Fourier-transform theory is Plancherel's theorem,† which states that if $X(f)$ is of integrable-square on the whole line, $-\infty < f < \infty$; i.e., if

$$\int_{-\infty}^{\infty} |X(f)|^2\,df < \infty \tag{6-10}$$

then there is a function $x(t)$ which is also of integrable-square on the whole line which is related to $X(f)$ by the equations

$$x(t) = \underset{A\to\infty}{\text{l.i.m.}} \int_{-A}^{A} X(f)e^{j\omega t}\,df \tag{6-11a}$$

and

$$X(f) = \underset{A\to\infty}{\text{l.i.m.}} \int_{-A}^{A} x(t)e^{-j\omega t}\,dt \tag{6-11b}$$

Under the condition of Eq. (6-10), an analogue of Parseval's theorem holds; i.e.,

$$\int_{-\infty}^{\infty} |X(f)|^2\,df = \int_{-\infty}^{\infty} |x(t)|^2\,dt \tag{6-12}$$

We shall speak of $x(t)$ and $X(f)$ as a *Fourier-transform pair*. If in addition $X(f)$ is absolutely integrable, $x(t)$ is given by Eq. (6-8); and if $x(t)$ is absolutely integrable, $X(f) = \hat{X}(f)$ is given by Eq. (6-9). As with Fourier series, we shall commonly write Fourier transforms as in Eqs. (6-8) and (6-9), even in some cases where the correct mathematical terminology requires a limit-in-the-mean.

6-2. The Spectral Density of a Periodic Function

Our chief concern in this chapter is with the function called the *power spectral density* (or simply *spectral density*), which gives the distribution in frequency of the power of a signal or a noise, and with its relation to the autocorrelation function. The situation where one considers a signal which is a known function of time is different from that where one considers a signal or noise which is represented as a sample function of a random process. Although there are strong analogies, the two must be treated separately. We shall begin by discussing the very simplest case, the power spectral density of a single periodic function.

Let $x(t)$ be a periodic function of period T, which has finite "energy" per period, i.e., which satisfies the condition of Eq. (6-4). Then by Parseval's theorem the time average of its energy, i.e., its power, is equal

† Titchmarsh (II, Art. 3.1).

to a sum of terms, each term associated with one frequency in its Fourier-series expansion.   Each term, in fact, can be interpreted as the time average of the energy of one particular component frequency.   Thus we are led to define a function $\mathbb{S}(f)$, the *power spectral density*, by

$$\mathbb{S}(f) = \sum_{n=-\infty}^{\infty} |a_n|^2 \, \delta(f - nf_o) \qquad f_o = \frac{1}{T} \qquad (6\text{-}13)$$

Hence $\mathbb{S}(f)$ consists of a series of impulses at the component frequencies of $x(t)$, each impulse having a strength equal to the power in that component frequency, and clearly is a measure of the distribution of the power in $x(t)$.   The total power in $x(t)$ is

$$\int_{-\infty}^{\infty} \mathbb{S}(f) \, df = \sum_{n=-\infty}^{\infty} |a_n|^2 = \frac{1}{T} \int_0^T |x^2(t)| \, dt \qquad (6\text{-}14)$$

The *power spectrum* $G(f)$ of $x(t)$ is defined to be

$$G(f) = \int_{-\infty}^{f} \mathbb{S}(\eta) \, d\eta \qquad (6\text{-}15)$$

The power spectrum of a periodic function is a staircase function having jumps at the harmonic frequencies.

It follows from Eq. (6-13) that all information about the phases of the various frequency components which is present in the Fourier-series expansion of $x(t)$ is lost in $\mathbb{S}(f)$, for two functions $x(t)$ and $x'(t)$ with Fourier coefficients of the same magnitude but different phases have the same spectral density.   Next, we note that if $x(t)$ is real, its Fourier coefficients a_n and a_{-n} are complex conjugates and hence $|a_n|^2 = |a_{-n}|^2$. Thus, if $x(t)$ is real, $\mathbb{S}(f)$, as given by Eq. (6-13), is an even function of frequency.   Finally, $\mathbb{S}(f)$ is nonnegative.

It was shown in Example 4-8.1 that the correlation function for the Fourier series given by Eq. (6-2) is

$$\mathfrak{R}(\tau) = \sum_{n=-\infty}^{\infty} |a_n|^2 e^{+jn\omega_0\tau} \qquad (6\text{-}16)$$

If we take the Fourier transform of $\mathfrak{R}(\tau)$, we get

$$\int_{-\infty}^{+\infty} \mathfrak{R}(\tau)e^{-j\omega\tau} \, d\tau = \int_{-\infty}^{\infty} \sum_{-\infty}^{\infty} |a_n|^2 e^{+jn\omega_0\tau} \, e^{-j\omega\tau} \, d\tau$$

$$= \sum_{n=-\infty}^{\infty} |a_n|^2 \, \delta(f - nf_o) = \mathbb{S}(f) \qquad (6\text{-}17)$$

Thus we have the fact that for a periodic function $x(t)$, its power spectral density and its correlation function are a Fourier-transform pair. Complete information about the frequency distribution of average energy of $x(t)$ is therefore contained in its correlation function.

6-3. Spectral Density of a Periodic Random Process

A wide-sense stationary random process with sample functions $x(t)$ is said to be *periodic* with period T if its correlation function $R(\tau)$ is periodic with period T. From this it follows (see Prob. 2) that the random variables x_t and x_{t+T} are equal with probability one for any choice of t. If all the sample functions of a random process are periodic, or even if all except a set which occurs with probability zero are periodic, the process is periodic in the sense defined above.

Let a wide-sense stationary periodic process of period T have sample functions $x(t)$. Suppose, for convenience, that the mean value $E(x_t)$ is zero. Then if the sample functions were periodic and could be expanded in Fourier series we would have

$$x(t) = \sum_{n=-\infty}^{\infty} x_n e^{jn\omega_0 t} \qquad \omega_0 = \frac{2\pi}{T} \qquad (6\text{-}18)$$

where

$$x_n = \frac{1}{T}\int_0^T x(t)e^{-jn\omega_0 t}\, dt \qquad (6\text{-}19)$$

For different sample functions, Eq. (6-19) yields in general different values for x_n; that is, as explained in Art. 4-7, if the whole ensemble of sample functions is considered, Eq. (6-19) defines x_n as a random variable. It follows from the theorem quoted in Art. 4-7 that the integral in Eq. (6-19) exists with probability one, and it can be shown that something like Eq. (6-18) is true, precisely,

$$x_t = \underset{N\to\infty}{\text{l.i.m.}} \sum_{n=-N}^{N} x_n e^{jn\omega_0 t} \qquad (6\text{-}20)$$

where l.i.m. is in the statistical sense, as in Art. 4-6. Actually, Eq. (6-20) is a special case of a result demonstrated in the next section. These expressions are like Eqs. (6-2) and (6-3), but now the *Fourier coefficients* x_n are random variables instead of numbers.

For most calculations which are made with random Fourier series, i.e., series like that in Eq. (6-18), it is extremely convenient if the series possesses a double orthogonality, that is, if not only the time functions of the nth and mth terms $n \neq m$ are orthogonal but in addition $E(x_n x_m^*) = 0$. We shall now show that the requirement that the random process be periodic guarantees that x_n and x_m, $n \neq m$, are uncorrelated.   Using Eq. (6-19) we get

$$E(x_n x_m^*) = \frac{1}{T^2} E\left[\int_0^T \int_0^T x(t)x^*(s)e^{-jn\omega_0 t}e^{jm\omega_0 s}\,ds\,dt\right]$$

$$= \frac{1}{T^2} \int_0^T \int_0^T R(t - s)e^{j\omega_0(ms-nt)}\,ds\,dt \tag{6-21}$$

where, since $R(\tau)$ is periodic, it can be written

$$R(\tau) = \sum_{n=-\infty}^{\infty} b_n e^{jn\omega_0 \tau} \tag{6-22}$$

for all τ.   Substituting from Eq. (6-22) in Eq. (6-21), where it will be noted the argument of $R(\tau)$ ranges from $-T$ to T, which is two periods of $R(\tau)$, then gives

$$E(x_n x_m^*) = \frac{1}{T^2} \int_0^T \int_0^T \sum_{k=-\infty}^{\infty} b_k e^{jk\omega_0(t-s)} e^{j\omega_0(ms-nt)}\,ds\,dt$$

$$= \frac{1}{T^2} \sum_{k=-\infty}^{\infty} b_k \int_0^T e^{j\omega_0(m-k)s}\,ds \int_0^T e^{j\omega_0(k-n)t}\,dt$$

$$= b_n \qquad \text{if } n = m$$
$$= 0 \qquad \text{if } n \neq m \tag{6-23}$$

Equation (6-23) states not only that x_n and x_m are uncorrelated for $n \neq m$ but also, in conjunction with Eq. (6-22), that the nth Fourier coefficient of the correlation function $R(\tau)$ is equal to the variance of the nth (random) Fourier coefficient of $x(t)$.   This is the statistical analogue of the fact that for a periodic function $x(t)$, the nth Fourier coefficient of its correlation function $R(\tau)$ is equal to the absolute value of the square of the nth Fourier coefficient of $x(t)$.

Now the energy in a noise signal represented by a random process is most reasonably defined to be the statistical average of the energies in the sample functions.   Thus we have for the energy in a noise signal over a time interval T

$$E\left(\int_0^T |x(t)|^2\,dt\right) = E\left[\int_0^T \sum_{n=-\infty}^{\infty} x_n e^{jn\omega_0 t} \sum_{k=-\infty}^{\infty} x_k^* e^{-jk\omega_0 t}\,dt\right]$$

$$= E\left[\sum_{n=-\infty}^{\infty} |x_n|^2 T\right] = T \sum_{n=-\infty}^{\infty} b_n \tag{6-24}$$

The time average of the energy in the signal is $\sum_{n=-\infty}^{\infty} b_n$.   By analogy with our definition in Art. 6-2, we define the *power spectral density* $S(f)$ for a periodic stationary process to be

$$S(f) = \sum_{n=-\infty}^{\infty} b_n \delta(f - nf_o) \qquad f_o = \frac{1}{T} \tag{6-25}$$

where

$$b_n = \frac{1}{T} \int_0^T R(t) e^{-in\omega_o t} \, dt$$

Here $b_n = E(|x_n|^2)$ is the statistical average of the power in the components of the sample functions of frequency $f = n/T$. The Fourier transform of the correlation function is

$$\int_{-\infty}^{+\infty} R(\tau) e^{-j\omega\tau} \, d\tau = \int_{-\infty}^{\infty} \sum_{n=-\infty}^{\infty} b_n e^{+in\omega_o\tau} e^{-j\omega\tau} \, d\tau$$

$$= \sum_{n=-\infty}^{\infty} b_n \delta(f - nf_o) = S(f) \tag{6-26}$$

Hence, as in Art. 6-2, we have observed that the power spectral density is the Fourier transform of the autocorrelation function, even though the correlation function of Art. 6-2 was a time average and the correlation function of this section is a statistical average. We shall, in fact, *define* the spectral density as the transform of the correlation function for a general wide-sense stationary process, and the result of this section serves as one heuristic justification for such a definition. Before discussing this, however, we shall consider in the next section the problem of representing a general, nonperiodic random process over a finite interval by a series of orthogonal functions with random coefficients.

6-4. Orthogonal Series Expansion of a Random Process

Trigonometrical Series Expansions. A nonperiodic function $f(t)$ of a real variable may be expanded, of course, in a Fourier series of complex exponentials, or sines and cosines, which represents $f(t)$ over any specified finite interval $a \leq t \leq b$. Such an expansion is not unique, because the series may be chosen so as to converge to any periodic function which agrees with $f(t)$ on $a \leq t \leq b$ and the coefficients of any such series are not influenced by the behaviour of $f(t)$ outside $a \leq t \leq b$. For a random process the situation is similar, but there are complications when the correlation between coefficients is considered.

Suppose now that we consider a wide-sense stationary process, as in the preceding section, but with no requirement of periodicity. We shall expand this process formally in a Fourier series which is to represent the process on the interval $a \leq t \leq b$; i.e.,

$$x(t) = \sum_{n=-\infty}^{\infty} x_n e^{jn\omega_o t} \qquad \omega_o = \frac{2\pi}{b-a} \qquad (6\text{-}27a)$$

where

$$x_n = \frac{1}{b-a} \int_a^b x(t) e^{-jn\omega_o t}\, dt \qquad (6\text{-}27b)$$

In Art. 6-3 we showed that if the process is periodic the random Fourier coefficients x_n and x_m, $n \neq m$, are uncorrelated. It can be shown, conversely, that in order for the coefficients to be uncorrelated, the process must be periodic.[†] The fact that periodicity (or its lack), which seems to be a condition on the behavior of the process outside $a \leq t \leq b$, can influence the coefficients of an expansion intended only for $a \leq t \leq b$ arises because we are concerned only with wide-sense stationary random processes. The requirement of stationarity limits the structure the process can have within one interval to be statistically similar to the structure it can have in any other interval of the same length.

Since the given process is not necessarily periodic, the correlation function cannot be written so simply in terms of the variances of the Fourier coefficients of $x(t)$, as in Eq. (6-22). We have now

$$R_x(\tau) = E[x(t+\tau)x^*(t)]$$
$$= \sum_{n=-\infty}^{+\infty} \sum_{k=-\infty}^{+\infty} E(x_n x_k^*) \exp[j(n-k)\omega_o t + jn\omega_o \tau] \qquad (6\text{-}28)$$

which is valid only when t and $t+\tau$ are in the same interval of length $(b-a)$. This expression cannot be simplified essentially for the general case.

Since the Fourier-series expansion of Eqs. (6-27a and b) is relatively simple to use when the coefficients are uncorrelated and has been used often with this property assumed, it is of interest to note that, in fact, as the length of the interval $a \leq t \leq b$ approaches infinity, the normalized correlation between different coefficients does approach zero.[‡] That is, let $T = b - a$; then

$$\lim_{T \to \infty} TE(x_n x_m^*) = 0 \qquad (6\text{-}29a)$$

when n and m are constant and $n \neq m$. Further,

$$\lim_{T \to \infty} TE(|x_n|^2) = \int_{-\infty}^{+\infty} R_x(u) \exp(-j2\pi f_n u)\, du = S_x(f_n) \qquad (6\text{-}29b)$$

[†] See Root and Pitcher (I, Theorem I).
[‡] Cf. Root and Pitcher (I).

when $n \to \infty$ as $T \to \infty$ in such fashion that $nf_o = f_n$ remains constant, and where $S_x(f_n)$ is the spectral density† of the random process, evaluated at $f = f_n$.

To show this, we have, as in Eq. (6-21),

$$E(x_n x_m^*) = \frac{1}{T^2} \int_a^b \int_a^b R_x(t - s) \exp[j\omega_o(ms - nt)] \, ds \, dt$$

Letting $u = t - s$, this becomes

$$E(x_n x_m^*) = \frac{1}{T^2} \int_0^T \int_{-s}^{T-s} R_x(u) \exp\{j\omega_o[ms - n(u + s)]\} \, du \, ds$$

and letting $v = s/T$,

$$T E(x_n x_m^*) = \int_0^1 \exp[j2\pi(m - n)v] \, dv \int_{-vT}^{T(1 - v)} R_x(u) \exp\left(\frac{j2\pi nu}{T}\right) du$$

When $m \neq n$ and when m and n are constant, the inner integral approaches $\int_{-\infty}^{+\infty} R_x(u) \, du =$ constant for every $v \neq 0$ as $T \to \infty$, and the entire integral vanishes in the limit. When $m = n$ and $n \to \infty$ as $T \to \infty$ in such fashion that $nf_o = f_n$ remains constant, the entire integral approaches

$$\int_{-\infty}^{+\infty} R_x(u) \exp(-j2\pi f_n u) \, du$$

Equations (6-29a and b) then follow.

For certain applications it is more convenient to expand a wide-sense stationary real random process in a sine–cosine Fourier series than in an exponential series. In such cases we can write

$$x(t) = \frac{x_{c0}}{2} + \sum_{n=1}^{\infty} (x_{cn} \cos n\omega_o t + x_{sn} \sin n\omega_o t) \qquad (6\text{-}30a)$$

where $\omega_o = 2\pi/(b - a)$, where x_{cn} is twice the real part of x_n:

$$x_{cn} = 2\Re(x_n) = \frac{2}{(b - a)} \int_a^b x(t) \cos n\omega_o t \, dt \qquad (6\text{-}30b)$$

and where x_{sn} is twice the imaginary part of x_n:

$$x_{sn} = 2\Im(x_n) = \frac{2}{(b - a)} \int_a^b x(t) \sin n\omega_o t \, dt \qquad (6\text{-}30c)$$

† Cf. Eq. (6-63).

In a manner similar to that used to prove Eqs. (6-29a and b), it may be shown that, where $T = (b - a)$ and m and n are constant,

$$\lim_{T \to \infty} TE(x_{cn}x_{cm}) = 0 \qquad \text{for } m \neq n \qquad (6\text{-}31a)$$

$$\lim_{T \to \infty} TE(x_{sn}x_{sm}) = 0 \qquad \text{for } m \neq n \qquad (6\text{-}31b)$$

and

$$\lim_{T \to \infty} TE(x_{cn}x_{sm}) = 0 \qquad \text{all } m, n \qquad (6\text{-}31c)$$

Further,

$$\lim_{T \to \infty} TE(x_{cn}{}^2) = \lim_{T \to \infty} TE(x_{sn}{}^2) = 2S_x(f_n) \qquad (6\text{-}31d)$$

where $n \to \infty$ as $T \to \infty$ in such fashion that $nf_o = f_n$ remains constant. The sine–cosine Fourier coefficients therefore also become uncorrelated as the length of the expansion interval increases without limit.

Expansion in Orthogonal Series with Uncorrelated Coefficients (Karhunen–Loéve Theorem).† As we pointed out above, a nonperiodic random process cannot be written as a trigonometric Fourier series with uncorrelated random coefficients. However, it turns out that if the term *Fourier series* is extended—as it often is—to include any series of orthogonal functions $\phi_n(t)$ with coefficients properly determined, then nonperiodic processes do have a Fourier-series expansion with uncorrelated coefficients. Let us now make this statement more precise and then show how the expansion is determined.

An expansion for $x(t)$ on an interval from a to b of the form

$$x(t) = \sum_n \sigma_n x_n \phi_n(t) \qquad a \leq t \leq b \qquad (6\text{-}32)$$

where

$$\int_a^b \phi_n(t) \phi_m^*(t)\, dt = 1 \qquad \text{if } m = n \qquad (6\text{-}33a)$$
$$= 0 \qquad \text{if } m \neq n$$

$$E(x_n x_m^*) = 1 \qquad \text{if } m = n \qquad (6\text{-}33b)$$
$$= 0 \qquad \text{if } m \neq n$$

and the σ_n are real or complex numbers will be called an *orthogonal expansion* of the random process on the given interval. The "equality" in Eq. (6-32) is to mean precisely that for every t, $a \leq t \leq b$

$$x_t = \operatorname*{l.i.m.}_{N \to \infty} \sum_{n=1}^{N} \sigma_n x_n \phi_n(t)$$

In Art. 6-3 it was shown that for a stationary periodic process such an expansion is given with $\phi_n(t) = \dfrac{1}{T} e^{jn\omega_o t}$ and $\sigma_n x_n$ equal to the corresponding random Fourier coefficients. In this section it has been pointed out

† See Loéve (I, p. 478).

that if the condition of periodicity is dropped, Eq. (6-33b) is no longer true with $\phi_n(t) = \dfrac{1}{T}\, e^{jn\omega_0 t}$ but is only a limiting relation as $T \to \infty$. It is true, however, that any random process—even a nonstationary one—with a continuous correlation function has an orthogonal expansion with some set of functions $\phi_n(t)$. To see how the functions $\phi_n(t)$ and the numbers σ_n are determined, suppose that Eqs. (6-32) and (6-33) are satisfied for some set of functions $\phi_n(t)$, some set of numbers σ_n, and some set of random variables x_n. Then

$$R(t,s) = E(x_t x_s^*) = E\left[\sum_n \sigma_n x_n \phi_n(t) \sum_k \sigma_k^* x_k^* \phi_k^*(s)\right]$$

$$= \sum_n |\sigma_n|^2 \phi_n(t)\phi_n^*(s) \qquad a \le t,\, s \le b \tag{6-34}$$

Using $R(t,s)$, as given by Eq. (6-34), we have

$$\int_a^b R(t,s)\phi_k(s)\, ds = \sum_n |\sigma_n|^2 \phi_n(t)\int_a^b \phi_n^*(s)\phi_k(s)\, ds$$

or
$$\int_a^b R(t,s)\phi_k(s)\, ds = |\sigma_k|^2 \phi_k(t) \tag{6-35}$$

In the language of integral equations, the numbers $|\sigma_k|^2$ must be the characteristic values and the functions $\phi_k(t)$ must be the characteristic functions† of the integral equation

$$\int_a^b R(t,s)\phi(s)\, ds = \lambda\phi(t) \qquad a \le t \le b \tag{6-36}$$

Conversely, we can construct an orthogonal expansion, valid over any given interval $a \le t \le b$, for a random process with a continuous correlation function by using for the σ's and $\phi(t)$'s of Eq. (6-32) the positive square roots of characteristic values and the characteristic functions of Eq. (6-36). Let $|\sigma_n|^2$ be the non-zero characteristic values of Eq. (6-36) (all of which are positive), with those characteristic values of multiplicity $r > 1$ indexed with r different numbers. Take σ_n as the positive square root of $|\sigma_n|^2$. Let $\{\phi_n(t)\}$ be a set of orthonormal characteristic functions of Eq. (6-36) with $\phi_n(t)$ corresponding to $|\sigma_n|^2$. Let the random variables x_n be given by

$$\sigma_n x_n = \int_a^b x(t)\phi_n^*(t)\, dt \tag{6-37}$$

Then
$$\sigma_n \sigma_m E(x_n x_m^*) = E\left[\int_a^b x(t)\phi^*(t)\, dt \int_a^b x^*(s)\phi_m(s)\, ds\right]$$

$$= \int_a^b \int_a^b R(t,s)\phi_n^*(t)\phi_m(s)\, dt\, ds$$

$$= \int_a^b \phi_n^*(t)|\sigma_m|^2 \phi_m(t)\, dt$$

$$= \sigma_n^2 \qquad \text{if } n = m$$

$$= 0 \qquad \text{if } n \ne m \tag{6-38}$$

† See Appendix 2, Art. A2-2.

Thus Eq. (6-33*b*) is satisfied. In addition, Eq. (6-33*a*) is satisfied because the $\phi_n(t)$ were chosen orthonormal. It remains to show that Eq. (6-32) is satisfied in the sense that x_t is the limit in the mean of partial sums on the right. That is, we must show that

$$\lim_{N \to \infty} E\left[\left| x(t) - \sum_{n=1}^{N} \sigma_n x_n \phi_n(t) \right|\right] = 0 \qquad (6\text{-}39)$$

If we let
$$x_N(t) = \sum_{n=1}^{N} \sigma_n x_n \phi_n(t) \qquad (6\text{-}40)$$

then a direct calculation shows that

$$E[x(t)x_N^*(t)] = E[x^*(t)x_N(t)] = E[x_N(t)x_N^*(t)]$$
$$= \sum_{n=1}^{N} \sigma_n{}^2 \phi_n(t) \phi_n^*(t) \qquad (6\text{-}41)$$

Hence

$$E[|x(t) - x_N(t)|^2] = R(t,t) - \sum_{n=1}^{N} \sigma_n{}^2 \phi_n(t) \phi_n^*(t) \qquad (6\text{-}42)$$

But by Mercer's theorem† the last term on the right converges to $R(t,t)$ as $N \to \infty$. Hence Eq. (6-39) holds and the demonstration is completed. The energy in a noise signal over the time interval $a \leq t \leq b$ is

$$E\left(\int_a^b |x(t)^2|\, dt \right) = E\left[\int_0^T \sum_{n=0}^{\infty} \sigma_n x_n \phi_n(t) \sum_{k=0}^{\infty} \sigma_k x_k^* \phi_k^*(t)\, dt \right]$$
$$= E\left[\sum_{n=0}^{\infty} \sigma_n{}^2 |x_n|^2 \right] = \sum_{n=0}^{\infty} \sigma_n{}^2 \qquad (6\text{-}43)$$

a result which is a generalization of Eq. (6-24). Thus a "spectral decomposition" of power can always be made with respect to the functions $\phi_n(t)$. This general treatment of orthogonal expansions over a finite interval includes what was done for periodic processes as a special case, where the characteristic functions of Eq. (6-36) become complex exponentials (or sines and cosines).

The expansion given by Eq. (6-32) is very useful in certain theoretical problems; practically, its usefulness is severely limited by two facts: procedures for finding solutions of integral equations of the form of Eq. (6-36) are not known in general,‡ and decomposition of the signal or its

---

† See Appendix 2, Art. A2-2.

‡ However, when the Fourier transform of the correlation function is a rational function, the integral equation can be solved. See Prob. 5 of this chapter and Art. A2-3 of Appendix 2.

power with respect to a set of orthonormal functions (the $\phi_n(t)$) which are not sines and cosines has not the simple engineering interpretation that a decomposition in frequency has—for example, in filter design. We close this section with an example† in which the characteristic functions can be found as solutions to an associated differential equation.

Example 6-4.1. In the Example 4-5.1 we found the autocorrelation function of a "random telegraph wave" $y(t)$ to be

$$R_y(\tau) = \frac{1}{4}\{1 + e^{-|2\alpha\tau|}\}$$

If we write

$$y(t) = \frac{1}{2} + x(t) \tag{6-44}$$

then $x(t)$ has mean zero and autocorrelation function

$$R_x(\tau) = \frac{1}{4}e^{-|2\alpha\tau|} \tag{6-45}$$

Let us now find the orthogonal expansion for $x(t)$ for $-A \leq t \leq A$. We have to find the characteristic values and functions of the integral equation

$$\frac{1}{4}\int_{-A}^{A} e^{-|2\alpha(u-v)|}\phi(v)\,dv = \mu\phi(u) \qquad -A \leq u \leq A \tag{6-46}$$

Making the substitutions $t = 2\alpha u$, $s = 2\alpha v$, $T = 2\alpha A$, and $f(t) = \phi(u)$, $\lambda = 8\alpha\mu$, Eq. (6-46) becomes

$$\int_{-T}^{T} e^{-|t-s|}f(s)\,ds = \lambda f(t) \qquad -T \leq t \leq T \tag{6-47}$$

We can solve Eq. (6-47) by finding a linear differential equation which $f(t)$ must satisfy and then substituting the general solution of the differential equation back in Eq. (6-47) to determine λ. Equation (6-47) can be written

$$\lambda f(t) = \int_{-T}^{t} e^{s-t}f(s)\,ds + \int_{t}^{T} e^{t-s}f(s)\,ds$$

Differentiating this twice we have

$$\lambda f'(t) = -\int_{-T}^{t} e^{s-t}f(s)\,ds + \int_{t}^{T} e^{t-s}f(s)\,ds$$

and

$$\lambda f''(t) = \int_{-T}^{T} e^{-|t-s|}f(s)\,ds - 2f(t)$$

Hence

$$\lambda f''(t) + 2f(t) = \lambda f(t)$$

or

$$f''(t) + \frac{2-\lambda}{\lambda}f(t) = 0 \tag{6-48}$$

Thus in order for $f(t)$ to satisfy the integral equation (6-47), it must satisfy the linear homogeneous differential equation (6-48). We shall substitute a general solution of Eq. (6-48) back in Eq. (6-47), considering separately the subcases: $\lambda = 0, 0 < \lambda < 2$, $\lambda = 2, 2 < \lambda$. Let us look first at the subcase $\lambda > 2$. Then

$$-1 < \frac{2-\lambda}{\lambda} < 0$$

and let us write

$$-a^2 = \frac{2-\lambda}{\lambda} \qquad 0 < a^2 < 1$$

† This example illustrates a procedure discussed in Appendix 2, Art. A2-3.

The differential equation becomes

$$f''(t) - a^2 f(t) = 0$$

which has the general solution

$$f(t) = c_1 e^{at} + c_2 e^{-at}$$

Substituting this expression in the left side of the integral equation (6-47), performing the indicated integrations and collecting terms yields

$$e^{at}\left[\frac{c_1}{a+1} - \frac{c_1}{a-1}\right] + e^{-at}\left[\frac{c_2}{a+1} - \frac{c_2}{a-1}\right]$$
$$+ e^{-t}\left[\frac{-c_1 e^{-(a+1)T}}{a+1} + \frac{c_2 e^{(a-1)T}}{a-1}\right]$$
$$+ e^{t}\left[\frac{-c_2 e^{-(a+1)T}}{a+1} + \frac{c_1 e^{(a-1)T}}{a-1}\right] \tag{6-49}$$

For $f(t)$ to satisfy the integral equation we must have the coefficients of the terms in e^t and e^{-t} equal to zero; i.e.,

$$\frac{c_1 e^{-aT}}{a+1} = \frac{c_2 e^{aT}}{a-1} \tag{6-50}$$

and

$$\frac{c_2 e^{-aT}}{a+1} = \frac{c_1 e^{aT}}{a-1} \tag{6-51}$$

Adding Eqs. (6-50) and (6-51) gives

$$(c_1 + c_2)\frac{e^{-aT}}{a+1} = (c_1 + c_2)\frac{e^{aT}}{a-1}$$

If $c_1 + c_2 \neq 0$, this equation cannot be satisfied for any a meeting the condition $0 < a^2 < 1$. Hence, if this equation is to be satisfied, we must have $c_1 = -c_2$. But substituting this condition into Eq. (6-50) gives

$$\frac{1-a}{1+a} = e^{2aT}$$

which cannot be satisfied by any a for which $0 < a^2 < 1$. The conclusion is that the integral equation cannot be satisfied for any $\lambda > 2$.

Let us consider next the subcase $0 < \lambda < 2$.

Then

$$\frac{2-\lambda}{\lambda} > 0$$

and if we write

$$b^2 = \frac{2-\lambda}{\lambda} \qquad 0 < b^2 < \infty$$

the solution of the differential equation is

$$f(t) = c_1 e^{jbt} + c_2 e^{-jbt}$$

By replacing jb by a, this is of the same form as the solution in the first case, and, as there, we have that Eqs. (6-50) and (6-51) must hold. It may readily be verified that if $c_1 \neq \pm c_2$, these equations cannot hold. Thus we have two subcases $c_1 = c_2$ and $c_1 = -c_2$. If $c_1 = c_2$, then we must have

$$\frac{a-1}{a+1} = e^{2aT}$$

or, in terms of b,

$$b \tan bT = 1$$

Call the non-zero solutions of this equation b_n (there are infinitely many b_n). Then if we replace b_n by λ_n from the definition of b, we find that the right side of Eq. (6-49) reduces to $\lambda f(t)$, so the integral equation is satisfied. If $c_1 = -c_2$, similar considerations will show that the integral equation is satisfied if b is a solution of

$$b \cot bT = 1$$

Summarizing, we find that a set of characteristic functions and their corresponding characteristic values for Eq. (6-47) are

$$f_n(t) = c \cos b_n t \qquad \lambda_n = \frac{2}{1 + b_n{}^2} \qquad (6\text{-}52)$$

where b_n satisfies $b \tan bT = 1$

and

$$\hat{f}_n(t) = c \sin \hat{b}_n t \qquad \hat{\lambda}_n = \frac{2}{1 + \hat{b}_n{}^2} \qquad (6\text{-}53)$$

where $\hat{b}_n$ satisfies $\hat{b} \cot \hat{b}T = +1$.

We have not yet considered the possibility that there may be characteristic values $\lambda = 2$ and $\lambda = 0$. In fact, there are not, as will be shown directly (Prob. 4). Hence Eqs. (6-52) and (6-53) give a complete set of characteristic values and functions.

The centered random telegraph wave $x(t)$ may now be written, according to the theorem of this section,

$$x(t) = \sum_n \sqrt{\mu_n} x_n \phi_n(t) + \sum_n \sqrt{\hat{\mu}_n} \hat{x}_n \hat{\phi}_n(t) \qquad (6\text{-}54)$$

where the x_n and $\hat{x}_n$ are mutually uncorrelated random variables with mean zero and variance one, and where

$$\mu_n = \frac{1}{4\alpha(1 + b_n{}^2)} \qquad (6\text{-}55a)$$

$$\hat{\mu}_n = \frac{1}{4\alpha(1 + \hat{b}_n{}^2)} \qquad (6\text{-}55b)$$

$$\phi_n(t) = \frac{1}{\sqrt{A + (\sin 4\alpha b_n A)/(4\alpha b_n)}} \cos 2\alpha b_n t \qquad (6\text{-}55c)$$

$$\hat{\phi}_n(t) = \frac{1}{\sqrt{A - (\sin 4\alpha \hat{b}_n A)/(4\alpha \hat{b}_n)}} \sin 2\alpha \hat{b}_n t \qquad (6\text{-}55d)$$

and b_n and $\hat{b}_n$ are solutions, respectively, of

$$b \tan 2\alpha A b = 1$$
$$\hat{b} \cot 2\alpha A \hat{b} = 1$$

6-5. Spectral Density for an Arbitrary Function

We have considered so far in this chapter the spectral analysis of random processes over a finite interval. In the remainder of the chapter we shall consider the problem for the interval $-\infty \leq t \leq \infty$. Again, we shall introduce the material by looking first at the power spectrum of a single function. Suppose a signal is given by a function $x(t)$, which is of integrable square; then $x(t)$ has a Fourier transform $X(f)$, and Eqs. (6-11) and (6-12) hold. The total energy in the signal is

$$\int_{-\infty}^{\infty} |x(t)|^2 \, dt = \int_{-\infty}^{\infty} |X(f)|^2 \, df$$

and the time average of energy, or power, is

$$\lim_{N \to \infty} \frac{1}{2T} \int_{-T}^{T} |x(t)|^2 \, dt = 0 \qquad (6\text{-}56)$$

Obviously it is useless to discuss power spectral density in this case; the density must be zero for all frequencies, since it must integrate to give the power. In other words, for signals of finite total energy it is the energy which must be analyzed, not the power.

Often it is a convenient fiction to suppose that a signal continues indefinitely. Then, although we usually require that it have finite energy over any finite interval, it may well have infinite energy over the infinite interval. In mathematical language, we want to treat functions $x(t)$, for which

$$\lim_{T \to \infty} \frac{1}{2T} \int_{-T}^{T} |x(t)|^2 \, dt < \infty \qquad (6\text{-}57)$$

The class of functions satisfying (6-57) includes the functions of integrable square, but it is of course a much larger class, and generally a function satisfying (6-57) will not have a Fourier transform. It is appropriate to discuss spectral decompositions of power of such a function $x(t)$, but we cannot proceed as we did in considering periodic functions over a finite interval because we do not have the Fourier transform of $x(t)$—the analogue of the Fourier series used there—with which to work. However, we can start from the autocorrelation function $\Re(\tau)$ and consider its Fourier transform.

Suppose the limit

$$\lim_{T \to \infty} \frac{1}{2T} \int_{-T}^{T} x(t)x^*(t - \tau) \, dt = \Re(\tau) \qquad (6\text{-}58)$$

exists for all τ and hence that (6-57) is satisfied. Then since†

$$\left| \int_{-T}^{T} x(t)x^*(t - \tau) \, dt \right|^2 \leq \int_{-T}^{T} |x(t)|^2 \, dt \int_{-T}^{T} |x(t - \tau)|^2 \, dt$$

it follows that

$$|\Re(\tau)| \leq \lim_{T \to \infty} \left\{ \frac{1}{2T} \int_{-T}^{T} |x(t)|^2 \, dt \, \frac{1}{2T} \int_{-T}^{T} |x(t - \tau)|^2 \, dt \right\}^{\frac{1}{2}}$$

$$= \lim_{T \to \infty} \frac{1}{2T} \int_{-T}^{T} |x(t)|^2 \, dt = \Re(0) \qquad (6\text{-}59)$$

† By the Schwartz inequality.

Thus if $x(t)$ has an autocorrelation function $\mathfrak{R}(\tau)$ it has an "average power," which is equal to $\mathfrak{R}(0)$; and $\mathfrak{R}(\tau)$ is bounded by $\mathfrak{R}(0)$. It will be recalled that a similar result was demonstrated for the autocorrelation function of a random process in Art. 4-5.

Let us now *define* the power spectral density $\mathcal{S}(f)$ of the function $x(t)$ to be the Fourier transform of the autocorrelation function

$$\mathcal{S}(f) = \int_{-\infty}^{\infty} \mathfrak{R}(\tau)e^{-j\omega\tau}\,d\tau \qquad \omega = 2\pi f \tag{6-60}$$

Then the inverse transform relation† gives

$$\mathfrak{R}(\tau) = \int_{-\infty}^{\infty} \mathcal{S}(f)e^{j\omega\tau}\,df \tag{6-61}$$

In Art. 6-2 we defined the spectral density of a periodic function satisfying the inequality in Eq. (6-4). Since such a function also satisfies the condition of Eq. (6-58), we have here redefined its spectral density. There is no inconsistency, of course, because Eq. (6-60) was a derived property in terms of the first definition. It follows immediately from Eq. (6-61) that the spectral density integrates to the average power of $x(t)$

$$\mathfrak{R}(0) = \int_{-\infty}^{\infty} \mathcal{S}(f)\,df = \lim_{T \to \infty} \frac{1}{2T} \int_{-\infty}^{\infty} |x(t)|^2\,dt \tag{6-62}$$

The study of the spectrum, spectral density, and the autocorrelation function of a single function is usually called *generalized harmonic analysis*.‡ We shall not discuss it further here, because, since we shall regard the signal and noise waves in the problems to be investigated later as sample functions of random processes, it is the harmonic analysis of random processes we shall need. Nevertheless there is a strong connection§ between the generalized harmonic analysis of functions and the harmonic analysis of wide-sense stationary random processes, which was developed later. The theorems in the two subjects are often analogues of each other. For ergodic stationary processes the spectrum defined from each sample function from Eq. (6-60) is with probability one the same as the spectrum of the random process.

---

† Equations (6-60) and (6-61) are correct as written if $\int_{-\infty}^{\infty} |\mathfrak{R}(\tau)|\,d\tau < \infty$, which is often satisfied in practice. In general, the spectral density $\mathcal{S}(f)$ may not exist and Eq. (6-61) has to be written as a Fourier–Stieltjes transform and Eq. (6-60) has to be written in an integrated form. See Doob (II, Chap. 11, Art. 3). By the use of impulse functions, the domain of application of Eqs. (6-60) and (6-61) may be extended formally to include most practical problems.

‡ See Wiener (I) or Wiener (II, Chap. 4).

§ For a brief discussion of this point see Doob (I).

6-6. Spectral Analysis of a Wide-sense Stationary Random Process†

We shall define the power spectral density of a wide-sense stationary random process, as of a single function, to be the Fourier transform of the autocorrelation function; thus

$$S(f) = \int_{-\infty}^{\infty} R(\tau) \exp(-j\omega\tau) \, d\tau \qquad \omega = 2\pi f \qquad (6\text{-}63)$$

Then
$$R(\tau) = \int_{-\infty}^{\infty} S(f) \exp(j\omega\tau) \, df \qquad (6\text{-}64)$$

The correlation function $R(\tau)$ is of course given by

$$R(\tau) = E[x(t)x^*(t - \tau)]$$

but if the process is ergodic, it is also given with probability one by

$$R(\tau) = \lim_{T\to\infty} \frac{1}{2T} \int_{-T}^{T} x(t)x^*(t - \tau) \, dt$$

Examples. Before investigating some of the properties of the spectral density function, we shall consider two simple examples.

Example 6-6.1. The Random Telegraph Wave. The random telegraph wave described in Example 4-5.1 has the autocorrelation function

$$R_x(\tau) = \tfrac{1}{4}[1 + \exp(-2a|\tau|)]$$

Then

$$
\begin{aligned}
S(f) &= \tfrac{1}{4} \int_{-\infty}^{\infty} [1 + \exp(-2a|\tau|)] \exp(-j2\pi f\tau) \, d\tau \\
&= \tfrac{1}{4}\delta(f) + \tfrac{1}{4} \int_{0}^{\infty} \exp(-2a\tau - 2j\pi f\tau) \, d\tau + \tfrac{1}{4} \int_{-\infty}^{0} \exp(2a\tau - 2j\pi f\tau) \, d\tau \\
&= \frac{1}{4} \left\{ \delta(f) + \frac{a}{a^2 + \pi^2 f^2} \right\} \qquad (6\text{-}65)
\end{aligned}
$$

The impulse function, which gives the power spectral density a spike at zero frequency, reflects the presence of a d-c component in the random telegraph wave. If the wave is centered about the zero voltage line, then the spike disappears.

Example 6-6.2. White Noise. It is often convenient to make the simplifying assumption in dealing with band-pass filter problems that the power spectral density of a signal is constant over a finite range of frequencies and zero outside; i.e., for $f_2 > f_1 \geq 0$,

---

† If a wide-sense stationary random process with sample functions $x(t)$ is "chopped off" at $t = -T$ and $t = T$ and expanded in a Fourier series over the interval from $-T$ to T, the random coefficients have an asymptotic orthogonality as $T \to \infty$, as was explained in Art. 6-4. This may suggest that there is a limiting relation in which the process is given as a kind of Fourier transform of another random process with sample functions $y'(f)$ replacing the random coefficients at a discrete set of frequencies in the Fourier series and with the property that the random variables y'_{f_1} and y'_{f_2} are uncorrelated for any f_1 and f_2. In fact, such a limit relation can be proved, although it is necessary to phrase the result in terms of a random process with sample functions $y(f)$, which is conceptually the integral of the y' process.

For discussions of this representation, which is called the *spectral representation* of the random process, see Doob (II, Chap. XI, Art. 4), or Bartlett (I, Art. 6-2).

$$S(f) = N_o \qquad -f_2 \leq f \leq -f_1, \qquad f_1 \leq f \leq f_2$$
$$= 0 \qquad \text{otherwise}$$

Then
$$R(\tau) = \int_{-\infty}^{\infty} S(f) \exp(j2\pi f\tau)\, df$$

$$= N_o \int_{-f_2}^{-f_1} \exp(j2\pi f\tau)\, df + N_o \int_{f_1}^{f_2} \exp(j2\pi f\tau)\, df$$

$$= \frac{N_o}{\pi\tau} [\sin 2\pi f_2\tau - \sin 2\pi f_1\tau]$$

$$= \frac{2N_o}{\pi\tau} \sin 2\pi\tau\, \Delta f \cos 2\pi\tau f_o \qquad (6\text{-}66)$$

where
$$f_o = \frac{f_1 + f_2}{2} \qquad \Delta f = \frac{f_2 - f_1}{2}$$

In the hypothetical limiting case for which $S(f)$ is constant for all frequencies, $R(\tau)$ becomes $N_0\delta(\tau)$. A random process with constant spectral density is usually called *white noise*.

The most commonly appearing examples of power spectral density, at least in applications to noise in electric circuits, are those of the spectral density of the output from a lumped-parameter linear circuit which is excited with white noise. The spectral density in such a case is a rational function.

Properties of the Spectral Density and Correlation Function. An interesting characterization of the class of continuous correlation functions of stationary processes is as follows. Let a wide-sense stationary process be given with sample functions $x(t)$. Let N be any positive integer, let $t_1, \ldots, t_N$ be any set of N values of the parameter t, and let $z_1, \ldots, z_N$ be any set of N complex numbers; then

$$E\left(\left|\sum_{m=1}^{N} x_{t_m} z_m\right|^2\right) = E\left(\sum_{m=1}^{N} x_{t_m} z_m \sum_{k=1}^{N} x_{t_k}^* z_k^*\right)$$

$$= \sum_{m,k=1}^{N} R(t_m - t_k) z_m z_k^* \qquad (6\text{-}67)$$

But since this is the average of a nonnegative random variable, it is nonnegative; i.e.,

$$\sum_{m,k=1}^{N} R(t_m - t_k) z_m z_k^* \geq 0 \qquad (6\text{-}68)$$

We have already pointed out in Art. 4-5 that

$$R(-\tau) = R^*(\tau) \qquad (6\text{-}69)$$

Thus the correlation function of any wide-sense stationary process satisfies Eqs. (6-68) and (6-69); conversely, it can be shown† that any con-

---

† Doob (I, p. 519). It should be pointed out that the class of characteristic functions of a random variable is exactly the same as the class of normalized autocorrelation functions of a possibly complex stationary random process.

tinuous function that satisfies these two equations is the correlation function of some wide-sense stationary process. Functions which satisfy Eqs. (6-68) and (6-69) are called *non-negative definite*.

An easy verification, using the criterion of non-negative definiteness, will now show that if $S(f)$ is any integrable nonnegative function whatever, its Fourier transform is a correlation function. Thus there is no restriction on what kind of function can be a power spectral density beyond the obvious requirements that it must be nonnegative and it must be integrable (finite power). However, it should be noted that the power spectral density of a *real* random process must be an even function of frequency. This follows from Eq. (6-63) upon observing that for a real process $R(\tau) = R(-\tau)$.

It is of some interest to demonstrate that the Fourier transform of any correlation function $R(\tau)$ must be nonnegative; indeed, the definition of power spectral density given by Eq. (6-63) would be physically meaningless if this were not so. The calculation by which we show this nonnegativeness will be of use for a different purpose later in this section. As in Eq. (6-67), for any $T > 0$,

$$E\left[\left|\int_0^T x(t) \exp(-2\pi jft)\, dt\right|^2\right] \tag{6-70}$$

$$= \int_0^T \int_0^T R(t-s) \exp[-2\pi j(t-s)f]\, ds\, dt \geq 0$$

Letting $t - s = \tau$, the double integral in Eq. (6-70) becomes

$$\int_0^T \int_\tau^T R(\tau) \exp(-2\pi j\tau f)\, dt\, d\tau + \int_{-T}^0 \int_0^{(T+\tau)} R(\tau) \exp(-2\pi j\tau f)\, dt\, d\tau$$

If the integration on t is performed and the whole expression is divided by T, we get

$$\frac{1}{T} \int_0^T R(\tau)(T-\tau) \exp(-2\pi j\tau f)\, d\tau + \frac{1}{T} \int_{-T}^0 R(\tau)(T+\tau) \exp(-2\pi j\tau f)\, d\tau$$

$$= \int_{-T}^T \left(1 - \frac{|\tau|}{T}\right) R(\tau) \exp(-2\pi j\tau f)\, d\tau = \int_{-\infty}^\infty R_T(\tau) \exp(-2\pi j\tau f)\, d\tau \geq 0 \tag{6-71}$$

where
$$R_T(\tau) = \left(1 - \frac{|\tau|}{T}\right) R(\tau) \qquad \text{for } |\tau| \leq T$$
$$= 0 \qquad \text{for } |\tau| > T$$

Now as $T \to \infty$, $R_T(\tau) \to R(\tau)$ for every τ and (if $R(\tau)$ is absolutely integrable) the integral of Eq. (6-71) approaches

$$\int_{-\infty}^\infty R(\tau) \exp(-2\pi j\tau f)\, d\tau = S(f)$$

Since the approximating integrals are nonnegative, the limit is non-negative; i.e., $S(f) \geq 0$.

If $R(\tau)$ is absolutely integrable, so that its Fourier transform exists in a strict sense, then by the so-called *Riemann–Lebesgue lemma*,[†]

$$\lim_{f \to \infty} S(f) = \lim_{f \to \infty} \int_{-\infty}^{\infty} R(\tau) \exp(-j\omega\tau) \, d\tau = 0 \qquad \omega = 2\pi f \qquad (6\text{-}72)$$

In this case $R(\tau)$ cannot be periodic, for a periodic function different from zero cannot be absolutely integrable. If $R(\tau)$ can be written as the sum of a periodic part plus a part which is integrable, then $S(f)$ will consist of a sum of impulse functions spaced at regular intervals plus a term which goes to zero as the frequency goes to infinity.

Estimation of Spectral Density and Spectrum. A type of problem which frequently arises in experimental work is the approximate determination of the spectral density or autocorrelation function of a (supposedly stationary) random process when only certain sample values or pieces of sample functions are known. This is really a problem of statistical estimation. What is required in each case is some function of the observations—that is, a random variable—which converges in some sense to the true value of $S(f_o)$ or $R(\tau_o)$, whichever is being sought. We cannot treat this rather difficult matter adequately here;[‡] we shall, however, demonstrate one important negative result.

An often used and intuitively "natural" way to estimate spectral density is by means of a function called the periodogram, which will be defined below. Nevertheless, it has been found that, despite its appeal, the periodogram gives a very questionable estimate of spectral density, as we shall now show.

If we put

$$X_T(f) = \int_0^T x(t) \exp(j\omega t) \, dt \qquad \omega = 2\pi f \qquad (6\text{-}73)$$

then for a particular sample function $x(t)$ the function

$$S(f,T) = \frac{|X_T(f)|^2}{T} \qquad (6\text{-}74)$$

which is called the *periodogram*, gives a frequency decomposition of the power of $x(t)$ over the interval $0 \leq t \leq T$. This might suggest that one could let $T \to \infty$ and have the random variables $S(f,T)$ approach the spectral density $S(f)$. In general this procedure is not justifiable, since it fails for a large class of examples, in particular for real gaussian random

[†] Titchmarsh (II, Art. 1.8).
[‡] See Grenander (I), and Doob (II, Chaps. X and XI, Arts. 7).

processes. It is always true that

$$\lim_{T \to \infty} E[S(f,T)] = S(f) \tag{6-75}$$

but the variance of $S(f,T)$ may not approach zero.
From Eqs. (6-70) and (6-71) we have

$$E[S(f,T)] = \int_{-\infty}^{\infty} R_T(\tau) \exp(-2\pi j f \tau) \, d\tau \tag{6-76}$$

from which Eq. (6-75) follows. Now if the process is real

$$E[S^2(f,T)] = \frac{1}{T^2} E \left[\int_0^T \int_0^T \int_0^T \int_0^T x(t)x(s)x(u)x(v) \right. \tag{6-77}$$

$$\left. \exp[-j\omega(t + u - s - v)] \, dt \, ds \, du \, dv \right]$$

If in addition the process is gaussian, it can be shown† that

$$E[x(t)x(s)x(u)x(v)] = R(t - s)R(u - v) + R(t - u)R(s - v)$$
$$+ R(t - v)R(s - u) \tag{6-78}$$

Equation (6-77) then becomes

$$E[S^2(f,T)] = \frac{1}{T^2} \int_0^T \int_0^T \int_0^T \int_0^T [R(t - s)R(u - v) + R(t - u)R(s - v)$$
$$+ R(t - v)R(s - u)] \exp[-j\omega(t + u - s - v)] \, dt \, ds \, du \, dv$$
$$= 2\{E[S(f,T)]\}^2$$
$$+ \frac{1}{T^2} \left| \int_0^T \int_0^T R(t - u) \exp[j\omega(t + u)] \, dt \, du \right|^2 \tag{6-79}$$

Hence $\qquad\qquad E[S^2(f,T)] \geq 2\{E[S(f,T)]\}^2$

and $\qquad\qquad \sigma^2[S(f,T)] = E[S^2(f,T)] - \{E[S(f,T)]\}^2$
$$\geq \{E[S(f,T)]\}^2 \tag{6-80}$$

Thus the variance of $S(f,T)$ does not approach zero as $T \to \infty$ for any value of f for which $S(f) > 0$. This means that as $T \to \infty$ the random variable $S(f,T)$ does not converge (in the mean) to the value $S(f)$ at any f where $S(f) > 0$.

6-7. Cross-spectral Densities

Frequently we have to consider a random process which is the sum of two random processes. Suppose, for example, we have $z(t) = x(t) + y(t)$, then the autocorrelation function of the sum process is

$$R_z(t, t - \tau) = E[(x_t + y_t)(x_{t-\tau}^* + y_{t-\tau}^*)]$$
$$= R_x(t, t - \tau) + R_y(t, t - \tau) + R_{xy}(t, t - \tau) + R_{yx}(t, t - \tau)$$

† See Chap. 8, Prob. 2.

If the $x(t)$ and $y(t)$ processes are wide-sense stationary and the cross-correlations are stationary—that is, functions of τ alone—then the $z(t)$ process is necessarily wide-sense stationary, and we have

$$R_z(\tau) = R_x(\tau) + R_y(\tau) + R_{xy}(\tau) + R_{yx}(\tau) \qquad (6\text{-}81)$$

and

$$S_z(f) = S_x(f) + S_y(f) + \int_{-\infty}^{\infty} R_{xy}(\tau) \exp(-j\omega\tau)\, d\tau$$
$$+ \int_{-\infty}^{\infty} R_{yx}(\tau) \exp(-j\omega\tau)\, d\tau \qquad (6\text{-}82)$$

It should be noted that the sum process may be wide-sense stationary and hence have a spectral density even when the $x(t)$ and $y(t)$ processes are not (e.g., let $u(t)$ and $v(t)$ be independent and stationary, each with mean zero and with identical correlation functions; then $x(t) = u(t) \cos t$ and $y(t) = v(t) \sin t$ are not wide-sense stationary but $z(t) = x(t) + y(t)$ is). Also, the $x(t)$ and $y(t)$ processes may be wide-sense stationary but not have stationary cross-correlations or a wide-sense stationary sum.† In such cases Eqs. (6-81) and (6-82) cannot be written, of course.

When the $x(t)$ and $y(t)$ processes have a stationary cross-correlation $R_{xy}(\tau)$, we call its Fourier transform $S_{xy}(f)$ a *cross-spectral density*. That is

$$S_{xy}(f) = \int_{-\infty}^{\infty} R_{xy}(\tau) \exp(-j\omega\tau)\, d\tau \qquad \omega = 2\pi f \qquad (6\text{-}83)$$

Then $\qquad\qquad R_{xy}(\tau) = \int_{-\infty}^{\infty} S_{xy}(f) \exp(j\omega\tau)\, df \qquad\qquad (6\text{-}84)$

Since $R_{xy}(\tau)$ is not necessarily an even function, $S_{xy}(f)$ is not necessarily real. The relation $R_{xy}(\tau) = R_{yx}^*(-\tau)$ gives, when a substitution is made in Eq. (6-83), the relation

$$S_{xy}(f) = S_{yx}^*(f) \qquad (6\text{-}85)$$

From Eqs. (6-82) and (6-85), a physical meaning can be assigned to the real part of the cross-spectral density. We have

$$S_{yx}(f) + S_{xy}(f) = S_z(f) - S_x(f) - S_y(f)$$

and hence, where $\Re$ denotes "real part of":

$$2\Re[S_{yx}(f)] = 2\Re[S_{xy}(f)] = S_{xy}(f) + S_{xy}^*(f) = S_z(f) - S_x(f) - S_y(f)$$

In words, the real part of $S_{xy}(f)$ or $S_{yx}(f)$ is one-half the power density at the frequency f which must be added to the sum of the power densities of $x(t)$ and $y(t)$ to give the power density of their sum. Obviously, if two stationary processes are uncorrelated, their cross-spectral density is zero at all frequencies and the spectral density of their sum is the sum of their spectral densities.

† See Prob. 11 for an example.

6-8. Problems

1. Find the spectral density $\mathcal{S}(f)$ for the function of Prob. 16, Chap. 4.

2. Show that if a wide-sense stationary random process has a periodic correlation function $R(\tau)$, then for any t, $E[|x_t - x_{t+T}|^2] = 0$ (and hence $x_t = x_{t+T}$ with probability one).

3. Find the spectral density $S(f)$ for the random process of Prob. 21, Chap. 4.

4a. Show that the integral equation

$$\int_{-T}^{T} \exp(-|t - s|) f(s)\, ds = \lambda f(t) \qquad -T \leq t \leq T$$

which appeared in the example of Art. 6-4 does not have the numbers 2 and 0 as characteristic values.

b. Find the characteristic values and functions of the equation

$$\int_{a}^{b} \exp(-a|t - s|) f(s)\, ds = \lambda f(t) \qquad a \leq t \leq b$$

5.† Show that if

$$R(t) = \int_{-\infty}^{\infty} \exp(j\omega t) \frac{P(-\omega^2)}{Q(-\omega^2)}\, df \qquad \omega = 2\pi f$$

where P is a polynomial of degree n in $(-\omega^2)$ and Q is a polynomial of degree m in $(-\omega^2)$, $n < m$, then any characteristic function $\phi(t)$ and its characteristic value λ of the integral equation

$$\int_{a}^{b} R(t - s)\phi(s)\, ds = \lambda\phi(t) \qquad a \leq t \leq b$$

must satisfy the linear homogeneous differential equation

$$\lambda Q\left(\frac{d^2}{dt^2}\right)\phi(t) = P\left(\frac{d^2}{dt^2}\right)\phi(t) \qquad a < t < b$$

6. Suppose $z(t) = x(t)y(t)$, where $x(t)$ and $y(t)$ are sample functions from independent stationary random processes for which

$$E[x_t] = m_x \qquad E[y_t] = m_y$$

and the autocorrelation functions of $x(t) - m_x$ and $y(t) - m_y$ are respectively

$$R_{x-m_x}(\tau) = \exp(-a|\tau|)$$
$$R_{y-m_y}(\tau) = \exp(-b|\tau|)$$

Find the autocorrelation function and spectral density of $z(t)$.

7. Let a random process have sample functions

$$x(t) = a \cos(2\pi f t + \phi)$$

where f and ϕ are independent random variables, ϕ is uniformly distributed over the interval $0 \leq \phi \leq 2\pi$, and f has a symmetric probability density $p(f)$. Show that the random process is wide-sense stationary and find its spectral-density function in terms of $p(f)$. Is this process periodic or not?

8a. Let

$$y(t) = x(t) \cos \omega_o t$$

† See Slepian (I, Appendix I).

where ω_o is a constant. Determine the time-average autocorrelation function $\mathcal{R}_y(\tau)$ and the spectral density $\mathcal{S}_y(f)$ in terms of $\mathcal{R}_x(\tau)$.

b. Let a random process have sample functions

$$y(t) = x(t) \cos (\omega_o t + \theta)$$

where ω_o is a constant, θ is a random variable uniformly distributed over the interval $0 \le \theta \le 2\pi$, and $x(t)$ is a wide-sense stationary random process which is independent of θ. Show that the $y(t)$ process is wide-sense stationary and determine its autocorrelation function and spectral density in terms of those for $x(t)$.

9.† Let a random process have sample functions

$$y(t) = a \cos [\omega_o t - \phi(t) + \theta]$$

where a and ω_o are constants, θ is a random variable uniformly distributed over the interval $0 \le \theta \le 2\pi$, and $\phi(t)$ is a stationary random process which is independent of θ. Show that $y(t)$ process is wide-sense stationary with autocorrelation function

$$R_y(t) = \frac{a^2}{2} \Re[\exp(j\omega_o t)E(\exp\{j[\phi(t_1) - \phi(t_2)]\})]$$

10. Let $\phi(t)$ in Prob. 9 be given by

$$\phi(t) = b \cos (\omega_m t + \theta')$$

where b and ω_m are constants and θ' is a random variable uniformly distributed over $0 \le \theta' \le 2\pi$. Using the result of Prob. 9, calculate the spectral-density function of $y(t)$. Hint: Use the identity

$$\exp(jz \cos \theta) = J_o(z) + \sum_{n=1}^{\infty} 2j^n J_n(z) \cos n\theta$$

11. Let $y(t)$ be as given in Prob. 8b. Let

$$w(t) = x(t) \cos [(\omega_o + \delta)t + \theta]$$

This represents $y(t)$ heterodyned up in frequency by an amount δ. Show that $w(t)$ is wide-sense stationary and find its autocorrelation function and spectral density in terms of those for $x(t)$. Show that the cross-correlations between $y(t)$ and $w(t)$ are not stationary, and show that $y(t) + w(t)$ is not wide-sense stationary. Show that if the heterodyning is done with random phase, i.e.,

$$w(t) = x(t) \cos [(\omega_o + \delta)t + \theta + \theta']$$

where θ' is uniformly distributed over $0 \le \theta' \le 2\pi$ and is independent of θ and $x(t)$, then $y(t) + w(t)$ is wide-sense stationary.

† See Middleton (III).

CHAPTER 7

SHOT NOISE

The usefulness of a thermionic vacuum tube is impaired in part by the fluctuations of its current due to the random emission of electrons from its heated cathode. In this chapter, we shall study the statistical properties of these random fluctuations, which are called *shot noise*. In particular, we shall study shot noise first in a temperature-limited diode, then in a space-charge-limited diode, and finally in space-charge-limited multi-electrode tubes. Our purposes here are twofold: (1) to obtain the statistical properties of shot noise so as to be able later to determine how it limits tube performance, and (2) to show how the statistical properties of a particular physical phenomenon may be derived. The second purpose is fully as important as the first.

7-1. Electronics Review

Before embarking on statistical derivations, it is perhaps worthwhile to review some of the equations which govern the operation of electron tubes. We shall restrict our discussion to tubes of the so-called "conventional receiving" type, whose electrode potential differences are of the order of several hundred volts or less and whose physical dimensions are small compared to a wave length at the frequencies of operation. The basic equations† for such tubes, as expressed in rationalized mks units, are

$$\mathbf{F}_m = m\mathbf{a} \tag{7-1}$$

relating the mechanical force $\mathbf{F}_m$ (in newtons) on an electron of mass m (in kilograms) to the resultant acceleration $\mathbf{a}$ (in meters per second) of the electron;

$$\mathbf{F}_e = -e\mathbf{E} \tag{7-2}$$

giving the electrical force $\mathbf{F}_e$ on an electron of charge e (in coulombs) due to the electric field intensity $\mathbf{E}$ (in volts per meter) acting on the electron; the gradient equation

$$\mathbf{E} = -\nabla V \tag{7-3}$$

† E.g., see Harman (I, Chaps. 5 and 6) or Spangenberg (I, Chaps. 8 and 16).

relating the electric field intensity to the electric potential V (in volts); the Poisson equation

$$\mathbf{\nabla}^2 V = -\frac{\rho}{\epsilon_o} \tag{7-4}$$

satisfied by the electric potential, where ρ is the space-charge density (in coulombs per cubic meter) and ϵ_o is the permittivity of free space (in farads per meter); and the equation

$$\mathbf{J} = -\rho\mathbf{v} \tag{7-5}$$

relating the current density $\mathbf{J}$ (in amperes per square meter) to the charge density and the charge velocity $\mathbf{v}$ (in meters per second).

The Parallel-plane Temperature-limited Diode. Let us now consider the specific case of the parallel-plane temperature-limited diode whose cathode–anode spacing is d (in meters) and whose cathode–anode potential difference is V_a. In a temperature-limited diode, the cathode–anode potential difference is so great that all electrons emitted by the cathode are pulled to the anode with such high velocities that space-charge effects are negligible; i.e., the Poisson equation reduces approximately to the Laplace equation

$$\mathbf{\nabla}^2 V = 0 \tag{7-6}$$

in this case. We shall assume for convenience that the cathode plane is that defined by $x = 0$ and the anode plane is that defined by $x = d$. Neglecting edge effects, the Laplace equation then becomes

$$\frac{\partial^2 V}{\partial x^2} = 0$$

and has the solution

$$V = V_a \left(\frac{x}{d}\right)$$

The electric field intensity is

$$\mathbf{E} = -\mathbf{i}\frac{\partial V}{\partial x} = -\mathbf{i}\frac{V_a}{d}$$

where $\mathbf{i}$ is the unit vector in the $+x$ direction. Equating the electrical and mechanical forces on an electron, we get

$$\frac{d^2x}{dt^2} = \frac{e}{m}\frac{V_a}{d}$$

as the differential equation governing the motion of an electron in the cathode–anode space of a temperature-limited diode. The velocity v and the position x of an electron at time t may be obtained by direct integration.   Thus

$$v = \left(\frac{eV_a}{md}\right) t \tag{7-7}$$

and

$$x = \left(\frac{eV_a}{md}\right) \frac{t^2}{2} \tag{7-8}$$

under the assumptions that since the diode is temperature-limited the
initial velocity of the electron is negligible compared with the velocity
of arrival at the anode and that the electron is emitted at $t = 0$.   The
transit time τ_a (i.e., the time required for an electron to traverse the
cathode–anode space) may be obtained by setting $x = d$ in Eq. (7-8)
and solving for $t = \tau_a$:

$$\tau_a = \left(\frac{2m}{eV_a}\right)^{\frac{1}{2}} d \tag{7-9}$$

The electron velocity and position can now be expressed in terms of this
transit time as

$$v = \left(\frac{2d}{\tau_a}\right)\left(\frac{t}{\tau_a}\right) = v_a\left(\frac{t}{\tau_a}\right) \tag{7-10}$$

and

$$x = d\left(\frac{t}{\tau_a}\right)^2 \tag{7-11}$$

where $v_a = (2d/\tau_a)$ is the electron velocity at the anode.

The current pulse induced in the anode circuit of the diode by the
flight of an electron through the cathode–anode space can be obtained
by finding the charge induced on the anode by the electron in motion
and taking the time derivative of this induced charge.   The charge q
induced on the anode may be found as follows: The energy U gained by
an electron moving through a potential difference V is

$$U = eV = eV_a \frac{x}{d}$$

This energy is equal to the amount of work W that must be done to
induce the charge q on the anode when the anode potential is V_a:

$$W = qV_a$$

On equating these energies and solving for q, we get

$$q = \frac{ex}{d}$$

The anode current pulse is therefore

$$i_e(t) = \frac{dq}{dt} = \frac{ev}{d} \tag{7-12}$$

during the flight of the electron and zero before the electron is emitted from the cathode and after its arrival at the anode. Thus

$$i_e(t) = \begin{cases} \dfrac{2e}{\tau_a}\left(\dfrac{t}{\tau_a}\right) & \text{for } 0 \le t \le \tau_a \\ 0 & \text{otherwise} \end{cases}$$

(7-13)

This current pulse is shown in Fig. 7-1.

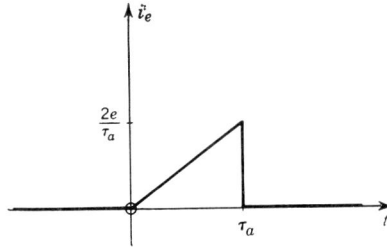

FIG. 7-1. An anode current pulse in a temperature-limited parallel-plane diode.

7-2. The Probability Distribution of Electron-emission Times†

In order to study the statistical properties of shot noise in a thermionic vacuum tube, we first need to determine the probability $P(K,\tau)$ that exactly K electrons are emitted from the tube's cathode during a time interval of length τ. It seems reasonable in the temperature-limited case to assume that the probability of emission of an electron during a given interval is statistically independent of the number of electrons emitted previously; that this probability varies as the length of the interval for short intervals, i.e., as $\Delta\tau \to 0$

$$P(1,\Delta\tau) = a\,\Delta\tau \tag{7-14}$$

where a is as yet an undetermined constant; and that the probability is negligible that more than one electron is emitted during such a short interval, i.e., approximately

$$P(0,\Delta\tau) + P(1,\Delta\tau) = 1 \tag{7-15}$$

for small $\Delta\tau$.

The probability that no electrons are emitted during an interval of length τ may be determined as follows: Consider an interval of length $\tau + \Delta\tau$ to be broken up into two subintervals, one of length τ and one of length $\Delta\tau$. Since the emission of an electron during $\Delta\tau$ is independent of the number of electrons emitted during τ, it follows that

$$P(0,\tau + \Delta\tau) = P(0,\tau)P(0,\Delta\tau)$$

If we substitute in this equation for $P(0,\Delta\tau)$ from Eqs. (7-14) and (7-15) we get for small $\Delta\tau$ the result

$$\frac{P(0,\tau + \Delta\tau) - P(0,\tau)}{\Delta\tau} = -aP(0,\tau)$$

† Cf. Feller (I, Art. 17.2) and Doob (II, Chap. VIII, Art. 4).

As $\Delta\tau \to 0$, this difference equation becomes the differential equation

$$\frac{dP(0,\tau)}{d\tau} = -aP(0,\tau) \qquad (7\text{-}16)$$

which has the solution

$$P(0,\tau) = \exp(-a\tau) \qquad (7\text{-}17)$$

where the boundary condition

$$P(0,0) = \lim_{\Delta\tau\to0} P(0,\Delta\tau) = 1 \qquad (7\text{-}18)$$

follows from Eqs. (7-14) and (7-15). Thus we have obtained a probability as the solution of a differential equation. This is an important technique.

Consider next the probability that K electrons are emitted during an interval of length $\tau + \Delta\tau$. Again we may break up that interval into two adjacent subintervals, one of length τ and the other of length $\Delta\tau$. If $\Delta\tau$ is small enough, there are only two possibilities during the subinterval of length $\Delta\tau$: either one electron is emitted during that interval, or none are. Therefore, for small $\Delta\tau$

$$P(K,\tau + \Delta\tau) = P(K - 1,\tau;1,\Delta\tau) + P(K,\tau;0,\Delta\tau)$$

Since again the emission of an electron during $\Delta\tau$ is independent of the number emitted during τ, it follows that

$$P(K,\tau + \Delta\tau) = P(K - 1,\tau)P(1,\Delta\tau) + P(K,\tau)P(0,\Delta\tau)$$

On substituting for $P(1,\Delta\tau)$ and $P(0,\Delta\tau)$, we find for small values of $\Delta\tau$ that

$$\frac{P(K,\tau + \Delta\tau) - P(K,\tau)}{\Delta\tau} + aP(K,\tau) = aP(K - 1,\tau)$$

Therefore as $\Delta\tau \to 0$ we obtain the differential equation

$$\frac{dP(K,\tau)}{d\tau} + aP(K,\tau) = aP(K - 1,\tau) \qquad (7\text{-}19)$$

as a recursion equation relating $P(K,\tau)$ to $P(K - 1,\tau)$. Since $P(K,0) = 0$, the solution of this first-order linear differential equation is†

$$P(K,\tau) = a \exp(-a\tau) \int_0^\tau \exp(at)P(K - 1,t)\, dt \qquad (7\text{-}20)$$

If now we take $K = 1$, we may use our previous result for $P(0,\tau)$ to obtain $P(1,\tau)$. This result can then be used in Eq. (7-20) to obtain

† E.g., see Courant (I, Chap. 6, Art. 3.1).

$P(2,\tau)$. Continuing this process of determining $P(K,\tau)$ from $P(K-1,\tau)$ gives

$$P(K,\tau) = \frac{(a\tau)^K \exp(-a\tau)}{K!} \tag{7-21}$$

for $K = 0, 1, 2, \ldots$

The probability that K electrons are emitted during a time interval of length τ is therefore given by the Poisson probability distribution.

The average number of electrons emitted during an interval of length τ is

$$E_\tau(K) = \sum_{K=0}^{\infty} \frac{K(a\tau)^K \exp(-a\tau)}{K!} = a\tau \tag{7-22}$$

since the possible number of electrons emitted during that interval ranges from zero to infinity. If we define $\bar{n} = E_\tau(K)/\tau$ as the average number of electrons emitted per second,† it follows that

$$a = \bar{n} \tag{7-23}$$

and that $P(K,\tau)$ may be expressed as

$$P(K,\tau) = \frac{(\bar{n}\tau)^K \exp(-\bar{n}\tau)}{K!} \tag{7-24}$$

Since the exponential tends to unity as $\bar{n}\Delta\tau \to 0$, for $K = 1$ and small $\bar{n}\Delta\tau$ this equation reduces approximately to

$$P(1,\Delta\tau) = \bar{n}\Delta\tau \tag{7-25}$$

which checks with Eq. (7-14). The probability that a single electron is emitted during a very short time interval is therefore approximately equal to the product of the average number of electrons emitted per second and the duration of that interval.

Independence of the Emission Times. Suppose that the interval $(t, t + \tau)$ is partitioned into M adjacent subintervals by the times $t = t_m$, where $m = 1, \ldots, M - 1$. Define $t_0 = t$, $t_M = t + \tau$, and

$$\tau_m = t_m - t_{m-1}$$

Then

$$\tau = \sum_{m=1}^{M} \tau_m$$

Consider now the probability $P(K_1,\tau_1; \ldots ;K_M,\tau_M|K,\tau)$ that if K elec-

† The assumption that a and hence $\bar{n}$ are constant for all time amounts to the assumption that the random process in question is stationary.

trons are emitted during the total interval of duration τ, then K_m electrons are emitted during the subinterval of duration τ_m, where

$$K = \sum_{m=1}^{M} K_m$$

From the definition of conditional probability

$$P(K_1,\tau_1; \ldots ; K_M,\tau_M | K,\tau) = \frac{P(K,\tau; K_1,\tau_1; \ldots ; K_M,\tau_M)}{P(K,\tau)}$$

Since the probability that K_M electrons are emitted during the last subinterval (that of length τ_M) is independent of the number of electrons emitted previously, we may write

$$P(K,\tau; K_1,\tau_1; \ldots ; K_M,\tau_M) = P(K,\tau; K_1,\tau_1; \ldots ; K_{M-1}, \tau_{M-1})P(K_M,\tau_M)$$

Continuing this process, it then follows that

$$P(K,\tau; K_1,\tau_1; \ldots ; K_M,\tau_M) = \prod_{m=1}^{M} P(K_m,\tau_m)$$

If now we use this result and the fact that the number of electrons emitted during a given interval has a Poisson distribution, we get

$$P(K_1,\tau_1; \ldots ; K_M,\tau_M | K,\tau)$$

$$= \left[\prod_{m=1}^{M} \frac{(\bar{n}\tau_m)^{K_m} \exp(-\bar{n}\tau_m)}{K_m!} \right] \bigg/ \frac{(\bar{n}\tau)^K \exp(-\bar{n}\tau)}{K!}$$

$$= \frac{K!}{\tau^K} \prod_{m=1}^{M} \frac{(\tau_m)^{K_m}}{K_m!} \tag{7-26}$$

Suppose next that $N \leq M$ of the K_m are unity and that the rest are zero, and let those subintervals for which $K_m = 1$ be indexed by n, where $n = 1, \ldots, N$. Here $K = N$, and we desire the probability that if N electrons are emitted during an interval of duration τ, then one is emitted in each of N nonoverlapping subintervals of duration τ_n, such that

$$\sum_{n=1}^{N} \tau_n \leq \tau$$

In this case, from Eq. (7-26)

$$P(K_1,\tau_1; \ldots ; K_M,\tau_M | N,\tau) = \frac{N!}{\tau^N} \prod_{n=1}^{N} \tau_n \tag{7-27}$$

It can be shown† that the same result may be obtained if we assume that

† Prob. 2 of this chapter.

the various electron-emission times are statistically independent random variables each with the uniform probability density function

$$p(t_n) = \begin{cases} \dfrac{1}{\tau} & \text{for } t \leq t_n \leq t + \tau \\ 0 & \text{otherwise} \end{cases} \qquad (7\text{-}28)$$

For this reason the Poisson process is sometimes called a "purely random" process.

7-3. Average Current through a Temperature-limited Diode

The total current flowing through a thermionic vacuum tube is the resultant of the current pulses produced by the individual electrons which pass through the tube. It was pointed out in Art. 7-1 that the space-charge effect of electrons in the cathode–anode space of a temperature-limited diode is negligible; hence there is effectively no interaction between the various electrons passing through such a diode. The total current, then, is simply the sum of the individual electronic current pulses, all of which have the same shape and differ only by translations in time due to the different emission times. If, therefore, K electrons are emitted in a particular temperature-limited diode during a time interval $(-T, +T)$ which is much larger than the transit time τ_a of an electron, we may write

$$I(t) = \sum_{k=1}^{K} i_e(t - t_k) \qquad \text{for } -T \leq t \leq T \qquad (7\text{-}29)$$

where $i_e(t)$ is the current pulse produced at the anode by an electron emitted at $t = 0$ and where t_k is the emission time of the kth electron emitted during the interval. This expression is not valid for values of t within τ_a of the left end of the given interval, but the end effects are usually negligible, since $2T >> \tau_a$.

Time Averaging. Let us now determine the time average $<I(t)>$ of the total current flowing through a diode. If K electrons are emitted by the given diode during the interval $(-T, +T)$, it follows from Eq. (7-29) and the definition of time average that

$$<I(t)> = \lim_{T \to \infty} \frac{1}{2T} \sum_{k=1}^{K} \int_{-T}^{+T} i_e(t - t_k) \, dt$$

Since all the current pulses have the same shape and differ only by translations in time,

$$\int_{-T}^{+T} i_e(t - t_k) \, dt = \int_{-T}^{+T} i_e(t) \, dt = e$$

where e is the charge of the electron. Each of the K terms of the above summation, therefore, has the same value e, and we obtain the hardly unexpected result that the time average of the total diode current is equal to the charge of an electron times the time average of the number of electrons passing through the diode per second; i.e.,

$$<I(t)> \ = e<n> \tag{7-30}$$

where

$$<n> \ = \lim_{T \to \infty} \frac{K}{2T} \tag{7-31}$$

Statistical Averaging.† It is of interest next to determine the statistical average of the total diode current. If we consider the particular diode studied above to be one member of an ensemble of temperature-limited diodes all having the same physical characteristics, its current waveform may be considered to be one sample function of an ergodic ensemble of diode current waveforms, as we have already assumed a statistically stationary system. The result presented in Eq. (7-30) for the time average of the diode current is, therefore, equal with probability one to the statistical average of the total diode current. However, we are not so much interested here in the result as in the method used to obtain that result.

Let us again consider the total diode current during the time interval $(-T,+T)$. The number K of electrons emitted during this interval is a random variable, as are the various emission times t_k. We may therefore obtain the statistical average $E(I_t)$ of the total diode current by averaging Eq. (7-29) with respect to the $K + 1$ random variables consisting of the K emission times t_k and K itself:

$$E(I_t) = \int_{-\infty}^{+\infty} \cdots \int_{-\infty}^{+\infty} \left[\sum_{k=1}^{K} i_e(t - t_k) \right]$$
$$p(t_1, \ldots ,t_K,K)\, dt_1 \cdots dt_K\, dK \tag{7-32}$$
$$= \int_{-\infty}^{+\infty} \cdots \int_{-\infty}^{+\infty} \left[\sum_{k=1}^{K} i_e(t - t_k) \right]$$
$$p(t_1, \ldots ,t_k|K)p(K)\, dt_1 \cdots dt_K\, dK$$

It was pointed out in Art. 7-2 that the emission times are uniformly distributed independent random variables. Therefore, using Eq. (7-28):

$$E(I_t) = \int_{-\infty}^{+\infty} p(K)\, dK \left[\sum_{k=1}^{K} \int_{-T}^{+T} \frac{dt_1}{2T} \cdots \int_{-T}^{+T} \frac{dt_K}{2T}\, i_e(t - t_k) \right]$$
$$= \int_{-\infty}^{+\infty} p(K)\, dK \left[\sum_{k=1}^{K} \frac{1}{2T} \int_{-T}^{+T} i_e(t - t_k)\, dt_k \right]$$

† Cf. Rice (I, Arts. 1.2 and 1.3).

As above, all the electronic current pulses have the same shape and integrate to e. Therefore all the K terms in the bracketed sum have the same value $e/2T$. Hence

$$E(I_t) = \frac{e}{2T} \int_{-\infty}^{+\infty} Kp(K) \, dK$$

From Art. 7-2 it follows that K has a Poisson distribution with a statistical average $\bar{n}2T$ where $\bar{n}$ is the statistical average of the number of electrons emitted per second. Thus, on defining $\bar{I} = E(I_t)$, we get

$$\bar{I} = e\bar{n} \tag{7-33}$$

which is the statistical equivalent of Eq. (7-30).

The process of statistical averaging with respect to the $K + 1$ random variables which consist of the K emission times t_k and K itself will be used several times in the remainder of this chapter.

7-4. Shot-noise Spectral Density for a Temperature-limited Diode

The autocorrelation function of the total diode current may be found by again considering the total current flowing during the interval $(-T, +T)$ to be a sum of electronic current pulses and performing the operation of statistical averaging as in Art. 7-3. Thus

$$R_I(\tau) = \int_{-\infty}^{+\infty} \cdots \int_{-\infty}^{+\infty} \left[\sum_{k=1}^{K} i_e(t - t_k) \right] \left[\sum_{j=1}^{K} i_e(t + \tau - t_j) \right]$$
$$p(t_1, \ldots, t_K, K) \, dt_1 \cdots dt_K \, dK \tag{7-34}$$

As before, the $(K + 1)$-fold joint probability density may be decomposed into product form because of the statistical independence of K and the various t_k:

$$R_I(\tau) = \int_{-\infty}^{+\infty} p(K) \, dK \left[\sum_{k=1}^{K} \sum_{j=1}^{K} \int_{-T}^{+T} \frac{dt_1}{2T} \cdots \right.$$
$$\left. \int_{-T}^{+T} \frac{dt_K}{2T} i_e(t - t_k) i_e(t + \tau - t_j) \right]$$

The K^2 terms of the double summation in this equation may be formed into two groups: the K terms for which $k = j$ and the $K^2 - K$ terms for which $k \neq j$. When $k = j$, the K-fold integral over the various t_k becomes

$$\int_{-T}^{+T} \frac{dt_1}{2T} \cdots \int_{-T}^{+T} \frac{dt_K}{2T} i_e(t - t_k) i_e(t + \tau - t_k)$$
$$= \frac{1}{2T} \int_{-\infty}^{+\infty} i_e(t) i_e(t + \tau) \, dt$$

for each value of k, since all the current pulses have the same shape (differing only by translations in time) and are all zero outside the interval $(-T, +T)$. When $k \neq j$, the integral over the t_k becomes

$$\int_{-T}^{+T} \frac{dt_1}{2T} \cdots \int_{-T}^{+T} \frac{dt_K}{2T} i_e(t - t_k) i_e(t + \tau - t_j)$$
$$= \frac{1}{2T} \int_{-T}^{+T} i_e(t - t_k) \, dt_k \frac{1}{2T} \int_{-T}^{+T} i_e(t + \tau - t_j) \, dt_j = \frac{e^2}{(2T)^2}$$

Substitution of these results into the above equation for $R_I(\tau)$ gives

$$R_I(\tau) = \frac{E_{2T}(K)}{2T} \int_{-\infty}^{+\infty} i_e(t) i_e(t + \tau) \, dt + \frac{e^2 E_{2T}(K^2 - K)}{(2T)^2}$$

Since K is a discrete random variable having a Poisson probability distribution $E_{2T}(K) = \bar{n}2T$ and

$$E_{2T}(K^2 - K) = \sum_{K=0}^{\infty} \frac{(K^2 - K)(\bar{n}2T)^K \exp(-\bar{n}2T)}{K!} = (\bar{n}2T)^2 \quad (7\text{-}35)$$

We therefore obtain the result that the autocorrelation function of the total diode current is

$$R_I(\tau) = \bar{n} \int_{-\infty}^{+\infty} i_e(t) i_e(t + \tau) \, dt + (e\bar{n})^2 \quad (7\text{-}36)$$

If now we define $i(t)$ to be the fluctuation of the total diode current about its mean value, i.e., the diode shot-noise current,

$$i(t) = I(t) - \bar{I} \quad (7\text{-}37)$$

we see that the autocorrelation function of the shot-noise current is

$$R_i(\tau) = \bar{n} \int_{-\infty}^{+\infty} i_e(t) i_e(t + \tau) \, dt \quad (7\text{-}38)$$

since $\bar{I} = e\bar{n}$. We shall evaluate $R_i(\tau)$ later for the particular case of a parallel-plane diode.

The Spectral Density. The spectral density $S_i(f)$ of the diode shot-noise current may be obtained by taking the Fourier transform of Eq. (7-38). Thus

$$S_i(f) = \bar{n} \int_{-\infty}^{+\infty} \exp(-j\omega\tau) \, d\tau \int_{-\infty}^{+\infty} i_e(t) i_e(t + \tau) \, dt$$

where $\omega = 2\pi f$. Letting $\tau = t' - t$

$$S_i(f) = \bar{n} \int_{-\infty}^{+\infty} i_e(t) \exp(j\omega t) \, dt \int_{-\infty}^{+\infty} i_e(t') \exp(-j\omega t') \, dt'$$

If now we define $G(f)$ to be the Fourier transform of the electronic current pulse $i_e(t)$,

$$G(f) = \int_{-\infty}^{+\infty} i_e(t) \exp(-j\omega t)\, dt$$

we obtain

$$S_i(f) = \bar{n}|G(f)|^2 \tag{7-39}$$

The spectral density of the total diode current is therefore

$$S_I(f) = \bar{n}|G(f)|^2 + (e\bar{n})^2\, \delta(f) \tag{7-40}$$

the impulse at zero frequency arising from the average component of the total diode current.

In many applications, the frequency of operation is small compared to the reciprocal of the transit time of the tube being used. In these cases, the spectral density of the shot-noise current may be assumed to be approximately equal to its value at zero frequency.† From the definition of $G(f)$, it follows that $G(0) = e$. From Eq. (7-39) it therefore follows that a low-frequency approximation for the spectral density of the shot-noise current generated by a temperature-limited diode is

$$S_i(f) = \bar{n}e^2 = e\bar{I} \tag{7-41}$$

This result is known as the *Schottky formula*.

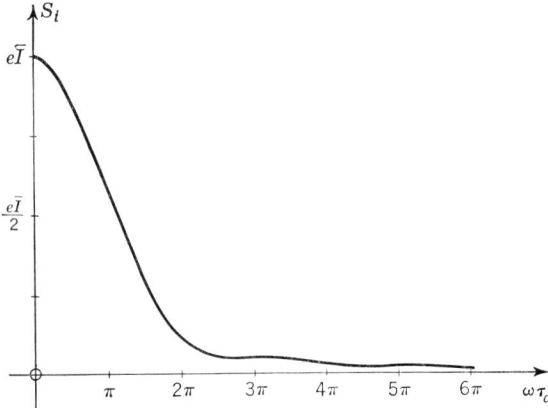

FIG. 7-2. Spectral density of shot noise in a temperature-limited parallel-plane diode.

The Parallel-plane Diode. Let us now apply these results to the case of a temperature-limited parallel-plane diode. In Art. 7-1 we found that the electronic current pulse in such a tube has the form of a triangle as given by Eq. (7-13). On substituting this into Eq. (7-38) we get

$$R_i(\tau) = \begin{cases} \dfrac{4e\bar{I}}{3\tau_a}\left(1 - \dfrac{3}{2}\dfrac{|\tau|}{\tau_a} + \dfrac{1}{2}\dfrac{|\tau|^3}{\tau_a{}^3}\right) & \text{when } -\tau_a \leq \tau \leq \tau_a \\ 0 & \text{otherwise} \end{cases} \tag{7-42}$$

† See, for example, Fig. 7-2.

as an expression for the autocorrelation function of the shot-noise current in such a tube. Taking the Fourier transform of $R_i(\tau)$ gives†

$$S_i(f) = e\bar{I}\,\frac{4}{(\omega\tau_a)^4}\,[(\omega\tau_a)^2 + 2(1 - \cos\omega\tau_a - \omega\tau_a\sin\omega\tau_a)] \qquad (7\text{-}43)$$

where $\omega = 2\pi f$, for the spectral density of the shot-noise current. This result is plotted in Fig. 7-2.

7-5. Shot-noise Probability Density for a Temperature-limited Diode‡

Since the total current flowing through a temperature-limited diode is a sum of statistically independent electronic current pulses, it might be expected from our previous discussion of the central limit theorem that the probability distribution of the total current would tend to become gaussian as the number of electrons flowing through the tube increased without limit. In fact, since approximately 6.25×10^{18} electrons per second flow through a tube per ampere of average current, we should be able to assume the total diode current to be a gaussian random process. However, rather than apply the central limit theorem directly, we shall derive a general expression for the probability distribution function of the total diode current and then study its limiting behavior as the average number of electrons passing through the tube increases without limit.

The General Form. As before, it is convenient to represent the total diode current flowing during the time interval $(-T, +T)$ as a function of the $K + 1$ random variables t_k and K. The probability density function $p(I_t)$ of the total diode current may be determined from the conditional probability density function $p(I_t|K)$ of the total current, subject to the hypothesis that K electrons have passed through the tube during the given interval, by use of the relation

$$p(I_t) = \sum_{K=0}^{\infty} p(I_t|K)P(K) \qquad (7\text{-}44)$$

which follows from Eqs. (3-20), (3-29a), and (3-33b). The conditional probability density $p(I_t|K)$ may most easily be obtained from its corresponding characteristic function $M_{(I_t|K)}(ju)$, where

$$M_{(I_t|K)}(ju) = E\left\{\exp\left[ju\sum_{k=1}^{K} i_e(t - t_k)\right]\right\}$$

† Cf. Rack (I, Part III).
‡ Cf. Rice (I, Arts. 1.4, 1.6, and 3.11).

The emission times t_k are statistically independent random variables, hence

$$M_{(I_t|K)}(ju) = \prod_{k=1}^{K} E\{\exp[jui_e(t - t_k)]\}$$

Since the emission times are uniformly distributed,

$$E\{\exp[jui_e(t - t_k)]\} = \frac{1}{2T} \int_{-T}^{+T} \exp[jui_e(t - t_k)] \, dt_k$$

All K terms of the product hence have the same form; we may therefore write

$$M_{(I_t|K)}(ju) = \left\{\frac{1}{2T} \int_{-T}^{+T} \exp[jui_e(t - \tau)] \, d\tau\right\}^K$$

The corresponding conditional probability density is

$$p(I_t|K) = \frac{1}{2\pi} \int_{-\infty}^{+\infty} \exp(-juI_t) \left\{\frac{1}{2T} \int_{-T}^{+T} \exp[jui_e(t - \tau)] \, d\tau\right\}^K du$$

Substituting this result into Eq. (7-44) and making use of the fact that K has a Poisson probability distribution, we get

$$p(I_t) = \frac{1}{2\pi} \int_{-\infty}^{+\infty} \exp(-juI_t - \bar{n}2T) \sum_{k=0}^{\infty} \frac{\left\{\bar{n} \int_{-T}^{+T} \exp[jui_e(t - \tau)] \, d\tau\right\}^K}{K!} \, du$$

The summation over K in this expression is a power-series expansion of an exponential function, hence

$$p(I_t) = \frac{1}{2\pi} \int_{-\infty}^{+\infty} \exp(-juI_t)$$
$$\exp\left\{-\bar{n}2T + \bar{n} \int_{-T}^{+T} \exp[jui_e(t - \tau)] \, d\tau\right\} du$$

If, for convenience, we now express $2T$ as the integral with respect to τ of unity over the interval $(-T, +T)$, we obtain

$$p(I_t) = \frac{1}{2\pi} \int_{-\infty}^{+\infty} \exp(-juI_t)$$
$$\exp\left(\bar{n} \int_{-T}^{+T} \left\{\exp[jui_e(t - \tau)] - 1\right\} d\tau\right) du \quad (7\text{-}45)$$

The electronic current pulse $i_e(t)$ is zero for values of t in excess of the transit time τ_a. Therefore

$$\exp[jui_e(t - \tau)] - 1 = 0 \quad (7\text{-}46)$$

for $|t - \tau| > \tau_a$ and for all t not within τ_a of the end points of the given interval. Both limits of the integral in Eq. (7-45) may hence be extended to infinity, and $t - \tau$ may be replaced by t', giving

$$p(I_t) = \frac{1}{2\pi} \int_{-\infty}^{+\infty} \exp(juI_t)$$

$$\exp\left(\bar{n} \int_{-\infty}^{+\infty} \{\exp[jui_e(t')] - 1\} \, dt'\right) du \quad (7\text{-}47)$$

as a general expression for the probability density function of the total current flowing through a temperature-limited diode. Note that this result is not a function of t, which confirms our previous assertion that the shot-noise process under study is a statistically stationary random process.

The Limiting Form. As it is not very evident from Eq. (7-47) just how $p(I_t)$ varies with I_t, let us now determine the limiting behavior of that result as the average number of electrons passing through the diode per second, $\bar{n}$, increases without limit. Equation (7-33) shows that the average of the total diode current increases without limit as $\bar{n} \to \infty$, and it follows from Eqs. (7-33) and (7-36) that the variance σ^2 of the total diode current does so also, since

$$\sigma^2 = R_I(0) - \bar{I}^2 = \bar{n} \int_{-\infty}^{+\infty} i_e^2(t) \, dt \quad (7\text{-}48)$$

To avoid such difficulties, we shall study the limiting behavior of the normalized random variable x_t corresponding to the diode current, where

$$x_t = \frac{I_t - \bar{I}}{\sigma} \quad (7\text{-}49)$$

rather than of I_t itself. The normalized diode current has a mean of zero and an rms value of unity, whatever the value of $\bar{n}$. The characteristic function of the normalized diode current is

$$M_{x_t}(jv) = E\left[\exp\left(jv\,\frac{I_t - \bar{I}}{\sigma}\right)\right] = \exp\left(-\frac{jv\bar{I}}{\sigma}\right) M_{I_t}\left(\frac{jv}{\sigma}\right) \quad (7\text{-}50)$$

where $M_{I_t}(ju)$ is the characteristic function of the total diode current, and is, from Eq. (7-47),

$$M_{I_t}(ju) = \exp\left(\bar{n} \int_{-\infty}^{+\infty} \{\exp[jui_e(t')] - 1\} \, dt'\right) \quad (7\text{-}51)$$

Therefore

$$M_{x_t}(jv) = \exp\left(-\frac{jv\bar{I}}{\sigma} + \bar{n} \int_{-\infty}^{+\infty} \left\{\exp\left[\frac{jvi_e(t')}{\sigma}\right] - 1\right\} dt'\right) \quad (7\text{-}52)$$

If now we expand in power series the argument of the integral in this equation and integrate the result term by term, we obtain

$$\bar{n} \int_{-\infty}^{+\infty} \left\{ \exp\left[\frac{jvi_e(t')}{\sigma}\right] - 1 \right\} dt'$$

$$= \frac{jv\bar{n}}{\sigma} \int_{-\infty}^{+\infty} i_e(t') \, dt' - \frac{v^2\bar{n}}{2\sigma^2} \int_{-\infty}^{+\infty} i_e^2(t') dt' - \frac{jv^3\bar{n}}{3!\sigma^3} \int_{-\infty}^{+\infty} i_e^3(t') \, dt'$$

$$+ \frac{v^4\bar{n}}{4!\sigma^4} \int_{-\infty}^{+\infty} i_e^4(t') \, dt' + \cdots$$

Substituting this result into Eq. (7-52) and using our previous result for $\bar{I}$ and σ^2 in terms of $\bar{n}$, it follows that

$$M_{x_t}(jv) = \exp\left[-\frac{v^2}{2} - \frac{jv^3\bar{n}}{3!\sigma^3} \int_{-\infty}^{+\infty} i_e^3(t') \, dt' \right.$$

$$\left. + \frac{v^4\bar{n}}{4!\sigma^4} \int_{-\infty}^{+\infty} i_e^4(t') \, dt' + \cdots \right]$$

Since σ varies as $(\bar{n})^{1/2}$, the term of the bracketed sum which contains v^k varies as $(\bar{n})^{-(k-2)/2}$ when $k \geq 3$. All the bracketed terms except the first therefore vanish as $\bar{n} \to \infty$; hence

$$\lim_{n \to \infty} M_{x_t}(jv) = \exp\left(-\frac{v^2}{2} \right) \tag{7-53}$$

The characteristic function, and hence the probability density, of the normalized diode current therefore becomes gaussian as the average number of electrons passing through the diode per second increases without limit.

Joint Probability Distribution. The various joint probability density functions which specify the statistical properties of the current flowing through a temperature-limited diode may be obtained from an analysis similar to that used above to determine $p(I_t)$. For example, in order to determine the joint probability density function $p(I_1, I_2)$ of the random variables $I_1 = I_{t+\tau}$ and $I_2 = I_t$, we may express the currents $I(t)$ and $I(t + \tau)$ as sums of electronic current pulses. Thus, assuming that K electrons are emitted during the interval $(-T, +T)$, we may write

$$I(t) = \sum_{k=1}^{K} i_e(t - t_k)$$

and

$$I(t + \tau) = \sum_{k=1}^{K} i_e(t + \tau - t_k)$$

where $-T \leq t \leq +T$ as before. Using the probability distributions of the emission times, we can determine first the conditional joint characteristic function

$$M_{(I_1,I_2|K)}(ju,jv) = E\left\{\exp\left[ju\sum_{k=1}^{K} i_e(t - t_k) + jv\sum_{k=1}^{K} i_e(t + \tau - t_k)\right]\right\}$$

and then determine the corresponding conditional joint probability density $p(I_1,I_2|K)$. The joint probability density $p(I_1,I_2)$ can be obtained by averaging $p(I_1,I_2|K)$ with respect to K. If we examine the limiting behavior of the result as $\bar{n} \to \infty$, we find that $p(I_1,I_2)$ tends to a two-dimensional gaussian probability density function. In a similar manner, it can also be shown that the Nth-order joint probability density function of the diode current tends to become gaussian as $\bar{n} \to \infty$ and hence that the diode current is a gaussian process.

7-6. Space-charge Limiting of Diode Current

Although the temperature-limited diode is of considerable theoretical interest, the space-charge-limited diode is more common in practice. The current flowing through the latter is limited by the action of the cloud of electrons in the cathode–anode space rather than by a limited supply of emitted electrons. In this section we shall derive the relation between the current flowing through a space-charge-limited diode and the voltage applied to it. In several of the following sections, we shall study the effect of space-charge limiting on shot noise. As before, we shall restrict our discussion to conventional receiving tubes as introduced in Art. 7-1.

Let us, for convenience, again consider a parallel-plane diode with cathode plane at $x = 0$, anode plane at $x = d$, a cathode voltage of zero, and an anode voltage V_a. Since the charge density in the cathode–anode space is not to be neglected here, the electric potential in that region must satisfy Poisson's equation

$$\frac{d^2V}{dx^2} = -\frac{\rho}{\epsilon_o} \tag{7-54}$$

The space-charge density at a point is a function of the current density flowing through the diode and the electron velocity at that point. The velocity v of an electron at a given point in the cathode–anode space is determined by its emission velocity v_o and the potential at that point, since the conservation of energy requires that

$$\frac{mv^2}{2} = \frac{mv_o{}^2}{2} + eV \tag{7-55}$$

Zero Emission Velocity.[†] Before discussing the case in which the emission velocity of an electron may have any positive value, let us first

[†] Cf. Harman (I, Art. 5.1).

study that in which all electrons have a zero emission velocity. Although the latter case is not very realistic physically, its analysis contains many of the steps of the more general case and yet is simple enough that the ideas involved are not obscured by the details.

Since the emission velocity of each electron is zero, each electron has, from Eq. (7-55), the velocity

$$v(x') = \left[\frac{2eV(x')}{m}\right]^{\frac{1}{2}}$$

at the plane $x = x'$ in the cathode–anode space. All electrons at x' have the same velocity; the charge density is, therefore, from Eq. (7-5),

$$-\rho = \frac{J}{v} = J\left(\frac{m}{2eV}\right)^{\frac{1}{2}}$$

and hence Poisson's equation may be written in the form

$$\frac{d^2V}{dx^2} = \frac{J}{\epsilon_o}\left(\frac{m}{2eV}\right)^{\frac{1}{2}} \qquad (7\text{-}56)$$

On multiplying both sides of Eq. (7-56) by $2dV/dx$ and integrating with respect to x, we get

$$\left(\frac{dV}{dx}\right)^2 = \frac{4J}{\epsilon_o}\left(\frac{mV}{2e}\right)^{\frac{1}{2}} \qquad (7\text{-}57)$$

since $V = 0$ and $dV/dx = 0$ at $x = 0$. The second condition results from differentiating Eq. (7-55) to obtain

$$mv\frac{dv}{dx} = e\frac{dV}{dX}$$

and using the fact (from above) that $v = 0$ at $x = 0$. Next, taking the square root of both sides of Eq. (7-57), integrating with respect to x and solving for the current density, gives

$$J = \frac{4\epsilon_o}{9}\left(\frac{2e}{m}\right)^{\frac{1}{2}}\frac{V_a^{\frac{3}{2}}}{d^2} \qquad (7\text{-}58)$$

as $V = V_a$ at $x = d$. This result is known as the *Langmuir–Child* equation and shows that the current density varies as the three-halves power of the applied voltage. Because we are dealing with a parallel-plane diode, the total current is simply the product of the current density and the anode area.

Distributed Emission Velocities.† Although the zero-emission-velocity case is attractive in its simplicity, the situation in an actual thermionic

† Cf. Langmuir (I).

diode is far more complicated. In such a diode, the emission velocity of an electron may take on any positive value and is in fact a random variable having the *Maxwell* or *Rayleigh* probability density:

$$p(v_o) = \begin{cases} \dfrac{mv_o}{kT_c} \exp\left(- \dfrac{mv_o^2}{2kT_c}\right) & \text{for } 0 \le v_o \\ 0 & \text{otherwise} \end{cases} \tag{7-59}$$

where k is Boltzmann's constant (1.38×10^{-23} joules per degree Kelvin) and T_c is the cathode temperature (in degrees Kelvin).†

Since the space charge in the cathode–anode region is due to the electrons in that region and since the charge of an electron is negative,

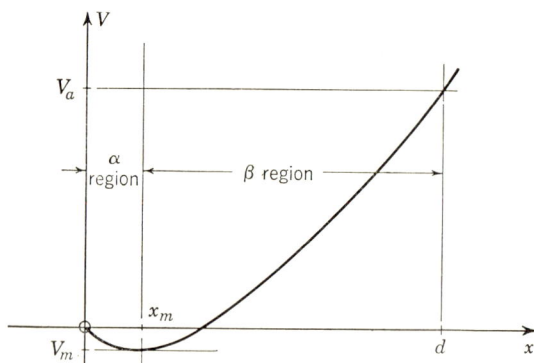

FIG. 7-3. The potential distribution in a space-charge-limited parallel-plane diode.

Poisson's equation (7-54) shows that the slope of the potential distribution in the cathode–anode region must increase with distance from the cathode. If the slope itself is positive, the corresponding electric field will exert an accelerating force on an electron and thus increase its velocity. If the slope is negative, an electron will experience a decelerating force and be slowed down. Therefore, if the space charge is to limit the flow of electrons to the anode, there must be a potential minimum between the cathode and the anode, and the potential distribution must have a form as shown in Fig. 7-3. Only those electrons having emission velocities great enough to enable them to overcome the decelerating field between the cathode and the potential minimum will be able to reach the anode. Since, from Eq. (7-55), the velocity v_m of an electron at the potential minimum is

$$v_m = \left(v_o^2 + \frac{2eV_m}{m}\right)^{1/2} \tag{7-60}$$

where V_m is the minimum potential (and is negative), the critical value

† Lindsay (I, Chap. V, Sec. 5), or van der Ziel (I, Art. 11.2).

v_c of emission velocity (i.e., the emission velocity of an electron which just comes to rest at the potential minimum) is

$$v_c = \left(-\frac{2eV_m}{m}\right)^{1/2} \tag{7-61}$$

If $v_o > v_c$, the electron will pass on to the anode; if $v_o < v_c$, the electron will return to the cathode.

Since for any reasonable value of cathode current there will be a very large number of electrons emitted per second, any measured velocity distribution will probably differ but little from the Maxwell distribution.† The emission current increment due to electrons with emission velocities in the range $(v_o, v_o + dv_o)$ therefore is

$$dJ(v_o) = J_S p(v_o)\, dv_o = J_S \left(\frac{mv_o}{kT_c}\right) \exp\left(-\frac{mv_o^2}{2kT_c}\right) dv_o \tag{7-62}$$

where J_S is the saturation or total value of emission current density. The current density J flowing past the potential minimum is then

$$J = J_S \int_{v_c}^{\infty} \left(\frac{mv_o}{kT_c}\right) \exp\left(-\frac{mv_o^2}{2kT_c}\right) dv_o = J_S \exp\left(\frac{eV_m}{kT_c}\right) \tag{7-63}$$

where the lower limit is v_c, since only those electrons with $v_o > v_c$ contribute to the anode current.

In order to be able to solve Poisson's equation, (7-54), we need to express the charge density in the plane at x in terms of the current density there. The space-charge cloud is composed of electrons having a distribution of velocities. The contribution $d\rho(v_o)$ to the space-charge density at x by electrons emitted with velocities in the range $(v_o, v_o + dv_o)$ is

$$-d\rho(v_o) = \frac{dJ(v_o)}{v(v_o)} \tag{7-64}$$

where $v(v_o)$ is the velocity at x of an electron which is emitted with a velocity v_o. Substitution in this expression for $dJ(v_o)$ from Eq. (7-62) gives

$$-d\rho(v_o) = \frac{J_S m v_o}{v(v_o)kT_c} \exp\left(-\frac{mv_o^2}{2kT_c}\right) dv_o$$

which becomes

$$-d\rho(v_o) = \frac{Jm}{kT_c} \exp\left[\frac{e(V-V_m)}{kT_c}\right] \exp\left[\frac{-mv^2(v_o)}{2kT_c}\right] \frac{v_o\, dv_o}{v(v_o)}$$

on substituting from Eq. (7-55) for v_o in the exponential and for J_S from

† Cf. Art. 5-5, in particular Eq. (5-32).

Eq. (7-63).   Since V is constant at x, it follows from Eq. (7-55) that

$$v\, dv = v_o\, dv_o$$

Hence,

$$-d\rho = \frac{Jm}{kT_c} \exp\left[\frac{e(V - V_m)}{kT_c}\right] \exp\left(\frac{-mv^2}{2kT_c}\right) dv \qquad (7\text{-}65)$$

The calculations are simplified if we now introduce a normalized potential energy

$$\eta = \frac{e(V - V_m)}{kT_c} \qquad (7\text{-}66)$$

and a normalized velocity

$$u = \left(\frac{m}{2kT_c}\right)^{\frac{1}{2}} v \qquad (7\text{-}67)$$

The equation above for the incremental charge density at x due to electrons having velocities in the range $(v, v + dv)$ can then be rewritten as

$$-d\rho = J\left(\frac{2m}{kT_c}\right)^{\frac{1}{2}} e^{\eta} e^{-u^2}\, du \qquad (7\text{-}68)$$

The total charge density at x can be obtained by integrating Eq. (7-68) over the appropriate range of velocities.

Suppose that the plane x is located in the region between the potential minimum and the anode; we shall denote this region as the β region. The only electrons in the β region are those whose emission velocities exceed the critical value, i.e., those for which v_o satisfies the inequalities

$$\frac{mv_c^2}{2} = -eV_m \leq \frac{mv_o^2}{2} < +\infty$$

The correspondng inequalities for the electron velocity at the plane x are, from Eq. (7-55),

$$e(V - V_m) \leq \frac{mv^2}{2} < +\infty$$

or, in terms of the normalized velocity at x,

$$\frac{e(V - V_m)}{kT_c} = \eta \leq u^2 < +\infty$$

The total charge density at a plane in the β region is therefore

$$-\rho_\beta = J\left(\frac{2m}{kT_c}\right)^{\frac{1}{2}} e^{\eta} \int_{\sqrt{\eta}}^{+\infty} e^{-u^2}\, du$$

The integral in this equation can be evaluated in terms of the erf function

$$\text{erf } z = \frac{2}{\sqrt{\pi}} \int_0^z e^{-u^2} \, du \tag{7-69}$$

where erf $(+\infty) = 1$.   The equation for ρ_β may be rewritten as

$$-\rho_\beta = J \left(\frac{\pi m}{2kT_c}\right)^{\frac{1}{2}} e^\eta \left[\frac{2}{\sqrt{\pi}} \int_0^\infty e^{-u^2} \, du - \frac{2}{\sqrt{\pi}} \int_0^{\sqrt{\eta}} e^{-u^2} \, du\right]$$

hence
$$-\rho_\beta = J \left(\frac{\pi m}{2kT_c}\right)^{\frac{1}{2}} e^\eta [1 - \text{erf } \sqrt{\eta}] \tag{7-70}$$

Suppose now that the plane x is located in the region between the cathode and the potential minimum; we shall denote this region as the α region.   There are now two classes of electrons passing the plane x: those passing x from the cathode going towards the potential minimum, and those passing x in the opposite direction returning to the cathode. The velocities at x of the electrons in the first class satisfy the inequalities

$$0 \le \frac{mv^2}{2} < +\infty$$

and hence
$$0 \le u^2 < +\infty$$

The electrons in the second class have emission velocities great enough to reach x but not great enough to pass the potential minimum.   For these,

$$0 \le \frac{mv^2}{2} \le e(V - V_m)$$

Hence
$$0 \le u^2 \le \eta$$

Since the charge densities due to these two classes of electrons both have the same sign, the total charge density at x is

$$-\rho_\alpha = J \left(\frac{2m}{kT_c}\right)^{\frac{1}{2}} e^\eta \left[\int_0^{+\infty} e^{-u^2} \, du + \int_0^{\sqrt{\eta}} e^{-u^2} \, du\right]$$

Therefore, using Eq. (7-69),

$$-\rho_\alpha = J \left(\frac{\pi m}{2kT_c}\right)^{\frac{1}{2}} e^\eta [1 + \text{erf } \sqrt{\eta}] \tag{7-71}$$

Let us now substitute the expressions for space-charge density obtained above into the Poisson equation.   Thus

$$\frac{d^2V}{dx^2} = \frac{J}{\epsilon} \left(\frac{\pi m}{2kT_c}\right)^{\frac{1}{2}} e^\eta [1 \pm \text{erf } \sqrt{\eta}] \tag{7-72}$$

where the positive sign corresponds to values of x in the α region and the negative sign corresponds to values of x in the β region.   On multiplying both sides of Eq. (7-72) by $2dV/dx$ and integrating from x_m to x, we get

$$\left(\frac{dV}{dx}\right)^2 = \frac{2J}{\epsilon}\left(\frac{\pi m}{2kT_c}\right)^{1/2} \int_{V_m}^{V} e^{\eta}[1 \pm \text{erf } \sqrt{\eta}] \, dV$$

since $(dV/dx) = 0$ at $x = x_m$. From the definition of η, it follows that

$$d\eta = \frac{e}{kT_c} dV$$

Hence

$$\left(\frac{d\eta}{dx}\right)^2 = \frac{Je}{\epsilon} \frac{(2\pi m)^{1/2}}{(kT_c)^{3/2}} \int_0^{\eta} e^{\eta}[1 \pm \text{erf } \sqrt{\eta}] \, d\eta \qquad (7\text{-}73)$$

In terms of a new space variable

$$\xi = \left(\frac{Je}{\epsilon}\right)^{1/2} \frac{(2\pi m)^{1/4}}{(kT_c)^{3/4}} (x - x_m) \qquad (7\text{-}74)$$

Eq. (7-73) can be rewritten as

$$\left(\frac{d\eta}{d\xi}\right)^2 = \int_0^{\eta} e^{\eta}[1 \pm \text{erf } \sqrt{\eta}] \, d\eta$$

$$= e^{\eta} - 1 \pm \int_0^{\eta} e^{\eta} \text{ erf } \sqrt{\eta} \, d\eta$$

The second integral can be integrated by parts giving the result

$$\left(\frac{d\eta}{d\xi}\right)^2 = \phi(\eta) \qquad (7\text{-}75)$$

where

$$\phi(\eta) = e^{\eta} - 1 \pm \left(e^{\eta} \text{ erf } \sqrt{\eta} - \frac{2}{\sqrt{\pi}} \sqrt{\eta}\right) \qquad (7\text{-}76)$$

The problem has thus been reduced to one of integration as the solution of Eq. (7-75) is the integral

$$\xi = \int_0^{\eta} \frac{d\eta}{[\phi(\eta)]^{1/2}} \qquad (7\text{-}77)$$

Unfortunately, a closed-form solution of Eq. (7-77) in terms of elementary functions is not possible and numerical integration is required. The results of such a numerical solution are tabulated in Langmuir (I).

An approximate solution of Eq. (7-77) can, however, be readily obtained under the assumption that η is large compared with unity. In this case one can show that Eq. (7-76) reduces approximately to

$$\phi_\beta(\eta) = 2\left(\frac{\eta}{\pi}\right)^{1/2} - 1$$

for values of x in the β region. Equation (7-77) then becomes approximately

$$\xi_\beta = \left(\frac{\pi}{4}\right)^{\frac14} \int_0^\eta \eta^{-\frac14}\left(1 - \frac{1}{2}\sqrt{\frac{\pi}{\eta}}\right)^{-\frac12} d\eta$$

A power-series expansion of the bracketed term gives

$$\left(1 - \frac{1}{2}\sqrt{\frac{\pi}{\eta}}\right)^{-\frac12} = 1 + \frac{1}{4}\sqrt{\frac{\pi}{\eta}} + \cdots$$

On substituting this expansion into our expression for ξ_β and integrating, we get

$$\xi_\beta = \frac{\pi^{\frac14}2^{\frac36}\eta^{\frac34}}{3}\left[1 + \frac{3}{4}\left(\frac{\pi}{\eta}\right)^{\frac12} + \cdots\right] \tag{7-78}$$

The corresponding expression for current density in terms of anode potential can be obtained by substituting in this result for $\xi(x = d)$ from Eq. (7-74) and for $\eta(V = V_a)$ from Eq. (7-66) and then solving for J. Thus

$$J = \frac{4\epsilon_o}{9}\left(\frac{2e}{m}\right)^{\frac12}\frac{(V_a - V_m)^{\frac32}}{(d - x_m)^2}\left\{1 + \frac{3}{2}\left[\frac{\pi k T_c}{e(V_a - V_m)}\right]^{\frac12} + \cdots\right\} \tag{7-79}$$

which, under the assumption that $\eta >> 1$, reduces to

$$J = \frac{4\epsilon_o}{9}\left(\frac{2e}{m}\right)^{\frac12}\frac{(V_a - V_m)^{\frac32}}{(d - x_m)^2} \tag{7-80}$$

If we compare this result with that obtained for the zero-emission-velocity case, we see that the space-charge-limited thermionic diode with distributed electron-emission velocities acts (when its anode potential is large compared with kT_c/e) as though its cathode were located at the potential minimum and each of its electrons had a zero emission velocity.

7-7. Shot Noise in a Space-charge-limited Diode

Fluctuations in emission of electrons from a heated cathode give rise to fluctuations in the current flowing through a space-charge-limited diode as well as in a temperature-limited diode. However, even though the electron-emission times in a space-charge-limited diode are statistically independent random variables, the arrival of an electron at the anode is certainly dependent upon the previously emitted electrons. For, as discussed in the previous section, whether or not a given electron reaches the anode depends upon whether or not its emission velocity is sufficient to enable it to pass the potential minimum, and the depth of the potential minimum depends upon the electrons previously emitted.

The effect of space-charge limiting on the shot-noise spectral density is not too difficult to determine qualitatively if we restrict our discussion to frequencies which are small compared to the reciprocal of the average transit time: Suppose that the rate of emission of electrons increases momentarily. The added electrons will increase the space charge present, hence the depth of the potential minimum as well as the critical value of the emission velocity will be increased. Thus, although more electrons are emitted, the proportion of emitted electrons which reach the anode is reduced. On the other hand, if there is a momentary decrease in the rate of electron emission, the depth of the potential minimum will decrease and the proportion of emitted electrons reaching the anode will be increased. The effect of space-charge limiting is, therefore, to smooth out the current fluctuations and hence to reduce the shot noise to a value below that in a temperature-limited diode with the same average current. This fluctuation-smoothing action of the space charge is essentially the same phenomenon as that which limits the diode current to a value below saturation.

Quantitative determination of the effect of space-charge limiting on the low-frequency spectral density of shot noise is rather complicated, and we shall not attempt it here. However, the steps involved are as follows:[†] Given that the electron-emission-velocity distribution is Maxwellian, the analysis of the preceding section determines the anode current density. Suppose now that there is an increase in emission of electrons with initial velocities in the range $(v_o, v_o + dv_o)$ and hence an increase $\delta J_o(v_o)$ in emission current density. The new value of anode current density may be obtained, as in the preceding section, by determining the space-charge density present, substituting in Poisson's equation, and solving that equation. If the increase in emission is small enough, the differential equation relating the normalized potential energy η and the space variable ξ will differ from Eq. (7-78) only by the addition of a perturbation term to $\phi(\eta)$. However, a numerical solution is again required, though an approximate solution can again be obtained which is valid for large η. For small changes, the resulting increase in anode current density δJ will be proportional to the increase in emission current density; i.e.,

$$\delta J = \gamma(v_o)\,\delta J_o(v_o)$$

where $\gamma(v_o)$ is called the *linear reduction factor* and is a function of v_o.

The actual distribution of electron-emission velocities fluctuates about the Maxwell distribution from instant to instant, giving rise to emission current fluctuations. The basic assumptions in an analysis of shot noise are that the emission current is temperature-limited and that any *fixed* division of the emission current is temperature-limited. The spectral density of the emission-current-density fluctuations caused by variations

† Cf. Rack (I) or Thompson, North, and Harris (I, pp. 75–125, Part II, "Diodes and Negative-grid Triodes").

in electron emission in the fixed velocity range $(v_o, v_o + dv_o)$ is therefore $e \, dJ_o(v_o)$. Since a fluctuation $\delta J_o(v_o)$ in emission current density produces a fluctuation $\gamma(v_o) \delta J_o(v_o)$ in anode current density, the spectral density $S_{\delta J}(f)$ of the resultant anode-current-density fluctuations is

$$S_{\delta J}(f) = \gamma^2(v_o) e \, dJ_o(v_o)$$

The spectral density $S_j(f)$ of the total fluctuation of the anode current density may be obtained by summing up this result over all velocity groups; thus

$$S_j(f) = e \int_0^\infty \gamma^2(v_o) \, dJ_o(v_o) = e \int_0^\infty \gamma^2(v_o) J \, sp(v_o) \, dv_o$$

where $p(v_o)$ is the velocity distribution given by Eq. (7-59) and $dJ(v_o)$ is given by Eq. (7-62). Using Eq. (7-63), this may be rewritten as

$$S_j(f) = eJ \int_0^\infty \gamma^2(v_o) \exp\left(-\frac{eV_m}{kT_c}\right) p(v_o) \, dv_o$$

The spectral density of the anode current fluctuations is given by this result multiplied by the anode area. Thus we see that the effect of space-charge limiting on the low-frequency spectral density of the anode current fluctuations may be accounted for by multiplying the Schottky formula, Eq. (7-41), by a *space-charge-smoothing factor* Γ^2:

$$S_i(f) = e\bar{I}\Gamma^2 \qquad (7\text{-}81)$$

where Γ^2 is given by the above integral and $0 \le \Gamma^2 \le 1$. A plot of Γ^2 vs. η is given in Fig. 7-4. This plot is based upon North's results, which assume that the diode current is well below its saturation value. An asymptotic expression for Γ^2 obtained by North is

$$\Gamma^2 = \frac{9(1 - \pi/4)}{\eta} = \frac{9(1 - \pi/4)kT_c}{e(V_a - V_m)} \qquad (7\text{-}82)$$

This result, also plotted in Fig. 7-4, is approximately valid for values of η large compared with unity but small enough that the diode still operates in the space-charge-limited region (and not in the temperature-limited region). The asymptotic form of Γ^2 may also be expressed as

$$\Gamma^2 = 3\left(1 - \frac{\pi}{4}\right)\frac{2kT_c g_d}{e\bar{I}} \qquad (7\text{-}83)$$

where g_d is the dynamic conductance of the diode;

$$g_d = \frac{\partial I}{\partial V_a} = \frac{3}{2}\left(\frac{I}{V_a - V_m}\right) \qquad (7\text{-}84)$$

following from Eq. (7-80).

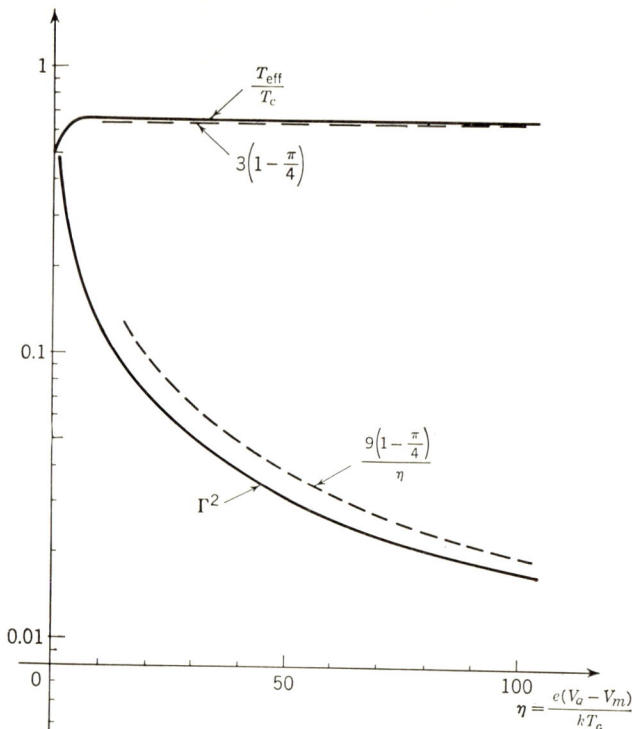

FIG. 7-4. Space-charge-reduction factor and effective temperature versus normalized potential energy for a space-charge-limited diode. (*Adapted from Fig. 5, Part II, of Thompson, North, and Harris (I). By permission of RCA.*)

Substituting Eq. (7-83) into the modified Schottky formula, Eq. (7-81), we find

$$S_i(f) = 2kT_{eff}g_d \qquad (7\text{-}85)$$

where
$$T_{eff} = 3\left(1 - \frac{\pi}{4}\right)T_c = 0.644T_c \qquad (7\text{-}86)$$

Equation (7-86) gives an asymptotic expression for T_{eff}; a plot of T_{eff}/T_c vs. η based on North's results is given in Fig. 7-4. In Art. 9-4 we shall see that a conductance g operating at a temperature T generates a gaussian thermal noise current which has a low-frequency spectral density

$$S(f) = 2kTg$$

From the noise viewpoint, then, a space-charge-limited diode acts as though it were a conductance g_d operating at a temperature equal to 0.644 times the cathode temperature.

7-8. Shot Noise in Space-charge-limited Triodes and Pentodes

It is beyond our scope to give a comprehensive account of the noise properties of multielectrode thermionic vacuum tubes, whether for shot or other types of noise. The reader is referred to other sources† for such studies. We do, however, wish to derive some low-frequency shot-noise results for the triode and the pentode.

The High-mu Triode. Neglecting the noise induced in the grid circuit by the random fluctuations of the electron current passing through the grid plane, a reasonably valid low-frequency analysis of the shot noise generated in a negative-grid high-mu triode may be made as follows: If it is assumed that no electrons are intercepted by the grid of the triode, and that all electrons are emitted from the cathode with zero initial velocity, it may be shown that‡ the anode current density of a parallel-plane negative-grid triode can be expressed as

$$J = \frac{4\epsilon_o}{9}\left(\frac{2e}{m}\right)^{\frac{1}{2}} \frac{[\sigma(V_g + V_p/\mu)]^{\frac{3}{2}}}{d_{cg}^{\ 2}} \tag{7-87}$$

where V_g is the cathode-grid potential difference of the triode, V_p is the cathode–anode potential difference, μ is the amplification factor, and σ is a geometrical factor which has the form

$$\sigma = \frac{1}{1 + \frac{1}{\mu}\left(\frac{d_{cp}}{d_{cg}}\right)^{\frac{4}{3}}} \tag{7-88}$$

where d_{cg} is the cathode-grid spacing of the triode, and d_{cp} is the cathode–anode spacing. For common receiving tubes, $0.5 \leq \sigma \leq 1$.

A comparison of Eq. (7-87) with the equivalent expression for the parallel-plane diode, Eq. (7-58), shows that the behavior of the triode can be expressed in terms of that of an *equivalent diode* which has a cathode–anode spacing equal to the cathode-grid spacing of the triode and a cathode–anode potential difference V_a given by the equation

$$V_a = \sigma\left(V_g + \frac{V_p}{\mu}\right) \tag{7-89}$$

The results of the previous article can now be applied to the equivalent diode in order to obtain the noise properties of the triode. For example, the low-frequency spectral density of the anode-current shot noise generated in a triode can be expressed from Eq. (7-85) as

$$S_{i_T}(f) = 2k(0.644T_c)g_{eq} \tag{7-90}$$

† E.g., van der Ziel (I, Chaps. 5, 6, 14, and 15).
‡ See Spangenberg (I, Art. 8.5).

where g_{eq} is the dynamic conductance of the equivalent diode. Since

$$g_{eq} = \frac{\partial I}{\partial V_a} = \frac{\partial I}{\partial V_g}\left(\frac{\partial V_g}{\partial V_a}\right) = g_m \frac{1}{\sigma}$$

where g_m is the dynamic transconductance of the triode, we may rewrite Eq. (7-90) in terms of the triode parameters as

$$S_{i_T}(f) = 2k\left(\frac{0.644T_c}{\sigma}\right)g_m \qquad (7\text{-}91)$$

The Pentode.† Let us now determine the spectral density of the shot noise present in the anode current of a pentode. We shall assume throughout that the voltages, with respect to the cathode, of all electrodes are maintained constant and that the control grid is biased negatively to the extent that it does not intercept any electrons. The region between the cathode and the screen grid then acts as a space-charge-limited triode, and our triode results above apply to the shot noise generated in the cathode current stream of the pentode. However, an additional effect takes place in a pentode as a result of the division of cathode current between the screen grid and the anode. If the screen grid is fine enough, whether or not a particular electron is intercepted by it is independent of what happens to other electrons; the random division of cathode current between the screen grid and the anode generates an additional noise called *partition noise*.

In order to study partition noise, it is most convenient to represent the cathode current (before it reaches the screen grid) as a sum of electronic current pulses in a manner similar to that used in Arts. 7-3 and 7-4. Thus, assuming that K electronic current pulses occur during the time interval $(-T, +T)$, we write for the cathode current,

$$I_c(t) = \sum_{k=1}^{K} i(t - t_k) \qquad (7\text{-}92)$$

where the "emission" times t_k correspond, say, to the times the electrons pass the potential minimum. The t_k may be assumed to be uniformly distributed; i.e.,

$$p(t_k) = \begin{cases} \dfrac{1}{2T} & \text{for } -T \le t_k \le +T \\ 0 & \text{otherwise} \end{cases} \qquad (7\text{-}93)$$

but are not independent random variables because of the action of the potential minimum. Since the interception of an electron by the screen

† Cf. Thompson, North, and Harris (I, Part III, "Multicollectors," pp. 126–142).

grid is independent from one electron to another, we can write for the anode current,

$$I_a(t) = \sum_{k=1}^{K} Y_k i(t - t_k) \tag{7-94}$$

where the Y_k are independent random variables, each of which may assume either of the two possible values: unity, with probability p; and zero, with probability $(1 - p)$.

Since the Y_k and t_k are independent random variables, the average value of the anode current is

$$\bar{I}_a = E \left[\sum_{k=1}^{K} Y_k i(t - t_k) \right]$$

$$= pE \left[\sum_{k=1}^{K} i(t - t_k) \right] = p\bar{I}_c \tag{7-95}$$

therefore,

$$p = \frac{\bar{I}_a}{\bar{I}_c} \quad \text{and} \quad (1 - p) = \frac{\bar{I}_2}{\bar{I}_c} \tag{7-96}$$

where $I_2(t) = I_c(t) - I_a(t)$ is the screen-grid current.

The autocorrelation function of the anode current is

$$R_a(\tau) = E \left[\sum_{k=1}^{K} Y_k i(t - t_k) \sum_{j=1}^{K} Y_j i(t + \tau - t_j) \right]$$

$$= E \left[\sum_{k=1}^{K} \sum_{j=1}^{K} \overline{Y_k Y_j} i(t - t_k) i(t + \tau - t_j) \right] \tag{7-97}$$

where $\overline{Y_k Y_j}$ is the average of $Y_k Y_j$. Since the Y_k are independent random variables,

$$\overline{Y_k Y_j} = \begin{cases} \overline{Y_k^2} = p & \text{when } k = j \\ \overline{Y_k Y_j} = p^2 & \text{when } k \neq j \end{cases} \tag{7-98}$$

Therefore, using these results in Eq. (7-97), it follows that

$$R_a(\tau) = pE \left[\sum_{k=1}^{K} i(t - t_k) i(t + \tau - t_k) \right]$$

$$+ p^2 E \left[\sum_{\substack{k=1 \\ (k \neq j)}}^{K} \sum_{j=1}^{K} i(t - t_k) i(t + \tau - t_j) \right]$$

This becomes, on adding and subtracting p^2 times the first term,

$$R_a(\tau) = p(1 - p)E\left[\sum_{k=1}^{K} i(t - t_k)i(t + \tau - t_k)\right]$$

$$+ p^2 E\left[\sum_{k=1}^{K}\sum_{j=1}^{K} i(t - t_k)i(t + \tau - t_j)\right] \quad (7\text{-}99)$$

Since all current pulses have (approximately) the same shape,

$$E\left[\sum_{k=1}^{K} i(t - t_k)i(t + \tau - t_k)\right]$$

$$= \int_{-\infty}^{+\infty} p(K)\,dK \sum_{k=1}^{K}\frac{1}{2T}\int_{-T}^{+T} i(t - t_k)i(t + \tau - t_k)\,dt_k$$

$$= \bar{n}\int_{-\infty}^{+\infty} i(t)i(t + \tau)\,dt \quad (7\text{-}100)$$

where $\bar{n} = \bar{K}/2T$ is the average number of electrons passing the potential minimum per second. The autocorrelation function of the anode current is therefore

$$R_a(\tau) = p(1 - p)\bar{n}\int_{-\infty}^{+\infty} i(t)i(t + \tau)\,dt + p^2 R_c(\tau) \quad (7\text{-}101)$$

since the second term in Eq. (7-99) is simply p^2 times the autocorrelation function $R_c(\tau)$ of the cathode current.

The spectral density of the anode current can now be obtained by taking the Fourier transform of Eq. (7-101). From a development paralleling that in the section on spectral density in Art. 7-4, it follows that the low-frequency approximation for the Fourier integral of the first term in Eq. (7-101) is $p(1 - p)e\bar{I}_c$. Since the cathode current stream is space-charge limited, its shot-noise spectral density has a low-frequency approximation $e\bar{I}_c\Gamma^2$. Using Eqs. (7-96), we can then express the low-frequency spectral density of the anode current as

$$S_a(f) = \frac{\bar{I}_a\bar{I}_2}{\bar{I}_c^{\,2}}\,e\bar{I}_c + \left(\frac{\bar{I}_a}{\bar{I}_c}\right)^2[e\bar{I}_c\Gamma^2 + \bar{I}_c^{\,2}\delta(f)]$$

the impulse term being due to the average value of the cathode current. The shot-noise portion $S_{a'}(f)$ of the low-frequency spectral density of the anode current is therefore

$$S_{a'}(f) = e\bar{I}_a\left[\frac{\bar{I}_2}{\bar{I}_c} + \frac{\bar{I}_a}{\bar{I}_c}\Gamma^2\right] \quad (7\text{-}102)$$

which may be rewritten as

$$S_{a'}(f) = e\bar{I}_a\left[\Gamma^2 + \frac{\bar{I}_2}{\bar{I}_c}(1 - \Gamma^2)\right] \quad (7\text{-}103)$$

Since $0 \leq \Gamma^2 \leq 1$ and $\bar{I}_2 + \bar{I}_a\Gamma^2 \leq \bar{I}_c$, it follows from Eqs. (7-102) and (7-103) that

$$e\bar{I}_a\Gamma^2 \leq S_{a'}(f) \leq e\bar{I}_a \qquad (7\text{-}104)$$

i.e., a space-charge-limited pentode generates more anode current noise than does a space-charge-limited diode but less than a temperature-limited diode when all have the same average value of anode current.

The results presented above are based upon the assumption that the screen grid voltage was maintained constant with respect to the cathode. If this condition is not satisfied, a correction term must be added to account for a correlation between screen grid and anode noise currents.

7-9. Problems

1.† Suppose that the emission times t_n are statistically independent random variables, that the probability that one electron is emitted during the short time interval $\Delta\tau$ is $a\,\Delta\tau < <1$, and that the probability that more than one electron is emitted during that interval is zero. Derive Eq. (7-17) by breaking the interval $(0,\tau)$ up into M subintervals of length $\Delta\tau = \tau/M$, finding the probability that no electron is emitted in any one of the subintervals, and letting $M \to \infty$.

2. Assuming that the emission times t_n are statistically independent random variables, each with the probability density

$$p(t_n) = \begin{cases} \dfrac{1}{\tau} & \text{for } t \leq t_n \leq t + \tau \\ 0 & \text{otherwise} \end{cases} \qquad (7\text{-}105)$$

derive Eq. (7-27).

3. Suppose that the electronic current pulse in a temperature-limited diode had the shape

$$i_e(t) = \begin{cases} ae^{-\alpha t} & \text{for } 0 \leq t \\ 0 & \text{for } t < 0 \end{cases} \qquad (7\text{-}106)$$

What would be the autocorrelation function and spectral density of the resultant shot noise?

4. The random process $x(t)$ is a linear superposition of identical pulses $i(t - t_k)$, where

$$\int_{-\infty}^{+\infty} i(t)\,dt = 0$$

The emission times t_k are statistically independent and have a uniform probability density, as in Eq. (7-105). The probability that K emission times occur in an interval of length τ is given by the Poisson probability distribution equation (7-24).

A new random process $y(t)$ is defined by

$$y(t) = x(t)x(t + \delta) \qquad (7\text{-}107)$$

where δ is fixed. Show that

$$R_y(\tau) = R_x{}^2(\tau) + R_x{}^2(\delta) + R_x(\delta + \tau)R_x(\delta - \tau) \qquad (7\text{-}108)$$

† Cf. Rice (I, Sec. 1.1).

5. Suppose that a parallel-plane thermionic diode with a Maxwell distribution of electron-emission velocities is operated with its anode potential V_a sufficiently negative with respect to the cathode that the minimum value of potential in the cathode–anode space occurs at the anode.

Show that the anode current I_a is given by

$$I_a = I_s \exp\left(\frac{eV_a}{kT_c}\right) \tag{7-109}$$

where I_s is the saturation value of emission current, that the space-charge-smoothing factor is unity, and hence that the low-frequency spectral density of the anode current fluctuations is given by

$$S_i(f) = e\bar{I}_a = 2k\left(\frac{T_c}{2}\right)g_d \tag{7-110}$$

where $g_d = \partial I_a/\partial V_a$ is the diode conductance.

6. Suppose that an electron entering an electron multiplier produces n electrons out with a probability p_n and that the arrival time of an input electron has a uniform probability density, as in Eq. (7-105). Show that

$$\bar{n} = \frac{\bar{I}_o}{\bar{I}_i}$$

where I_o and I_i are the output and input currents, respectively, of the electron multiplier. Show further that the low-frequency spectral density of the output-current fluctuations is given by

$$S_o(f) = \sigma^2(n)e\bar{I}_i + \bar{n}^2 S_i(f)$$

where $S_i(f)$ is the low-frequency spectral density of the input current fluctuations.

THE GAUSSIAN PROCESS

We saw in Chap. 5 that the probability distribution of a sum of independent random variables tends to become gaussian as the number of random variables being summed increases without limit. In Chap. 7 we saw that the shot noise generated in a thermionic vacuum tube is a gaussian process, and we shall observe in Chap. 9 that the voltage fluctuation produced by thermal agitation of electrons in a resistor is also a gaussian process. The gaussian random variable and the gaussian random process are therefore of special importance and warrant special study. We shall investigate some of their properties in this chapter.

8-1. The Gaussian Random Variable

A real random variable x having the probability density function

$$p(x) = \frac{1}{\sqrt{2\pi}} e^{-x^2/2} \tag{8-1}$$

shown in Fig. 8-1 is a *gaussian random variable* and has the characteristic function

$$M_x(jv) = e^{-v^2/2} \tag{8-2}$$

The corresponding probability distribution function† is

$$P(x \leq X) = \frac{1}{\sqrt{2\pi}} \int_{-\infty}^{X} e^{-x^2/2} \, dx \tag{8-3}$$

An asymptotic expansion‡ for $P(x \leq X)$ valid for large values of X is

$$P(x \leq X) = 1 - \frac{e^{-X^2/2}}{\sqrt{2\pi}\, X} \left[1 - \frac{1}{X^2} + \frac{1 \cdot 3}{X^4} - \cdots \right] \tag{8-4}$$

† Extensive tables of the gaussian $p(x)$ and $2P(\xi \leq x) - 1$ are given in National Bureau of Standards (I).
‡ Cf. Dwight (I, Eq. 586).

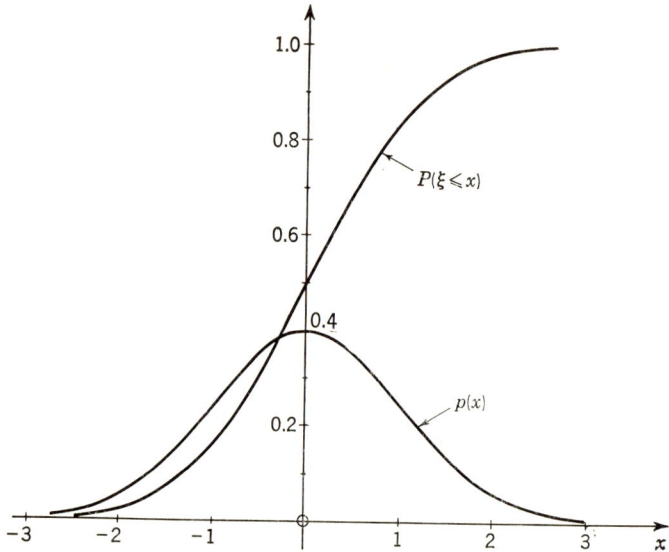

FIG. 8-1. Gaussian probability functions.

Since the gaussian probability density function defined above is an even function of x, the nth moment

$$E(x^n) = \frac{1}{\sqrt{2\pi}} \int_{-\infty}^{+\infty} x^n e^{-x^2/2} \, dx$$

is zero for odd values of n:

$$E(x^n) = 0 \qquad (n \text{ odd}) \tag{8-5a}$$

When $n \geq 2$ is even, we obtain

$$E(x^n) = 1 \cdot 3 \cdot 5 \cdots (n-1) \tag{8-5b}$$

either by direct evaluation of the integral,† or by successive differentiations of the gaussian characteristic function. In particular,

$$E(x) = 0$$

and

$$\sigma^2(x) = E(x^2) = 1 \tag{8-6}$$

Consider the new random variable

$$y = \sigma x + m \tag{8-7}$$

where x is a gaussian random variable as above. From Eqs. (8-6) and (8-7), it follows that

$$E(y) = m$$

and

$$\sigma^2(y) = \sigma^2 \tag{8-8}$$

† See for example Cramér (I, Art. 10.5) or Dwight [I, Eq. (861.7)].

The probability density function of y is, using Eq. (3-43),

$$p(y) = \frac{1}{\sqrt{2\pi}\,\sigma} \exp\left[-\frac{(y-m)^2}{2\sigma^2}\right] \tag{8-9}$$

The corresponding characteristic function is

$$M_y(jv) = \exp\left[jvm - \frac{v^2\sigma^2}{2}\right] \tag{8-10}$$

The random variable y is a *gaussian random variable with mean m and variance σ^2.*

The nth central moment of the random variable y is given by

$$\mu_n = E[(y-m)^n] = \sigma^n E(x^n)$$

It therefore follows, from Eqs. (8-5), that

$$\mu_n = 0 \qquad \text{for } n \text{ odd} \tag{8-11a}$$

and
$$\mu_n = 1 \cdot 3 \cdot 5 \, \cdots \, (n-1)\sigma^n \tag{8-11b}$$

when $n \geq 2$ is even.

8-2. The Bivariate Distribution

Suppose that x_1 and x_2 are statistically independent gaussian random variables with zero means and with variances $\sigma_1{}^2$ and $\sigma_2{}^2$, respectively. Their joint probability density function is, from Eqs. (3-40) and (8-9),

$$p(x_1, x_2) = \frac{1}{2\pi\sigma_1\sigma_2} \exp\left(-\frac{x_1{}^2}{2\sigma_1{}^2} - \frac{x_2{}^2}{2\sigma_2{}^2}\right) \tag{8-12}$$

Let us now define two random variables y_1 and y_2 in terms of x_1 and x_2 by the rotational transformation:

$$\begin{aligned} y_1 &= x_1 \cos\theta - x_2 \sin\theta \\ y_2 &= x_1 \sin\theta + x_2 \cos\theta \end{aligned} \tag{8-13}$$

The means of y_1 and y_2 are both zero, and the variances are

$$\begin{aligned} \mu_{20} &= E(y_1{}^2) = \sigma_1{}^2 \cos^2\theta + \sigma_2{}^2 \sin^2\theta \\ \mu_{02} &= E(y_2{}^2) = \sigma_1{}^2 \sin^2\theta + \sigma_2{}^2 \cos^2\theta \end{aligned} \tag{8-14}$$

The covariance is

$$\mu_{11} = E(y_1 y_2) = (\sigma_1{}^2 - \sigma_2{}^2) \sin\theta \cos\theta \tag{8-15}$$

which in general is not zero.

The joint probability density function of y_1 and y_2 can now be obtained from that of x_1 and x_2. On solving Eq. (8-13) for x_1 and x_2 in terms of y_1 and y_2, we get

$$x_1 = y_1 \cos \theta + y_2 \sin \theta$$
$$x_2 = -y_1 \sin \theta + y_2 \cos \theta$$

i.e., the inverse of a rotation is a rotation. The Jacobian of this transformation is

$$|J| = \begin{vmatrix} \dfrac{\partial x_1}{\partial y_1} & \dfrac{\partial x_2}{\partial y_1} \\[2mm] \dfrac{\partial x_1}{\partial y_2} & \dfrac{\partial x_2}{\partial y_2} \end{vmatrix} = \begin{vmatrix} \cos \theta & -\sin \theta \\[1mm] \sin \theta & \cos \theta \end{vmatrix} = 1$$

and it therefore follows, using Eqs. (3-54) and (8-12), that

$$p(y_1,y_2) = \frac{1}{2\pi\sigma_1\sigma_2} \exp\left[-\frac{(y_1 \cos \theta + y_2 \sin \theta)^2}{2\sigma_1{}^2} - \frac{(-y_1 \sin \theta + y_2 \cos \theta)^2}{2\sigma_2{}^2} \right]$$

This result may be expressed in terms of the various second-order moments of y_1 and y_2 as

$$p(y_1,y_2) = \frac{1}{2\pi(\mu_{20}\mu_{02} - \mu_{11}{}^2)^{1/2}} \exp\left[\frac{-\mu_{02}y_1{}^2 + 2\mu_{11}y_1y_2 - \mu_{20}y_2{}^2}{2(\mu_{20}\mu_{02} - \mu_{11}{}^2)} \right] \quad (8\text{-}16)$$

The random variables y_1 and y_2 are said to have a *bivariate gaussian* probability density function. Hence two random variables y_1 and y_2 which have a bivariate gaussian probability density function, as in Eq. (8-16), can be transformed into two independent gaussian random variables by a rotation of coordinates.

The joint characteristic function of y_1 and y_2 can similarly be obtained from that of x_1 and x_2, and is

$$M(jv_1,jv_2) = \exp[-\tfrac{1}{2}(\mu_{20}v_1{}^2 + 2\mu_{11}v_1v_2 + \mu_{02}v_2{}^2)] \quad (8\text{-}17)$$

General Formulas. The random variables

$$Y_1 = \frac{y_1}{\mu_{20}{}^{1/2}} \quad \text{and} \quad Y_2 = \frac{y_2}{\mu_{02}{}^{1/2}} \quad (8\text{-}18)$$

are standardized random variables, since

$$E(Y_1) = 0 = E(Y_2)$$
$$\sigma^2(Y_1) = 1 = \sigma^2(Y_2) \quad (8\text{-}19)$$

Their joint probability density function is, from Eq. (8-18),

$$p(Y_1,Y_2) = \frac{1}{2\pi(1 - \rho^2)^{1/2}} \exp\left[-\frac{Y_1{}^2 - 2\rho Y_1 Y_2 + Y_2{}^2}{2(1 - \rho^2)} \right] \quad (8\text{-}20)$$

where ρ is the correlation coefficient of Y_1 and Y_2. Equation (8-20) gives the *general form* of the joint probability density function of two *gaussian* standardized random variables. The corresponding characteristic function is

$$M(jv_1, jv_2) = \exp[-\tfrac{1}{2}(v_1^2 + 2\rho v_1 v_2 + v_2^2)] \qquad (8\text{-}21)$$

which follows from Eq. (8-19).

The general gaussian joint probability density function of the two real random variables y_1 and y_2 with means m_1 and m_2, variances σ_1 and σ_2, and correlation coefficient ρ is

$$p(y_1, y_2) = \frac{\exp\left[-\dfrac{\sigma_2^2(y_1 - m_1)^2 - 2\sigma_1\sigma_2\rho(y_1 - m_1)(y_2 - m_2) + \sigma_1^2(y_2 - m_2)^2}{2\sigma_1^2\sigma_2^2(1 - \rho^2)}\right]}{2\pi\sigma_1\sigma_2(1 - \rho^2)^{1/2}} \qquad (8\text{-}22)$$

Their joint characteristic function is

$$M(jv_1, jv_2) = \exp[j(m_1 v_1 + m_2 v_2) - \tfrac{1}{2}(\sigma_1^2 v_1^2 + 2\sigma_1\sigma_2\rho v_1 v_2 + \sigma_2^2 v_2^2)] \qquad (8\text{-}23)$$

Dependence and Independence. Let Y_1 and Y_2 be standardized gaussian random variables. The conditional probability density function of Y_2 given Y_1 may be obtained by dividing Eq. (8-20) by $p(Y_1)$ and is

$$p(Y_2 | Y_1) = \frac{1}{[2\pi(1 - \rho^2)]^{1/2}} \exp\left[-\frac{(Y_2 - \rho Y_1)^2}{2(1 - \rho^2)}\right] \qquad (8\text{-}24)$$

The conditional probability density of Y_2 given Y_1 is therefore gaussian with mean ρY_1 and variance $(1 - \rho^2)$.

Suppose that Y_1 and Y_2 are uncorrelated, i.e., that $\rho = 0$. It follows from Eq. (8-20) that

$$p(Y_1, Y_2) = \frac{1}{2\pi} \exp\left(-\frac{Y_1^2 + Y_2^2}{2}\right) = p(Y_1)p(Y_2) \qquad (8\text{-}25)$$

Thus we see that if two standardized *gaussian* random variables are *uncorrelated*, they are also *statistically independent*. It follows from this that if any two gaussian random variables are uncorrelated, they are independent.

Linear Transformations.† It often becomes convenient to use a matrix notation‡ when dealing with multiple random variables. For example, let **y** be the column matrix of the two random variables y_1 and y_2:

$$\mathbf{y} = \begin{bmatrix} y_1 \\ y_2 \end{bmatrix} \qquad (8\text{-}26)$$

† Cf. Cramér (I, Arts. 22.6 and 24.4) and Courant and Hilbert (I, Chap. 1).
‡ See, for example, Hildebrand (I, Chap. I).

and let $\mathbf{v}$ be the column matrix of v_1 and v_2:

$$\mathbf{v} = \begin{bmatrix} v_1 \\ v_2 \end{bmatrix} \tag{8-27}$$

whose transpose is the row matrix $\mathbf{v}'$:

$$\mathbf{v}' = [v_1 \quad v_2]$$

Then since

$$\mathbf{v}'\mathbf{y} = v_1 y_1 + v_2 y_2$$

we may express the characteristic function of y_1 and y_2 as

$$M(jv_1, jv_2) = M(j\mathbf{v}) = E[\exp(j\mathbf{v}'\mathbf{y})] \tag{8-28}$$

In particular, if y_1 and y_2 are gaussian random variables, we can write, from Eq. (8-23),

$$M(j\mathbf{v}) = \exp\left(j\mathbf{m}'\mathbf{v} - \frac{1}{2}\mathbf{v}'\mathbf{\Lambda}\,\mathbf{v}\right) \tag{8-29}$$

where $\mathbf{m}$ is the column matrix of the means:

$$\mathbf{m} = \begin{bmatrix} m_1 \\ m_2 \end{bmatrix} \tag{8-30}$$

and $\mathbf{\Lambda}$ is the covariance matrix:

$$\mathbf{\Lambda} = \begin{bmatrix} \sigma_1{}^2 & \sigma_1\sigma_2\rho \\ \sigma_1\sigma_2\rho & \sigma_2{}^2 \end{bmatrix} \tag{8-31}$$

Suppose now that the random variables y_1 and y_2 have zero means and that they are transformed into the random variables z_1 and z_2 by the linear transformation

$$\begin{aligned} z_1 &= a_{11}y_1 + a_{12}y_2 \\ z_2 &= a_{21}y_1 + a_{22}y_2 \end{aligned} \tag{8-32}$$

This transformation may be expressed in matrix form as

$$\mathbf{z} = \mathbf{A}\mathbf{y} \tag{8-33}$$

where $\mathbf{A}$ is the transformation matrix

$$\mathbf{A} = \begin{bmatrix} a_{11} & a_{12} \\ a_{21} & a_{22} \end{bmatrix} \tag{8-34}$$

and $\mathbf{z}$ is the column matrix of z_1 and z_2.

Since y_1 and y_2 have zero means, so do z_1 and z_2:

$$E(z_1) = 0 = E(z_2)$$

The variances and covariance of z_1 and z_2 are

$$\sigma^2(z_1) = a_{11}{}^2\sigma_1{}^2 + 2a_{11}a_{12}\sigma_1\sigma_2\rho + a_{12}{}^2\sigma_2{}^2$$
$$\sigma^2(z_2) = a_{21}{}^2\sigma_1{}^2 + 2a_{21}a_{22}\sigma_1\sigma_2\rho + a_{22}{}^2\sigma_2{}^2 \qquad (8\text{-}35)$$
$$E(z_1z_2) = a_{11}a_{21}\sigma_1{}^2 + (a_{11}a_{22} + a_{21}a_{12})\sigma_1\sigma_2\rho + a_{12}a_{22}\sigma_2{}^2$$

Direct calculation then shows that the covariance matrix $\mathbf{\mu}$ of z_1 and z_2 may be expressed in terms of $\mathbf{\Lambda}$ by

$$\mathbf{\mu} = \mathbf{A}\mathbf{\Lambda}\mathbf{A}' \qquad (8\text{-}36)$$

The characteristic function of z_1 and z_2 may be written as

$$M_z(j\mathbf{u}) = E[\exp(j\mathbf{u}'\mathbf{z})]$$

from Eq. (8-28). This becomes, from Eq. (8-33),

$$M_z(j\mathbf{u}) = E[\exp(j\mathbf{u}'\mathbf{A}\,\mathbf{y})]$$
$$= E[\exp(j\mathbf{w}'\mathbf{y})] = M_y(j\mathbf{w}') \qquad (8\text{-}37)$$

where we have defined

$$\mathbf{w} = \mathbf{A}'\mathbf{u}$$

If y_1 and y_2 are gaussian random variables with zero means, their characteristic function is

$$M_y(j\mathbf{w}) = \exp(-\tfrac{1}{2}\mathbf{w}'\mathbf{\Lambda}\mathbf{w})$$

from Eq. (8-29). It follows from the definition of $\mathbf{w}$, and from Eq. (8-36), that

$$\mathbf{w}'\mathbf{\Lambda}\mathbf{w} = \mathbf{u}'\mathbf{A}\mathbf{\Lambda}\mathbf{A}'\mathbf{u} = \mathbf{u}'\mathbf{\mu}\mathbf{u} \qquad (8\text{-}38)$$

From this result, and from Eq. (8-37), we then see that the characteristic function of z_1 and z_2 is

$$M_z(j\mathbf{u}) = \exp(-\tfrac{1}{2}\mathbf{u}'\mathbf{\mu}\,\mathbf{u}) \qquad (8\text{-}39)$$

which is the characteristic function of a pair of gaussian random variables with a covariance matrix $\mathbf{\mu}$. Thus we have shown that *a linear transformation of a pair of gaussian random variables results in a pair of gaussian random variables.* The development leading to Eq. (8-16) is a special case of this result.

8-3. The Multivariate Distribution†

The multivariate joint probability density function of N standardized gaussian real random variables Y_n is defined by

$$p(Y_1, \ldots, Y_N) = \frac{\exp\left[-\dfrac{1}{2|\rho|}\displaystyle\sum_{n=1}^{N}\sum_{m=1}^{N}|\rho|_{nm}Y_nY_m\right]}{(2\pi)^{N/2}|\rho|^{\frac{1}{2}}} \qquad (8\text{-}40)$$

† Cf. Cramér (I, Chap. 24).

where $|\rho|_{nm}$ is the cofactor of the element ρ_{nm} in the determinant $|\rho|$, and where $|\rho|$ is the determinant of the correlation matrix

$$\varrho = \begin{bmatrix} \rho_{11} & \rho_{12} & \cdots & \rho_{1N} \\ \rho_{21} & \rho_{22} & \cdots & \rho_{2N} \\ \cdots & \cdots & \cdots & \cdots \\ \rho_{N1} & \rho_{N2} & \cdots & \rho_{NN} \end{bmatrix} \qquad (8\text{-}41)$$

in which

$$\rho_{nm} = E(Y_n Y_m) \qquad \text{and} \qquad \rho_{nn} = 1 \qquad (8\text{-}42)$$

The corresponding gaussian joint characteristic function is

$$M_Y(jv_1, \ldots, jv_N) = \exp\left(-\frac{1}{2}\sum_{n=1}^{N}\sum_{m=1}^{N}\rho_{nm}v_n v_m\right) \qquad (8\text{-}43)$$

which may be expressed in matrix form as

$$M_Y(j\mathbf{v}) = \exp(-\tfrac{1}{2}\mathbf{v}'\varrho\mathbf{v}) \qquad (8\text{-}44)$$

where $\mathbf{v}$ is the column matrix

$$\mathbf{v} = \begin{bmatrix} v_1 \\ v_2 \\ \cdot \\ \cdot \\ \cdot \\ v_N \end{bmatrix} \qquad (8\text{-}45)$$

The joint probability density function of N gaussian real random variables y_n with means m_n and variances σ_n^2 can be obtained from Eq. (8-40) and is

$$p(y_1, \ldots, y_N) = \frac{\exp\left[-\dfrac{1}{2|\Lambda|}\displaystyle\sum_{n=1}^{N}\sum_{m=1}^{N}|\Lambda|_{nm}(y_n - m_n)(y_m - m_m)\right]}{(2\pi)^{N/2}|\Lambda|^{\frac{1}{2}}}$$

$$(8\text{-}46)$$

where $|\Lambda|_{nm}$ is the cofactor of the element λ_{nm} in the determinant $|\Lambda|$ of the covariance matrix

$$\Lambda = \begin{bmatrix} \lambda_{11} & \lambda_{12} & \cdots & \lambda_{1N} \\ \lambda_{21} & \lambda_{22} & \cdots & \lambda_{2N} \\ \cdots & \cdots & \cdots & \cdots \\ \lambda_{N1} & \lambda_{N2} & \cdots & \lambda_{NN} \end{bmatrix} \qquad (8\text{-}47)$$

in which

$$\lambda_{nm} = E[(y_n - m_n)(y_m - m_m)] = \sigma_n\sigma_m\rho_{nm} \qquad (8\text{-}48)$$

The corresponding multivariate gaussian characteristic function is

$$M_y(jv_1, \ldots, jv_N) = \exp\left(j \sum_{n=1}^{N} v_n m_n - \frac{1}{2} \sum_{n=1}^{N} \sum_{m=1}^{N} \lambda_{nm} v_n v_m\right) \quad (8\text{-}49)$$

which may be expressed in matrix form as

$$M_y(j\mathbf{v}) = \exp(j\mathbf{m}'\mathbf{v} - \frac{1}{2}\mathbf{v}'\mathbf{\Lambda}\mathbf{v}) \quad (8\text{-}50)$$

where $\mathbf{m}$ is the matrix of the means:

$$\mathbf{m} = \begin{bmatrix} m_1 \\ m_2 \\ \cdot \\ \cdot \\ \cdot \\ m_N \end{bmatrix} \quad (8\text{-}51)$$

It should be noted that Eq. (8-50) contains Eq. (8-29) as a special case.

Suppose now that the N standardized gaussian random variables Y_n are uncorrelated, i.e., that

$$\rho_{nm} = \begin{cases} 1 & n = m \\ 0 & n \neq m \end{cases} \quad (8\text{-}52)$$

The joint probability density function of the Y_n then becomes, from Eq. (8-40),

$$p(Y_1, \ldots, Y_N) = \frac{\exp\left(-\frac{1}{2} \sum_{n=1}^{N} Y_n^2\right)}{(2\pi)^{N/2}} = \prod_{n=1}^{N} p(Y_n) \quad (8\text{-}53)$$

Therefore *if N gaussian random variables are all uncorrelated, they are also statistically independent.*

Suppose next that N gaussian random variables y_n, with zero means, are transformed into the N random variables z_n by the linear transformation

$$\begin{aligned} z_1 &= a_{11}y_1 + a_{12}y_2 + \cdots + a_{1N}y_N \\ z_2 &= a_{21}y_1 + a_{22}y_2 + \cdots + a_{2N}y_N \\ &\cdots\cdots\cdots\cdots\cdots\cdots\cdots\cdots\cdots \\ z_N &= a_{N1}y_1 + a_{N2}y_2 + \cdots + a_{NN}y_N \end{aligned} \quad (8\text{-}54)$$

Let $\mathbf{A}$ be the matrix of this transformation

$$\mathbf{A} = \begin{bmatrix} a_{11} & a_{12} & \cdots & a_{1N} \\ a_{21} & a_{22} & \cdots & a_{2N} \\ \cdots & \cdots & \cdots & \cdots \\ a_{N1} & a_{N2} & \cdots & a_{NN} \end{bmatrix} \quad (8\text{-}55)$$

We can then write

$$\mathbf{z} = \mathbf{Ay} \tag{8-56}$$

where $\mathbf{z}$ is the column matrix of the z's and $\mathbf{y}$ is the column matrix of the y's. It can then be shown by direct calculation that the covariance matrix $\mathbf{\mu}$ of the z's is related to the covariance matrix $\mathbf{\Lambda}$ of the y's by

$$\mathbf{\mu} = \mathbf{A\Lambda A'} \tag{8-57}$$

as in the bivariate case. The matrix argument used in the bivariate case to show that the linearly transformed variables are gaussian applies without change to the multivariate case. Hence the z_n are also gaussian random variables. That is, *a linear transformation of gaussian random variables yields gaussian random variables.*

8-4. The Gaussian Random Process

A random process is said to be a *gaussian random process* if, for every finite set of time instants t_n, the random variables $x_n = x_{t_n}$ have a gaussian joint probability density function. If the given process is real, the joint probability density function of the N random variables x_n is, from Eq. (8-46),

$$p(x_1, \ldots , x_N) = \frac{\exp\left[-\dfrac{1}{2|\Lambda|}\displaystyle\sum_{n=1}^{N}\sum_{m=1}^{N}|\Lambda|_{nm}(x_n - m_n)(x_m - m_m)\right]}{(2\pi)^{N/2}|\Lambda|^{\frac{1}{2}}} \tag{8-58}$$

where

$$m_n = E(x_n) = E(x_{t_n}) \tag{8-59}$$

and where Λ is the covariance matrix of elements

$$\begin{aligned} \lambda_{nm} &= E[(x_n - m_n)(x_m - m_m)] \\ &= R_x(t_n,t_m) - m_n m_m \end{aligned} \tag{8-60}$$

If the process in question is a wide-sense stationary random process, then $R_x(t_n,t_m) = R_x(t_n - t_m)$ and $m_n = m_m = m$. The covariances, and hence the joint probability densities, then become functions of the time differences $(t_n - t_m)$ and not of t_n and t_m separately. It therefore follows that *a wide-sense stationary gaussian random process is also stationary in the strict sense.*

If the given process is a complex random process, then the N complex random variables

$$z_n = z_{t_n} = x_n + jy_n$$

and the real random variables x_n and y_n together have a $2N$-dimensional gaussian joint probability density function.

Linear Transformations. Suppose that the random process having the sample functions $x(t)$ is a real gaussian random process and that the integral

$$y = \int_a^b x(t)a(t)\, dt \tag{8-61}$$

exists in the sense of Art. 4-7, where $a(t)$ is a continuous real function of t. We will now show that the random variable y is a real gaussian random variable.

Let the interval $a < t \le b$ be partitioned into N subintervals $t_{n-1} < t \le t_n$, $n = 1, \ldots, N$, where $t_0 = a$ and $t_N = b$, and consider the approximating sum

$$y_N = \sum_{n=1}^N x(t_n)a(t_n)\, \Delta t_n \tag{8-62}$$

where $\Delta t_n = t_n - t_{n-1}$.   The average value of the approximating sum is

$$E(y_N) = \sum_{n=1}^N E(x_{t_n})a(t_n)\, \Delta t_n$$

If the average of the random variable y exists, i.e., if

$$E(y) = \int_a^b E(x_t)a(t)\, dt < +\infty \tag{8-63}$$

and if the average of the random process $E(x_t)$ is a continuous function of t, then†

$$\lim_{N\to\infty} E(y_N) = E(y) \tag{8-64}$$

where all the Δt_n go to zero as $N \to \infty$.   The mean square of the approximating sum is

$$E(y_N{}^2) = \sum_{n=1}^N \sum_{m=1}^N E(x_{t_n}x_{t_m})a(t_n)a(t_m)\, \Delta t_n\, \Delta t_m$$

If the mean square value of y exists, i.e., if

$$E(y^2) = \int_a^b \int_a^b E(x_t x_s)a(t)a(s)\, dt\, ds < +\infty \tag{8-65}$$

and if the correlation $E(x_t x_s)$ is a continuous function of t and s, then

$$\lim_{N\to\infty} E(y_N{}^2) = E(y^2) \tag{8-66}$$

It therefore follows from Eqs. (8-64) and (8-66) that

$$\lim_{N\to\infty} \sigma^2(y_N) = \sigma^2(y) \tag{8-67}$$

† Courant (I, Vol. I, pp. 131ff.).

We can now show that y_N converges in the mean to y.   Consider

$$E[(y - y_N)^2] = E(y^2) - 2E(yy_N) + E(y_N^2) \tag{8-68}$$

We may express $E(yy_N)$ in the form

$$E(yy_N) = \sum_{n=1}^{N} \left[\int_a^b E(x_t x_{t_n}) a(t)\ dt \right] a(t_n)\ \Delta t_n$$

Since $E(x_t x_{t_n})$ is continuous in t_n, so is the term in square brackets. Hence, from Eq. (8-65),

$$\lim_{N \to \infty} E(yy_N) = \int_a^b a(s) \int_a^b E(x_t x_s) a(t)\ dt\ ds = E(y^2) \tag{8-69}$$

It then follows, on substituting from Eqs. (8-66) and (8-69) into Eq. (8-68), that

$$\lim_{N \to \infty} E[(y - y_N)^2] = 0 \tag{8-70}$$

Thus the approximating sums converge in the mean, and hence in probability, to the random variable y.   It further follows, from the results of Art. 4-6, that the probability distribution functions of the approximating sums converge to the probability distribution function of y.   It now remains to find the form of the probability distribution function of y.

Equation (8-62) defines y_N as a linear transformation of a set of gaussian random variables.   The approximating sum y_N is therefore a gaussian random variable and has the characteristic function, from Eq. (8-10),

$$M_{y_N}(jv) = \exp\left[jvE(y_N) - \frac{v^2 \sigma^2(y_N)}{2} \right]$$

It then follows from Eqs. (8-64) and (8-67) that the limiting form of this characteristic function is also gaussian:

$$\lim_{N \to \infty} M_{y_N}(jv) = \exp\left[jvE(y) - \frac{v^2 \sigma^2(y)}{2} \right] \tag{8-71}$$

Since we have already shown that the probability distribution functions of the approximating sums converge to the probability distribution function of y, and since the limiting form of the characteristic function of y_N is continuous in v about $v = 0$, it follows that:†

$$\lim_{N \to \infty} M_{y_N}(jv) = M_y(jv) \tag{8-72}$$

and hence, from Eq. (8-71), that y is a real gaussian random variable.

† Cramér (I, Art. 10.4).

If the random process having the sample functions $x(t)$ is a complex gaussian random process, the above argument may be applied separately to the real and imaginary parts of $x(t)$ to show that the random variable y defined by Eq. (8-61) is a complex gaussian random variable.

Under suitable integrability conditions, it can similarly be shown that if the random process having the sample functions $x(t)$ is a gaussian random process, and if the integral

$$y(t) = \int_{-\infty}^{+\infty} x(\tau)b(t,\tau)\ d\tau \qquad (8\text{-}73)$$

exists where $b(t,\tau)$ is a continuous function of t and τ, then the random process having the sample functions $y(t)$ is also a gaussian random process.

Orthogonal Expansions. Let $x(t)$ be a sample function from a stationary gaussian random process. It then follows, from Art. 6-4 and the preceding section, that we may represent that process over the interval $a < t \leq b$ by the Fourier series

$$x(t) = \sum_{n=-\infty}^{+\infty} x_n e^{jn\omega_0 t} \qquad \omega_0 = \frac{2\pi}{b-a} \qquad (8\text{-}74)$$

where the coefficients

$$x_n = \frac{1}{b-a} \int_a^b x(t)e^{-jn\omega_0 t}\ dt \qquad (8\text{-}75)$$

are complex gaussian random variables. Since these coefficients become uncorrelated in the limit as $(b-a) \to \infty$, they also become statistically independent.

It also follows from the results of Art. 6-4 that the gaussian random process having the sample function $x(t)$ may be represented whether stationary or not over the interval $[a,b]$ by the general Fourier series

$$x(t) = \sum_n \sigma_n x_n \phi_n(t) \qquad (8\text{-}76)$$

where the $|\sigma_n|^2$ and the $\phi_n(t)$ are the characteristic values and characteristic functions, respectively, of the integral equation

$$\int_a^b R(t,s)\phi(s)\ ds = \lambda\phi(t) \qquad (8\text{-}77)$$

and where the coefficients

$$x_n = \frac{1}{\sigma_n} \int_a^b x(t)\phi_n^*(t)\ dt \qquad (8\text{-}78)$$

are uncorrelated (and hence statistically independent) gaussian random variables.

8-5. The Narrow-band Gaussian Random Process[†]

A random process is said to be a *narrow-band random process* if the width Δf of the significant region of its spectral density is small compared to the center frequency f_c of that region. A typical narrow-band spectral density is pictured in Fig. 8-2. If a sample function of such a random process is examined with an oscilloscope, what appears to be

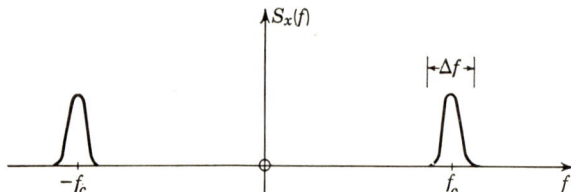

FIG. 8-2. A narrow-band spectral density.

a sinusoidal wave with slowly varying envelope (amplitude) and phase will be seen. That is, a sample function $x(t)$ of a narrow-band random process appears to be expressible in the form

$$x(t) = V(t) \cos [\omega_c t + \phi(t)] \tag{8-79}$$

where $f_c = \omega_c/2\pi$ is the mean frequency of the spectral band and where the *envelope* $V(t)$ and the *phase* $\phi(t)$ are slowly varying functions of time. Formally, such a representation need not be restricted to narrow-band waves; the concept of an envelope has significance, however, only when the variations of $V(t)$ and $\phi(t)$ are slow compared to those of $\cos \omega_c t$.

Envelope and Phase Probability Distributions. Let us now determine some of the statistical properties of the envelope and phase when the narrow-band random process in question is a stationary gaussian random process. To this end, it is convenient to represent the given process over the interval $0 < t \leq T$ by the Fourier series

$$x(t) = \sum_{n=1}^{\infty} (x_{cn} \cos n\omega_o t + x_{sn} \sin n\omega_o t) \tag{8-80}$$

where $\omega_o = 2\pi/T$,

$$x_{cn} = \frac{2}{T} \int_0^T x(t) \cos n\omega_o t \, dt \tag{8-81a}$$

and

$$x_{sn} = \frac{2}{T} \int_0^T x(t) \sin n\omega_o t \, dt \tag{8-81b}$$

It then follows from the results of Art. 6-4 and the preceding article that these coefficients are gaussian random variables which become uncorrelated as the duration of the expansion interval increases without limit. The mean frequency of the narrow spectral band may be introduced by

[†] Cf. Rice (I, Art. 3.7).

writing $n\omega_o$ in Eq. (8-80) as $(n\omega_o - \omega_c) + \omega_c$, where $\omega_c = 2\pi f_c$, and expanding the sine and cosine factors. In this way, we can obtain the expression

$$x(t) = x_c(t) \cos \omega_c t - x_s(t) \sin \omega_c t \qquad (8\text{-}82)$$

where we have defined

$$x_c(t) = \sum_{n=1}^{\infty} [x_{cn} \cos (n\omega_0 - \omega_c)t + x_{sn} \sin (n\omega_0 - \omega_c)t] \qquad (8\text{-}83a)$$

$$x_s(t) = \sum_{n=1}^{\infty} [x_{cn} \sin (n\omega_o - \omega_c)t - x_{sn} \cos (n\omega_o - \omega_c)t] \qquad (8\text{-}83b)$$

It then follows from Eqs. (8-79) and (8-82) that

$$x_c(t) = V(t) \cos \phi(t) \qquad (8\text{-}84a)$$

and

$$x_s(t) = V(t) \sin \phi(t) \qquad (8\text{-}84b)$$

and hence that

$$V(t) = [x_c^2(t) + x_s^2(t)]^{1/2} \qquad (8\text{-}85a)$$

and

$$\phi(t) = \tan^{-1} \left[\frac{x_s(t)}{x_c(t)}\right] \qquad (8\text{-}85b)$$

where $0 \leq V(t)$ and $0 \leq \phi(t) \leq 2\pi$. Since the only nonvanishing terms in the sums in Eqs. (8-83) are those for which the values of nf_o fall in the given narrow spectral band, the sample functions $x_c(t)$ and $x_s(t)$ have frequency components only in a band of width Δf centered on zero frequency. The frequency components of the envelope and phase are therefore confined to a similar region about zero frequency.

The random variables x_{ct} and x_{st}, which refer to the possible values of $x_c(t)$ and $x_s(t)$, respectively, are defined as sums of gaussian random variables, and hence are gaussian random variables. Their means are zero:

$$E(x_{ct}) = 0 = E(x_{st}) \qquad (8\text{-}86)$$

since the original process has a zero mean. The mean square of x_{ct} is, from Eq. (8-83a),

$$E(x_{ct}^2) = \sum_{n=1}^{\infty} \sum_{m=1}^{\infty} \begin{bmatrix} E(x_{cn}x_{cm}) \cos (n\omega_o - \omega_c)t \cos (m\omega_o - \omega_c)t \\ + E(x_{cn}x_{sm}) \cos (n\omega_o - \omega_c)t \sin (m\omega_o - \omega_c)t \\ + E(x_{sn}x_{cm}) \sin (n\omega_o - \omega_c)t \cos (m\omega_o - \omega_c)t \\ + E(x_{sn}x_{sm}) \sin (n\omega_o - \omega_c)t \sin (m\omega_o - \omega_c)t \end{bmatrix}$$

It then follows, on using the limiting properties of the coefficients, Eqs. (6-31), that as $T \to \infty$,

$$E(x_{ct}^2) = \lim_{T \to \infty} \sum_{n=1}^{\infty} E(x_{cn}^2)[\cos^2 (n\omega_o - \omega_c)t + \sin^2 (n\omega_o - \omega_c)t]$$

$$= 2 \int_0^{\infty} S_x(f) \, df = E(x_t^2)$$

where $S_x(f)$ is the spectral density of the gaussian random process. Similarly, it may be shown that

$$E(x_{st}{}^2) = E(x_t{}^2)$$

and hence, using Eq. (8-86), that

$$\sigma^2(x_{ct}) = \sigma^2(x_{st}) = \sigma_x{}^2 \qquad (8\text{-}87)$$

where $\sigma_x = \sigma(x_t)$. The covariance of x_{ct} and x_{st} is

$$E(x_{ct}x_{st}) = \sum_{n=1}^{\infty} \sum_{m=1}^{\infty} \begin{bmatrix} E(x_{cn}x_{cm}) \cos (n\omega_o - \omega_c)t \sin (m\omega_o - \omega_c)t \\ - E(x_{cn}x_{sm}) \cos (n\omega_o - \omega_c)t \cos (m\omega_o - \omega_c)t \\ + E(x_{sn}x_{cm}) \sin (n\omega_o - \omega_c)t \sin (m\omega_o - \omega_c)t \\ - E(x_{sn}x_{sm}) \sin (n\omega_o - \omega_c)t \cos (m\omega_o - \omega_c)t \end{bmatrix}$$

which becomes, as $T \to \infty$,

$$E(x_{ct}x_{st}) = \lim_{T \to \infty} \sum_{n=1}^{\infty} E(x_{cn}{}^2) \begin{bmatrix} \cos (n\omega_o - \omega_c)t \sin (n\omega_o - \omega_c)t \\ - \sin (n\omega_o - \omega_c)t \cos (n\omega_o - \omega_c)t \end{bmatrix}$$

Hence

$$E(x_{ct}x_{st}) = 0 \qquad (8\text{-}88)$$

The random variables x_{ct} and x_{st} are therefore independent gaussian random variables with zero means and variances $\sigma_x{}^2$. Their joint probability density then is, from Eq. (8-12),

$$p(x_{ct}, x_{st}) = \frac{1}{2\pi\sigma_x{}^2} \exp\left(-\frac{x_{ct}{}^2 + x_{st}{}^2}{2\sigma_x{}^2}\right) \qquad (8\text{-}89)$$

The joint probability density function of the envelope and phase random variables V_t and ϕ_t, respectively, may now be found from that of x_{ct} and x_{st}. From Eq. (8-84), the Jacobian of the transformation from x_{ct} and x_{st} to V_t and ϕ_t is seen to be

$$|J| = V_t$$

It therefore follows from Eqs. (3-54) and (8-89) that

$$p(V_t, \phi_t) = \begin{cases} \dfrac{V_t}{2\pi\sigma_x{}^2} \exp\left(-\dfrac{V_t{}^2}{2\sigma_x{}^2}\right) & \text{for } V_t \geq 0 \text{ and } 0 \leq \phi_t \leq 2\pi \\ 0 & \text{otherwise} \end{cases} \qquad (8\text{-}90)$$

The probability density function of V_t can be obtained by integrating this result over ϕ_t from 0 to 2π and is[†]

$$p(V_t) = \begin{cases} \dfrac{V_t}{\sigma_x{}^2} \exp\left(-\dfrac{V_t{}^2}{2\sigma_x{}^2}\right) & \text{for } V_t \geq 0 \\ 0 & \text{otherwise} \end{cases} \qquad (8\text{-}91)$$

† Cf. Chap. 3, Prob. 13.

This is the *Rayleigh* probability density function and is shown in Fig. 8-3. The probability density function of ϕ_t can be obtained by integrating Eq. (8-90) over V_t and is†

$$p(\phi_t) = \begin{cases} \dfrac{1}{2\pi} & \text{if } 0 \leq \phi_t \leq 2\pi \\ 0 & \text{otherwise} \end{cases} \tag{8-92}$$

The random phase angle is therefore uniformly distributed. It then follows from Eqs. (8-90), (8-91), and (8-92) that

$$p(V_t, \phi_t) = p(V_t)p(\phi_t)$$

and hence that V_t and ϕ_t are independent random variables. As we shall show later, however, the envelope and phase random processes, which

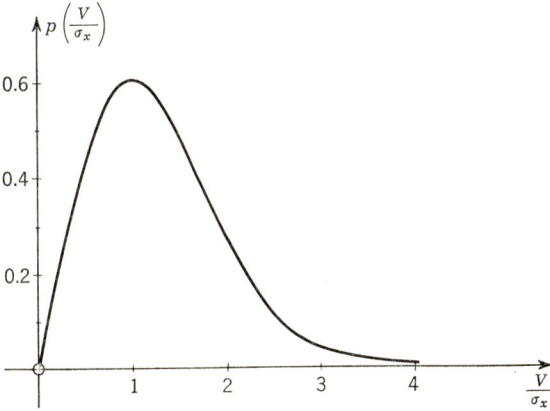

FIG. 8-3. The Rayleigh probability density function.

have the sample functions $V(t)$ and $\phi(t)$, respectively, are *not* independent random *processes*.

Joint Densities. Next let us determine the joint probability densities of $V_1 = V_t$ and $V_2 = V_{t-\tau}$ and of $\phi_1 = \phi_t$ and $\phi_2 = \phi_{t-\tau}$ from the joint probability density of the random variables $x_{c1} = x_{ct}$, $x_{s1} = x_{st}$, $x_{c2} = x_{c(t-\tau)}$, and $x_{s2} = x_{s(t-\tau)}$. It follows from Eqs. (8-83) that x_{c1}, x_{c2}, x_{s1}, and x_{s2} are gaussian; it remains only to find their covariances in order to determine their joint probability density function.

From Eq. (8-87), we have

$$\sigma^2(x_{c1}) = \sigma^2(x_{s1}) = \sigma^2(x_{c2}) = \sigma^2(x_{s2}) = \sigma_x^2 \tag{8-93}$$

and from Eq. (8-88),

$$E(x_{c1}x_{s1}) = 0 = E(x_{c2}x_{s2}) \tag{8-94}$$

† Cf. Chap. 3, Prob. 13.

The covariance of x_{c1} and x_{c2} is

$$R_c(\tau) = \sum_{n=1}^{\infty} \sum_{m=1}^{\infty} \begin{bmatrix} E(x_{cn}x_{cm}) \cos (n\omega_o - \omega_c)t \cos (m\omega_o - \omega_c)(t - \tau) \\ + E(x_{cn}x_{sm}) \cos (n\omega_o - \omega_c)t \sin (m\omega_o - \omega_c)(t - \tau) \\ + E(x_{sn}x_{cm}) \sin (n\omega_o - \omega_c)t \cos (m\omega_o - \omega_c)(t - \tau) \\ + E(x_{sn}x_{sm}) \sin (n\omega_o - \omega_c)t \sin (m\omega_o - \omega_c)(t - \tau) \end{bmatrix}$$

which becomes, using Eqs. (6-31), as $T \to \infty$,

$$R_c(\tau) = \lim_{T \to \infty} \sum_{n=1}^{\infty} E(x_{cn}{}^2) \cos (n\omega_o - \omega_c)\tau$$

Hence
$$R_c(\tau) = 2 \int_0^{\infty} S_x(f) \cos 2\pi(f - f_c)\tau \, df \qquad (8\text{-}95)$$

In a similar manner it may be shown that

$$R_s(\tau) = E(x_{s1}x_{s2}) = 2 \int_0^{\infty} S_x(f) \cos 2\pi(f - f_c)\tau \, df \qquad (8\text{-}96)$$
$$= R_c(\tau)$$

and that

$$R_{cs}(\tau) = E(x_{c1}x_{s2}) = -E(x_{s1}x_{c2})$$
$$= 2 \int_0^{\infty} S_x(f) \sin 2\pi(f - f_c)\tau \, df \qquad (8\text{-}97)$$

The covariance matrix of the random variables x_{c1}, x_{s1}, x_{c2}, and x_{s2} then is

$$\Lambda = \begin{bmatrix} \sigma_x{}^2 & 0 & R_c(\tau) & R_{cs}(\tau) \\ 0 & \sigma_x{}^2 & -R_{cs}(\tau) & R_c(\tau) \\ R_c(\tau) & -R_{cs}(\tau) & \sigma_x{}^2 & 0 \\ R_{cs}(\tau) & R_c(\tau) & 0 & \sigma_x{}^2 \end{bmatrix} \qquad (8\text{-}98)$$

The determinant of this matrix is

$$|\Lambda| = [\sigma_x{}^4 - R_c{}^2(\tau) - R_{cs}{}^2(\tau)]^2 \qquad (8\text{-}99)$$

The various cofactors are

$$\begin{aligned} \Lambda_{11} &= \Lambda_{22} = \Lambda_{33} = \Lambda_{44} = \sigma_x{}^2|\Lambda|^{1/2} \\ \Lambda_{12} &= \Lambda_{21} = \Lambda_{34} = \Lambda_{43} = 0 \\ \Lambda_{13} &= \Lambda_{31} = \Lambda_{24} = \Lambda_{42} = -R_c(\tau)|\Lambda|^{1/2} \\ \Lambda_{14} &= \Lambda_{41} = -\Lambda_{23} = -\Lambda_{32} = -R_{cs}(\tau)|\Lambda|^{1/2} \end{aligned} \qquad (8\text{-}100)$$

It then follows from Eq. (8-46) that the joint probability density function of x_{c1}, x_{s1}, x_{c2}, and x_{s2} is

$$\begin{aligned} &p(x_{c1},x_{s1},x_{c2},x_{s2}) \\ &= \frac{1}{4\pi^2|\Lambda|^{1/2}} \exp \left\{ -\frac{1}{2|\Lambda|^{1/2}} \begin{bmatrix} \sigma_x{}^2(x_{c1}{}^2 + x_{s1}{}^2 + x_{c2}{}^2 + x_{s2}{}^2) \\ -2R_c(\tau)(x_{c1}x_{c2} + x_{s1}x_{s2}) \\ -2R_{cs}(\tau)(x_{c1}x_{s2} - x_{s1}x_{c2}) \end{bmatrix} \right\} \end{aligned} \qquad (8\text{-}101)$$

where $|\Lambda|$ is given by Eq. (8-99).

The Jacobian of the transformation from x_{c1}, x_{s1}, x_{c2}, and x_{s2} to V_1, ϕ_1, V_2, and ϕ_2 is, from Eq. (8-84),

$$|J| = V_1 V_2$$

It therefore follows from Eqs. (3-54) and (8-101) that

$$p(V_1,\phi_1,V_2,\phi_2) = \begin{cases} \dfrac{V_1 V_2}{4\pi^2 |\Lambda|^{\frac{1}{2}}} \exp\left\{ -\dfrac{1}{2|\Lambda|^{\frac{1}{2}}} \begin{bmatrix} \sigma_x^2(V_1^2 + V_2^2) \\ - 2R_c(\tau)V_1V_2\cos(\phi_2 - \phi_1) \\ - 2R_{cs}(\tau)V_1V_2\sin(\phi_2 - \phi_1) \end{bmatrix} \right\} \\ \qquad \text{for } V_1,\, V_2 \geq 0 \quad \text{and} \quad 0 \leq \phi_1,\, \phi_2 \leq 2\pi \\ 0 \quad \text{otherwise} \end{cases}$$

(8-102)

The joint probability density function of V_1 and V_2 can now be obtained by integrating this result over ϕ_1 and ϕ_2. Since the exponential in Eq. (8-102) is periodic in ϕ_2, we get (where $\alpha = \phi_2 - \phi_1 - \tan^{-1}[R_{cs}(\tau)/R_c(\tau)]$)

$$\frac{1}{4\pi^2}\int_0^{2\pi} d\phi_1 \int_0^{2\pi} d\phi_2$$

$$\exp\left\{ -\frac{V_1 V_2}{|\Lambda|^{\frac{1}{2}}}[R_c(\tau)\cos(\phi_2 - \phi_1) + R_{cs}(\tau)\sin(\phi_2 - \phi_1)]\right\}$$

$$= \frac{1}{2\pi}\int_0^{2\pi} d\phi_1 \frac{1}{2\pi}\int_0^{2\pi} \exp\left\{ -\frac{V_1V_2[R_c^2(\tau) + R_{cs}^2(\tau)]^{\frac{1}{2}}}{|\Lambda|^{\frac{1}{2}}}\cos\alpha \right\} d\alpha$$

$$= I_0\left\{ \frac{V_1V_2[R_c^2(\tau) + R_{cs}^2(\tau)]^{\frac{1}{2}}}{|\Lambda|^{\frac{1}{2}}} \right\}$$

where $I_0(z)$ is the zero-order modified Bessel function of the first kind.†
The joint probability density function of V_1 and V_2 is therefore

$$p(V_1,V_2) = \begin{cases} \dfrac{V_1 V_2}{|\Lambda|^{\frac{1}{2}}} I_0\left\{ \dfrac{V_1V_2[R_c^2(\tau) + R_{cs}^2(\tau)]^{\frac{1}{2}}}{|\Lambda|^{\frac{1}{2}}} \right\} \exp\left[-\dfrac{\sigma_x^2(V_1^2 + V_2^2)}{2|\Lambda|^{\frac{1}{2}}} \right] \\ \qquad\qquad\qquad\qquad\qquad \text{where } V_1,\, V_2 \geq 0 \\ 0 \qquad\qquad\qquad\qquad\qquad \text{otherwise} \end{cases}$$

(8-103)

where $|\Lambda|$ is given by Eq. (8-99).
The joint probability density function of ϕ_1 and ϕ_2 may be gotten by integrating $p(V_1,\phi_1,V_2,\phi_2)$ over V_1 and V_2.‡ On defining

$$\beta = \frac{R_c(\tau)}{\sigma_x^2}\cos(\phi_2 - \phi_1) + \frac{R_{cs}(\tau)}{\sigma_x^2}\sin(\phi_2 - \phi_1)$$

(8-104)

we obtain from Eq. (8-102)

$$p(\phi_1,\phi_2) = \frac{1}{4\pi^2|\Lambda|^{\frac{1}{2}}}\int_0^\infty dV_1 \int_0^\infty dV_2\, V_1 V_2$$

$$\exp\left[-\frac{\sigma_x^2(V_1^2 + V_2^2 - 2V_1V_2\beta)}{2|\Lambda|^{\frac{1}{2}}} \right] \quad \text{for } 0 \leq \phi_1,\, \phi_2 \leq 2\pi.$$

† See, for example, Magnus and Oberhettinger (I, Chap. III, Arts. 1 and 5) or Watson (I, Arts. 3.7 and 3.71).
‡ Cf. MacDonald (I).

In order to facilitate evaluation of this double integral, let us introduce two new variables y and z by the relations

$$V_1{}^2 = \frac{|\Lambda|^{\frac{1}{2}}}{\sigma_x{}^2} \, ye^{2z}$$

and

$$V_2{}^2 = \frac{|\Lambda|^{\frac{1}{2}}}{\sigma_x{}^2} \, ye^{-2z} \tag{8-105}$$

The magnitude of the Jacobian of the transformation from V_1 and V_2 to y and z is

$$|J| = \frac{|\Lambda|^{\frac{1}{2}}}{\sigma_x{}^2}$$

The domain of y is $(0,+\infty)$ and that of z is $(-\infty,+\infty)$. After some manipulation we can therefore obtain

$$p(\phi_1,\phi_2) = \frac{|\Lambda|^{\frac{1}{2}}}{4\pi^2\sigma_x{}^4} \int_0^\infty \left[\int_{-\infty}^{+\infty} \exp(-y \cosh 2z) \, dz \right] ye^{\beta y} \, dy$$

$$= \frac{|\Lambda|^{\frac{1}{2}}}{4\pi^2\sigma_x{}^4} \int_0^\infty K_0(y)ye^{\beta y} \, dy$$

where $K_0(y)$ is the zero-order modified Hankel function.[†] It is known[‡] that the formula

$$\int_0^\infty e^{-ay}K_0(y) \, dy = \frac{\cos^{-1} a}{(1 - a^2)^{\frac{1}{2}}}$$

is valid for $a > -1$. Hence

$$\int_0^\infty e^{\beta y}K_0(y) \, dy = \frac{\pi - \cos^{-1} \beta}{(1 - \beta^2)^{\frac{1}{2}}}$$

is valid for $\beta < 1$. Differentiation of this result with respect to β gives

$$\int_0^\infty ye^{\beta y}K_0(y) \, dy = \frac{1}{(1 - \beta^2)} + \frac{\beta(\pi - \cos^{-1} \beta)}{(1 - \beta^2)^{\frac{3}{2}}}$$

The joint probability density function of ϕ_1 and ϕ_2 is therefore

$$p(\phi_1,\phi_2) = \begin{cases} \dfrac{|\Lambda|^{\frac{1}{2}}}{4\pi^2\sigma_x{}^4} \left[\dfrac{(1 - \beta^2)^{\frac{1}{2}} + \beta(\pi - \cos^{-1} \beta)}{(1 - \beta^2)^{\frac{3}{2}}} \right] \\ \qquad\qquad\qquad\qquad\qquad \text{where } 0 \leq \phi_1,\, \phi_2 \leq 2\pi \\ 0 \qquad\qquad\qquad\qquad\quad \text{otherwise} \end{cases} \tag{8.106}$$

where $|\Lambda|$ is given by Eq. (8-99) and β is given by Eq. (8-104).

---

† See, for example, Magnus and Oberhettinger (I, Chap. III, Arts. 1 and 5) or Watson (I, Arts. 3.7 and 6.22).

‡ See, for example, Magnus and Oberhettinger (I, Chap. III, Art. 7), or Watson (I, Art. 13.21).

Evaluation of Eqs. (8-102), (8-103), and (8-106), at $\phi_1 = \phi_2$, for example, shows that

$$p(V_1,\phi_1,V_2,\phi_2) \neq p(V_1,V_2)p(\phi_1,\phi_2) \tag{8-107}$$

The envelope and phase random processes are therefore *not* statistically independent.

8-6. Sine Wave Plus Narrow-band Gaussian Random Process†

For our final results of this chapter we shall derive expressions for the probability density functions of the envelope and phase-angle of the sum of a sine wave and a narrow-band gaussian random process.

Let $x(t)$ be a sample function of a stationary narrow-band gaussian random process and let

$$y(t) = P \cos(\omega_c t + \psi) + x(t) \tag{8-108}$$

where P is a constant and the random variable ψ is uniformly distributed over the interval $(0,2\pi)$ and independent of the gaussian random process. Using Eq. (8-82), we can write

$$y(t) = X_c(t) \cos \omega_c t - X_s(t) \sin \omega_c t \tag{8-109}$$
where $\qquad X_c(t) = P \cos \psi + x_c(t)$
and $\qquad X_s(t) = P \sin \psi + x_s(t) \tag{8-110}$

If we express $y(t)$ in terms of an envelope and phase,

$$y(t) = V(t) \cos[\omega_c t + \phi(t)] \tag{8-111}$$

it follows that
$$\begin{aligned} X_c(t) &= V(t) \cos \phi(t) \\ X_s(t) &= V(t) \sin \phi(t) \end{aligned} \tag{8-112}$$
and
and hence that
$$V(t) = [X_c{}^2(t) + X_s{}^2(t)]^{1/2} \tag{8-113}$$

As in Art. 8-5, the random variables x_{ct} and x_{st} are independent gaussian random variables with zero means and variance $\sigma_x{}^2$. The joint probability density function of X_{ct}, X_{st}, and ψ is therefore

$$\begin{aligned} p(X_{ct},X_{st},\psi) &= \frac{1}{2\pi} \frac{\exp\left[-\dfrac{(X_c - P \cos \psi)^2}{2\sigma_x{}^2}\right]}{(2\pi\sigma_x{}^2)^{1/2}} \frac{\exp\left[-\dfrac{(X_s - P \sin \psi)^2}{2\sigma_x{}^2}\right]}{(2\pi\sigma_x{}^2)^{1/2}} \\ &= \frac{1}{4\pi^2\sigma_x{}^2} \exp\left[-\frac{X_c{}^2 + X_s{}^2 + P^2 - 2P(X_c \cos \psi + X_s \sin \psi)}{2\sigma_x{}^2}\right] \end{aligned}$$

for $0 \leq \psi \leq 2\pi$.

† Cf. Rice (I, Art. 3.10) and Middleton (II, Art. 5).

Hence the joint probability density function of V_t, ϕ_t, and ψ is

$$p(V_t, \phi_t, \psi) = \begin{cases} \dfrac{V_t}{4\pi^2\sigma_x{}^2} \exp\left[-\dfrac{V_t{}^2 + P^2 - 2PV_t\cos(\phi-\psi)}{2\sigma_x{}^2}\right] & \\ \qquad\qquad\text{where } V_t \geq 0 \text{ and } 0 \leq \phi_t, \psi \leq 2\pi & \\ 0 \qquad\qquad \text{otherwise} & \end{cases} \quad (8\text{-}114)$$

We can now determine the probability density function of V_t by integrating this result over ϕ_t and ψ. Thus, for $V_t \geq 0$,

$$p(V_t) = \frac{V_t}{\sigma_x{}^2} \exp\left(-\frac{V_t{}^2 + P^2}{2\sigma_x{}^2}\right) \frac{1}{2\pi} \int_0^{2\pi} d\psi \frac{1}{2\pi} \int_{-\psi}^{2\pi-\psi} \exp\left(\frac{PV_t}{\sigma_x{}^2}\cos\theta\right) d\theta$$

where $\theta = \phi - \psi$. Since the exponential integrand is periodic in θ, we can integrate over the interval $(0, 2\pi)$ in θ, and so get

$$p(V_t) = \begin{cases} \dfrac{V_t}{\sigma_x{}^2} \exp\left(-\dfrac{V_t{}^2 + P^2}{2\sigma_x{}^2}\right) I_0\left(\dfrac{PV_t}{\sigma_x{}^2}\right) & \text{for } V_t \geq 0 \\ 0 & \text{otherwise} \end{cases} \quad (8\text{-}115)$$

for the probability density function of the envelope of the sum of a sine wave and a narrow-band gaussian random process. This result reduces to that given by Eq. (8-91) when $P = 0$.

An asymptotic series expansion† of the modified Bessel function, valid for large values of the argument, is

$$I_0(x) = \frac{e^x}{(2\pi x)^{1/2}}\left(1 + \frac{1}{8x} + \frac{9}{128x^2} + \cdots\right) \quad (8\text{-}116)$$

It therefore follows that when $PV_t \gg \sigma_x{}^2$ we get approximately

$$p(V_t) = \frac{1}{\sigma_x}\left(\frac{V_t}{2\pi P}\right)^{1/2} \exp\left[-\frac{(V_t - P)^2}{2\sigma_x{}^2}\right] \quad (8\text{-}117)$$

for $V_t \geq 0$.

Hence, when the magnitude P of the sine wave is large compared to σ_x and when V_t is near P, the probability density function of the envelope of the sum process is approximately gaussian.

The joint probability density function of the phase angles ϕ_t and ψ can be gotten from that of V_t, ϕ_t, and ψ by integrating over V_t. Thus, from Eq. (8-114),

$$p(\phi_t, \psi) = \frac{1}{4\pi^2\sigma_x{}^2} \int_0^\infty V_t \exp\left[-\frac{V_t{}^2 + P^2 - 2PV_t\cos(\phi-\psi)}{2\sigma_x{}^2}\right] dV_t$$

This becomes, on completing the square,

$$p(\phi_t, \psi) = \frac{1}{4\pi^2\sigma_x{}^2} \exp\left(-\frac{P^2\sin^2\theta}{2\sigma_x{}^2}\right) \int_0^\infty V_t \exp\left[-\frac{(V_t - P\cos\theta)^2}{2\sigma_x{}^2}\right] dV_t$$

† Dwight (I, Eq. 814.1).

where $\theta = \phi_t - \psi$. Hence, setting $u = (V_t - P \cos \theta)/\sigma_x$,

$$p(\phi_t,\psi) = \frac{1}{4\pi^2} \exp \left(- \frac{P^2 \sin^2 \theta}{2\sigma_x{}^2} \right) \int_{-P \cos \theta/\sigma_x}^{\infty} u \exp \left(\frac{-u^2}{2} \right) du$$

$$+ \frac{P \cos \theta}{4\pi^2 \sigma_x} \exp \left(- \frac{P^2 \sin^2 \theta}{2\sigma_x{}^2} \right) \int_{-P \cos \theta/\sigma_x}^{\infty} \exp \left(\frac{-u^2}{2} \right) du$$

The first integral evaluates to $\exp(-P^2 \cos^2 \theta/2\sigma_x{}^2)$. Since the integrand of the second is even, we therefore get

$$p(\phi_t,\psi) = \frac{\exp\left(- P^2/2\sigma_x{}^2\right)}{4\pi^2} + \frac{P \cos \theta}{4\pi^2 \sigma_x}$$
$$\exp \left(- \frac{P^2 \sin^2 \theta}{2\sigma_x{}^2} \right) \int_{-\infty}^{P \cos \theta/\sigma_x} \exp \left(\frac{-u^2}{2} \right) du \quad \text{(8-118)}$$

where $\theta = \phi_t - \psi$ and $0 \leq \phi_t, \psi \leq 2\pi$. The integral here is $(2\pi)^{\frac{1}{2}}$ times the probability distribution function of a gaussian random variable. When the amplitude of the sine wave is zero, Eq. (8-118) reduces to

$$p(\phi_t,\psi) = \frac{1}{4\pi^2} \qquad \text{when } P = 0 \qquad \text{(8-119)}$$

as it should.

An approximate expression for the joint probability density function of the phase angles ϕ_t and ψ can be obtained by using Eq. (8-4) in Eq. (8-118). Thus we get, approximately,

$$p(\phi_t,\psi) = \frac{P \cos (\phi_t - \psi)}{(2\pi)^{3/2}\sigma_x} \exp \left[- \frac{P^2 \sin^2 (\phi_t - \psi)}{2\sigma_x{}^2} \right]$$
$$\text{when } P \cos (\phi_t - \psi) >> \sigma_x \quad \text{(8-120)}$$

where $0 \leq \phi_t, \psi \leq 2\pi$.

We have tried in this and the preceding article to present a few of the significant statistical properties of a narrow-band gaussian random process. More extensive results may be found in the technical literature, particularly in the papers of Rice.†

8-7. Problems

1. Let x and y be independent gaussian random variables with means m_x and m_y, and variances $\sigma_x{}^2$ and $\sigma_y{}^2$, respectively. Let

$$z = x + y$$

a. Determine the characteristic function of z.
b. Determine the probability density function of z.

† E.g., Rice (I) and Rice (II).

2. Let x_1, x_2, x_3, and x_4 be real random variables with a gaussian joint probability density function, and let their means all be zero. Show that†

$$E(x_1x_2x_3x_4) = E(x_1x_2)E(x_3x_4) + E(x_1x_3)E(x_2x_4) + E(x_1x_4)E(x_2x_3) \quad (8\text{-}121)$$

3. Let $x(t)$ be a sample function of a stationary real gaussian random process with a zero mean. Let a new random process be defined with the sample functions

$$y(t) = x^2(t)$$

Show that

$$R_y(\tau) = R_x{}^2(0) + 2R_x{}^2(\tau) \quad (8\text{-}122)$$

4. Let x be a gaussian random variable with mean zero and unit variance. Let a new random variable y be defined as follows: If $x = x_o$, then

$$y = \begin{cases} x_o \text{ with probability } \frac{1}{2} \\ -x_o \text{ with probability } \frac{1}{2} \end{cases}$$

a. Determine the joint probability density function of x and y.
b. Determine the probability density function of y alone.

Note that although x and y are gaussian random variables, the joint probability density function of x and y is not gaussian.

5. Derive Eq. (8-17).

6. Let the random variables x_1 and x_2 have a gaussian joint probability density function. Show that if the random variables y_1 and y_2 are defined by a rotational transformation of x_1 and x_2 about the point $[E(x_1),E(x_2)]$, then y_1 and y_2 are *independent* gaussian random variables if the angle of rotation ϕ is chosen such that

$$\tan 2\phi = \frac{2\mu_{11}}{\mu_{20} - \mu_{02}} \quad (8\text{-}123)$$

7. Let the autocorrelation function of the stationary gaussian random process with sample functions $x(t)$ be expanded over the interval $\left[-\dfrac{T}{2}, +\dfrac{T}{2} \right]$ in a Fourier series

$$R(\tau) = \sum_{n=-\infty}^{+\infty} a_n e^{jn\omega_o \tau} \qquad \omega_o = \frac{2\pi}{T}$$

Let the independent gaussian random variables x_n, $(n = -\infty, \ldots, -1,0,+1, \ldots, +\infty)$ have zero means and unit variances, and consider the random process defined by the sample functions

$$y(t) = \sum_{n=-\infty}^{+\infty} b_n x_n e^{jn\omega_o t} \qquad \omega_o = \frac{2\pi}{T}$$

where the b_n are complex constants.

Show‡ that if the b_n are chosen so that

$$|b_n|^2 = a_n$$

then the random process defined by the sample functions $y(t)$ has the same multi-

† Cf. Chap. 7, Prob. 4.
‡ Cf. Root and Pitcher (I, Theorem 2).

variate probability density functions as does the process defined by the sample functions $x(t)$ so long as $0 \leq t \leq T/2$.

8. Let V_t be the envelope of a stationary narrow-band real gaussian random process. Show that

$$E(V_t) = \left(\frac{\pi}{2}\right)^{1/2} \sigma_x \qquad (8\text{-}124)$$

and

$$\sigma^2(V_t) = \left(2 - \frac{\pi}{2}\right) \sigma_x{}^2 \qquad (8\text{-}125)$$

where $\sigma_x{}^2$ is the variance of the gaussian random process.

9. Let $x(t)$ be a sample function of a stationary narrow-band real gaussian random process. Consider a new random process defined with the sample functions

$$y(t) = x(t) \cos \omega_o t$$

where $f_o = \omega_o/2\pi$ is small compared to the center frequency f_c of the original process but large compared to the spectral width of the original process. If we write

$$x(t) = V(t) \cos [\omega_c t + \phi(t)]$$

then we may define

$$y_L(t) = \frac{V(t)}{2} \cos [(\omega_c - \omega_o)t + \phi(t)]$$

to be the sample functions of the "lower sideband" of the new process, and

$$y_U(t) = \frac{V(t)}{2} \cos [(\omega_c + \omega_o)t + \phi(t)]$$

to be the sample functions of the "upper sideband" of the new process.

a. Show that the upper and lower sideband random processes are each stationary random processes even though their sum is nonstationary.

b. Show that the upper and lower sideband random processes are *not* statistically independent.

10. Let

$$x(t) = x_c(t) \cos \omega_c t - x_s(t) \sin \omega_c t$$

be a sample function of a stationary narrow-band real gaussian random process, where $f_c = \omega_c/2\pi$ is the center frequency of the narrow spectral band.

a. Show that

$$R_x(\tau) = R_c(\tau) \cos \omega_c \tau - R_{cs}(\tau) \sin \omega_c \tau \qquad (8\text{-}126)$$

where
$$R_c(\tau) = E[x_{ct} x_{c(t-\tau)}]$$
and
$$R_{cs}(\tau) = E[x_{ct} x_{s(t-\tau)}]$$

b. Show further that, on defining

$$R_\xi(\tau) = [R_c{}^2(\tau) + R_{cs}{}^2(\tau)]^{1/2} \qquad (8\text{-}127a)$$

and

$$\theta(\tau) = \tan^{-1} \left[\frac{R_{cs}(\tau)}{R_c(\tau)} \right] \qquad (8\text{-}127b)$$

where
$$R_\xi(0) = R_c(0) \qquad \text{and} \qquad \theta(0) = 0$$
we may write

$$R_x(\tau) = R_\xi(\tau) \cos [\omega_c \tau + \theta(\tau)] \qquad (8\text{-}128)$$

c. Show that if $S_x(f)$ is even about f_c for $f \geq 0$, then

$$R_{cs}(\tau) = 0, \qquad \theta(\tau) = 0, \qquad R_\xi(\tau) = R_c(\tau)$$

and hence
$$R_x(\tau) = R_c(\tau) \cos \omega_c \tau \qquad (8\text{-}129)$$

11. Let the spectral density of the random process of Prob. 10 be given by

$$S_x(f) = \frac{\sigma_x{}^2}{2(2\pi\sigma^2)^{\frac{1}{2}}} \left\{ \exp\left[-\frac{(f - f_c)^2}{2\sigma^2} \right] + \exp\left[-\frac{(f + f_c)^2}{2\sigma^2} \right] \right\}$$

where $\sigma << f_c$.

    a. Evaluate $R_c(\tau)$.
    b. Evaluate $R_{cs}(\tau)$.

12. Let $x(t)$ be a sample function of a stationary narrow-band real gaussian random process. Define σ_τ and $\lambda(\tau)$ by

$$\sigma_\tau e^{j\lambda(\tau)} = \frac{2}{\sigma_x{}^2} \int_0^\infty S_x(f)\, \exp[j(f - f_c)\tau]\, df$$

$$= \frac{R_c(\tau)}{\sigma_x{}^2} + j\, \frac{R_{cs}(\tau)}{\sigma_x{}^2} \tag{8-130}$$

where σ_τ is chosen to be real and nonnegative. Show that the autocorrelation function of the *envelope* of the narrow-band process is†

$$R_V(\tau) = \sigma_x{}^2[2E(\sigma_\tau) - (1 - \sigma_\tau{}^2)K(\sigma_\tau)]$$

$$= \frac{\pi}{2}\sigma_x{}^2\, {}_2F_1\left(-\frac{1}{2}, -\frac{1}{2};1;\sigma_\tau{}^2 \right) \tag{8-131}$$

where K and E are complete elliptic integrals of the first and second kind respectively and where ${}_2F_1$ is a hypergeometric function.‡

13. Let the sample functions of the random process of Prob. 12 be expressed in the form

$$x(t) = V(t)y(t)$$

where $V(t)$ is the envelope and where

$$y(t) = \cos[\omega_c t + \phi(t)]$$

Show that the autocorrelation function of the phase-modulated carrier $y(t)$ of the given narrow-band process is given by§

$$R_y(\tau) = \left[\frac{E(\sigma_\tau) - (1 - \sigma_\tau{}^2)K(\sigma_\tau)}{2\sigma_\tau} \right] \cos[\omega_c \tau + \lambda(\tau)]$$

$$= \frac{\pi\sigma_\tau}{8}\, {}_2F_1\left(\frac{1}{2},\frac{1}{2};2;\sigma_\tau{}^2 \right) \cos[\omega_c \tau + \lambda(\tau)] \tag{8-132}$$

14. Let $V(t)$ and $\phi(t)$ be the envelope and phase, respectively, of a sample function of a stationary narrow-band real gaussian process. Show, by evaluating Eqs. (8-102), (8-103), and (8-106) at $\phi_2 = \phi_1$, that

$$p(V_1,\phi_1,V_2,\phi_2) \neq p(V_1,V_2)p(\phi_1,\phi_2)$$

and hence that the pairs of random variables (V_1,V_2) and (ϕ_1,ϕ_2) are not statistically independent.

---

† Cf. Price (II) or Middleton (II, Arts. 6 and 7).
‡ Cf. Magnus and Oberhettinger (I, Chap. II, Art. 1).
§ Cf. Price (II).

CHAPTER 9

LINEAR SYSTEMS

The concepts of *random variable* and *random process* were developed in the preceding chapters, and some of their statistical properties were discussed. We shall use these ideas in the remaining chapters of this book to determine the effects of passing random processes through various kinds of systems. Thus, for example, we shall make such a study of linear systems in this and the two following chapters; in particular, we shall introduce linear-system analysis in this chapter, investigate some of the problems of noise in amplifiers in the next, and consider the optimization of linear systems in Chap. 11.

9-1. Elements of Linear-system Analysis

It is assumed that the reader is generally familiar with the methods of analysis of linear systems.† Nevertheless, we shall review here some of the elements of that theory.

The System Function. Suppose that $x(t)$ and $y(t)$, as shown in Fig. 9-1, are the input and output, respectively, of a fixed-parameter linear sys-

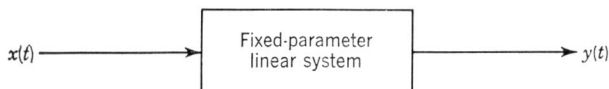

FIG. 9-1. The linear system.

tem. By *fixed parameter* we mean that if the input $x(t)$ produces the output $y(t)$, then the input $x(t + \tau)$ produces the output $y(t + \tau)$. By *linear* we mean that if the input $x_i(t)$ produces the output $y_i(t)$, then the input

$$x(t) = a_1x_1(t) + a_2x_2(t)$$

produces the output

$$y(t) = a_1y_1(t) + a_2y_2(t)$$

An example of such a system would be one governed by a set of linear differential equations with constant coefficients.

If the time function

$$x(t) = e^{j\omega t}$$

† E.g., as presented in Bode (I), Guillemin (I), or James, Nichols, and Phillips (I).

where ω is a real number, is applied for an infinitely long time to the input of a fixed-parameter linear system, and if an output exists after this infinite time, it is of the same form; i.e.,

$$y(t) = Ae^{j\omega t}$$

where A does not depend on t. For suppose that an input of the form $\exp(j\omega t)$ which has been present since $t = -\infty$ produces a well-defined output $y(t)$, which is called the *steady-state* response. Then, since the system is a fixed-parameter system, the input

$$x(t + t') = e^{j\omega(t+t')} = e^{j\omega t'}e^{j\omega t}$$

produces an output $y(t + t')$. However t' does not depend on t, and the system is linear; therefore, the input $\exp(j\omega t')\exp(j\omega t)$ gives the output $\exp(j\omega t')y(t)$. Hence

$$y(t + t') = e^{j\omega t'}y(t)$$

Upon setting $t = 0$, it follows that

$$y(t') = y(0)e^{j\omega t'}$$

is the system response to the input $\exp(j\omega t')$, thus proving our assertion with $A = y(0)$. If now the input has a complex amplitude $X(j\omega)$, that is, if

$$x(t) = X(j\omega)e^{j\omega t} \tag{9-1}$$

the output is

$$y(t) = AX(j\omega)e^{j\omega t}$$

which we write

$$y(t) = Y(j\omega)e^{j\omega t} \tag{9-2}$$

The ratio A of complex amplitudes of $y(t)$ and $x(t)$ is a function of ω, and we denote it henceforth by $H(j\omega)$; thus

$$H(j\omega) = \frac{Y(j\omega)}{X(j\omega)} \tag{9-3}$$

$H(j\omega)$ is called the *system function* of the fixed-parameter linear system. It should be noted that a fixed-parameter linear system may not have a well-defined output if it has been excited by an input of the form $\exp(j\omega t)$ for an infinitely long time, e.g., a lossless L-C circuit with resonant frequency ω. This fact is connected with the idea of stability, which we shall discuss briefly a little later on.

Suppose that the system input is a periodic time function whose Fourier series converges. Then we can write

$$x(t) = \sum_{n=-\infty}^{+\infty} \alpha(jn\omega_o) \exp(jn\omega_o t) \tag{9-4}$$

where

$$T_o = \frac{1}{f_o} = \frac{2\pi}{\omega_o} \tag{9-5}$$

is the fundamental period, and the complex coefficients $\alpha(jn\omega_o)$ are given by

$$\alpha(jn\omega_o) = \frac{1}{T_o} \int_{-T_o/2}^{T_o/2} x(t) \, \exp(-jn\omega_o t) \, dt \qquad (9\text{-}6)$$

It follows from Eq. (9-3) that

$$y_n(t) = Y(jn\omega_o) \exp(jn\omega_o t) = \alpha(jn\omega_o)H(jn\omega_o) \exp(jn\omega_o t)$$

is the system output in response to the input component

$$x_n(t) = \alpha(jn\omega_o) \exp(jn\omega_o t)$$

Since the system is linear, its total output in response to $x(t)$ is the sum of the component outputs $y_n(t)$. Thus

$$y(t) = \sum_{n=-\infty}^{+\infty} \alpha(jn\omega_o)H(jn\omega_o) \exp(jn\omega_o t) \qquad (9\text{-}7)$$

is the steady-state output of a fixed-parameter linear system when its input is a periodic function of time as given by Eq. (9-4).

Next, let us assume that the system input is a transient function of time for which a Fourier transform exists. Our series result above may be extended heuristically to this case as follows: Suppose that we replace the transient input $x(t)$ by a periodic input

$$\tilde{x}(t) = \sum_{n=-\infty}^{+\infty} \alpha(jn\omega_o) \exp(jn\omega_o t)$$

where the coefficients $\alpha(jn\omega_o)$ are determined from $x(t)$ by Eq. (9-6). The corresponding periodic output is, from Eq. (9-7),

$$y(t) = \sum_{n=-\infty}^{+\infty} \alpha(jn\omega_o)H(jn\omega_o) \exp(jn\omega_o t)$$

On multiplying and dividing both equations by T_o and realizing that ω_o is the increment in $n\omega_o$, we get

$$\tilde{x}(t) = \frac{1}{2\pi} \sum_{n=-\infty}^{+\infty} T_o\alpha(jn\omega_o) \exp(jn\omega_o t) \, \Delta(n\omega_o)$$

and

$$\tilde{y}(t) = \frac{1}{2\pi} \sum_{n=-\infty}^{+\infty} T_o\alpha(jn\omega_o)H(jn\omega_o) \exp(jn\omega_o t) \, \Delta(n\omega_o)$$

If now we let $T_o \to \infty$ (and let $n \to \infty$ while $\omega_o \to 0$ so that $n\omega_o = \omega$), it follows from Eq. (9-6) that

$$\lim_{T_o \to \infty} T_o \alpha(jn\omega_o) = \int_{-\infty}^{+\infty} x(t)e^{-j\omega t}\, dt = X(j\omega) \qquad (9\text{-}8)$$

where $X(j\omega)$ is the Fourier transform of the transient input, and hence that

$$\lim_{T_o \to \infty} \tilde{x}(t) = \frac{1}{2\pi}\int_{-\infty}^{+\infty} X(j\omega)e^{j\omega t}\, d\omega = x(t) \qquad (9\text{-}9)$$

and $$\lim_{T_o \to \infty} \tilde{y}(t) = y(t) = \frac{1}{2\pi}\int_{-\infty}^{+\infty} X(j\omega)H(j\omega)e^{j\omega t}\, d\omega \qquad (9\text{-}10)$$

The system output is thus given by the Fourier transform of

$$Y(j\omega) = X(j\omega)H(j\omega) \qquad (9\text{-}11)$$

i.e., the Fourier transform of the output of a fixed-parameter linear system in response to a transient input is equal to the Fourier transform of the input times the system function (when the integral in Eq. (9-10) converges).

Unit Impulse Response. A transient input of special importance is the unit impulse

$$x(t) = \delta(t) \qquad (9\text{-}12)$$

As shown in Appendix 1, the Fourier transform of this input is equal to unity for all ω:

$$X(j\omega) = 1 \qquad (9\text{-}13)$$

The Fourier transform of the corresponding output is, from Eq. (9-11),

$$Y(j\omega) = 1 \cdot H(j\omega) = H(j\omega) \qquad (9\text{-}14)$$

and the output $h(t)$ is, from Eq. (9-10),

$$h(t) = \frac{1}{2\pi}\int_{-\infty}^{+\infty} H(j\omega)e^{j\omega t}\, d\omega \qquad (9\text{-}15)$$

The *unit impulse response* $h(t)$ of a fixed-parameter linear system is therefore given by the Fourier transform of the system function $H(j\omega)$ of that system; conversely

$$H(j\omega) = \int_{-\infty}^{+\infty} h(t)e^{-j\omega t}\, dt \qquad (9\text{-}16)$$

If the unit impulse response is zero for negative values of t, i.e., if

$$h(t) = 0 \qquad \text{when } t < 0 \qquad (9\text{-}17)$$

the corresponding linear system is said to be *physically realizable*.

The Convolution Integral. The response of a fixed-parameter linear system to a transient input can be expressed in terms of the system's unit impulse response by substituting for the system function from Eq. (9-16) into Eq. (9-10). Thus we get

$$y(t) = \int_{-\infty}^{+\infty} h(\tau)\, d\tau \, \frac{1}{2\pi} \int_{-\infty}^{+\infty} X(j\omega)e^{j\omega(t-\tau)}\, d\omega$$

The integral over ω is the inverse Fourier transform of $X(j\omega)$ evaluated at the time $t - \tau$ (i.e., the input at $t - \tau$). Hence

$$y(t) = \int_{-\infty}^{+\infty} h(\tau)x(t - \tau)\, d\tau \qquad (9\text{-}18)$$

Alternatively, by expressing $X(j\omega)$ in Eq. (9-10) in terms of $x(t)$, or by simply changing variables in Eq. (9-18), we get

$$y(t) = \int_{-\infty}^{+\infty} x(\tau)h(t - \tau)\, d\tau \qquad (9\text{-}19)$$

These integrals, known as *convolution integrals*, show that a fixed-parameter linear system can be characterized by an integral operator. Since the system output is thus expressed as a mean of the past values of the input weighted by the unit impulse response of the system, the unit impulse response is sometimes called the *weighting function* of the corresponding linear system.

As the unit impulse response of a physically realizable linear system is zero for negative values of its argument, we can replace the infinite lower limit of the integral in Eq. (9-18) by zero:

$$y(t) = \int_{0}^{\infty} h(\tau)x(t - \tau)\, d\tau$$

Further, if we assume that the input was zero before $t = -a$ (as shown in Fig. 9-2), the upper limit may be changed from $+\infty$ to $t + a$. Thus we get

$$y(t) = \int_{0}^{t+a} h(\tau)x(t - \tau)\, d\tau \qquad (9\text{-}20)$$

for the output of a physically realizable linear system in response to an input starting at $t = -a$.

An intuitive feeling for the convolution integral may perhaps be obtained as follows: Consider the output of a physically realizable linear system at some specific instant of time $t_1 > -a$. Let the system input $x(t)$ be approximated over the interval $(-a, t_1)$ by a set of N nonoverlapping rectangular pulses of width $\Delta\tau$ (as shown in Fig. 9-2), where

$$N \, \Delta\tau = t_1 + a$$

Since the system is linear, the system output at t_1 is given by the sum of the outputs produced by the N previous rectangular input pulses.

Consider now the output at t_1 produced by the single input pulse occurring at the earlier time $(t_1 - n\,\Delta\tau)$, where $n \leq N$. As $\Delta\tau \to 0$, the width of the input pulse approaches zero (and $N \to \infty$) and the output due to this pulse approaches that which would be obtained from an impulse input applied at the same time and equal in area to that of the

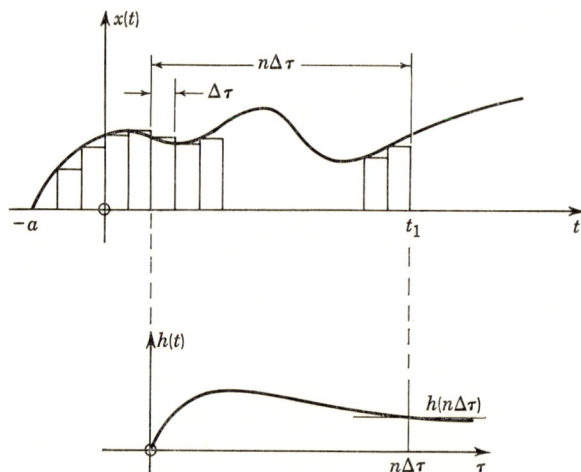

FIG. 9-2. The convolution integral.

given rectangular input pulse, i.e., to $x(t_1 - n\,\Delta\tau)\,\Delta\tau$. Since the system output at t_1 due to a unit impulse input at $(t_1 - n\,\Delta\tau)$ is $h(n\,\Delta\tau)$, the output at t_1 due to the given rectangular input pulse is approximately $h(n\,\Delta\tau)x(t_1 - n\,\Delta\tau)\,\Delta\tau$. The total output at t_1 is then approximately

$$y(t_1) = \sum_{n=1}^{N} h(n\,\Delta\tau)x(t_1 - n\,\Delta\tau)\,\Delta\tau$$

If now we let $\Delta\tau \to 0$ (and $n \to \infty$ so that $n\,\Delta\tau = \tau$), we obtain

$$y(t_1) = \int_0^{t_1+a} h(\tau)x(t_1 - \tau)\,d\tau$$

which is Eq. (9-20) evaluated at $t = t_1$.

A fixed-parameter linear system is said to be *stable* if every bounded input function produces a bounded output function. A stability requirement on the unit impulse response can be obtained as follows: It follows from Eq. (9-18) that

$$|y(t)| = \left| \int_{-\infty}^{+\infty} h(\tau)x(t - \tau)\,d\tau \right| \leq \int_{-\infty}^{+\infty} |h(\tau)|\,|x(t - \tau)|\,d\tau$$

If the input is bounded, there exists some positive constant A such that

$$|x(t)| \leq A < +\infty$$

for all t.   Hence, for such an input,

$$|y(t)| \leq A \int_{-\infty}^{+\infty} |h(\tau)| \, d\tau$$

for all t.   Therefore, if the unit impulse response is absolutely integrable, i.e., if

$$\int_{-\infty}^{+\infty} |h(\tau)| \, d\tau < +\infty$$

then the output is bounded and the system is stable.   On the other hand, it may be shown† that if $h(t)$ is not integrable, then the system is unstable.

It is often useful to extend the system function as we have defined it to a function of the complex variable $p = \alpha + j\omega$.   Let $H(p)$ be defined by the complex Fourier transform

$$H(p) = \int_{-\infty}^{\infty} h(t) e^{-(\alpha+j\omega)t} \, dt \tag{9-21}$$

in that region of the p plane where the integral converges.   Equation (9-21) reduces to our earlier expression for the system function, Eq. (9-16), when $\alpha = 0$.   In particular, if the filter is realizable and stable, i.e., if $h(t) = 0$ for $t < 0$ and $h(t)$ is integrable, then the integral in Eq. (9-21) converges uniformly for all $\alpha \geq 0$.   It follows that $H(p)$ is defined as a uniformly bounded function for all p with $\alpha \geq 0$ (i.e., $|H(p)| \leq A = $ constant for all $\alpha \geq 0$).   Under the same conditions it may be shown by standard methods‡ that $H(p)$ is in fact an analytic function for all p with $\alpha > 0$.   Thus $H(p)$ can have no poles on the $j\omega$ axis or in the right half p plane.   Conversely, it may be proved that if $H(p)$ is a bounded analytic function of p for $\alpha \geq 0$ and if $H(j\omega)$ satisfies an appropriate condition regulating its behavior for large ω, then the Fourier transform of $H(p)$ is a function of t which vanishes for $t < 0$; i.e., its Fourier transform is a realizable impulse response function.   In fact, if $H(j\omega)$ is of integrable square along the $j\omega$ axis, then

$$H(p) = \int_{0}^{\infty} h(t) e^{-(\alpha+j\omega)t} \, dt$$

(where the convergence of the integral is in the sense of convergence-in-the-mean) for some $h(t)$.§   For a more detailed discussion of $H(p)$ as

† James, Nichols, and Phillips (I, Secs. 2.8 and 5.3).
‡ Titchmarsh (I, Chap. II).
§ Paley and Wiener (I, Theorem VIII, p. 11).

a function of a complex variable, the reader is referred to other sources.†
Generally we shall consider only stable linear systems in the discussions
that follow.

Evaluation of Network-system Functions. In some of our later discussions, it will be necessary to consider multiple-input linear systems.
Let us therefore review some of the input–output relations for an N-input
single-output stable linear system. Since the system is linear, we may
express its output $y(t)$ as a sum of terms $y_n(t)$:

$$y(t) = \sum_{n=1}^{N} y_n(t) \tag{9-22}$$

where $y_n(t)$ is defined to be that part of the output which is produced
by the nth input $x_n(t)$; i.e., $y(t)$ is equal to $y_n(t)$ when all the inputs are
zero except the nth. The problem now is to determine $y_n(t)$ given $x_n(t)$.

Suppose that the nth input is a sinusoidal function of time $X_n(j\omega)$
$\exp(j\omega t)$ and that all other inputs are zero. We can then define a system function $H_n(j\omega)$, relating the output to the nth input, by

$$H_n(j\omega) = \frac{Y_n(j\omega)}{X_n(j\omega)} \tag{9-23}$$

where $Y_n(j\omega)$ is the complex amplitude of the resulting response. We
can also define a unit impulse response function $h_n(t)$ as the Fourier
transform of $H_n(j\omega)$:

$$h_n(t) = \frac{1}{2\pi} \int_{-\infty}^{+\infty} H_n(j\omega)e^{j\omega t}\, d\omega \tag{9-24}$$

This impulse response is the system output in response to a unit impulse
at the nth input when all the other inputs are zero.

A particular case of interest is that in which the system under discussion is an $N + 1$ terminal-pair electrical network as shown in Fig. 9-3.
It is desirable to evaluate the system functions H_n in terms of the more
familiar network impedances and admittances. Assuming sinusoidal
inputs, we may write

$$v_n(t) = V_n e^{j\omega t} \qquad \text{and} \qquad i_n(t) = I_n e^{j\omega t}$$

where generally V_n and I_n are complex functions of frequency. The
voltage and current relations at the $N + 1$ terminal pairs can then be

---

† See James, Nichols, and Phillips (I, Chap. 2) and Bode (I, Chap. VII). For a
mathematical discussion of the complex Fourier transform see Paley and Wiener (I),
especially the Introduction and Sec. 8.

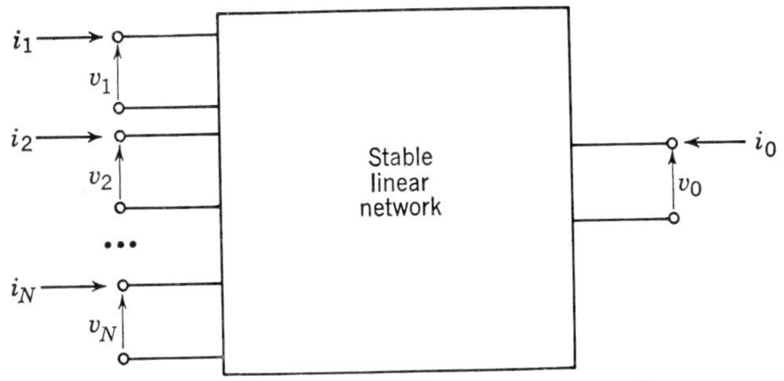

FIG. 9-3. The $N + 1$ terminal-pair network.

expressed by the $N + 1$ simultaneous equations:[†]

$$I_0 = y_{00}V_0 + y_{01}V_1 + \cdots + y_{0N}V_N$$
$$I_1 = y_{10}V_0 + y_{11}V_1 + \cdots + y_{1N}V_N$$
$$\cdots \cdots \cdots \cdots \cdots \cdots \cdots \cdots$$
$$I_N = y_{N0}V_0 + y_{N1}V_1 + \cdots + y_{NN}V_N \qquad (9\text{-}25)$$

where the coefficients y_{nk} are the so-called *short-circuit transfer admittances*. If we set all the input voltages to zero except V_k (i.e., if we short-circuit all inputs except the kth), Eqs. (9-25) become

$$I_0 = y_{0k}V_k$$
$$I_1 = y_{1k}V_k$$
$$\cdots \cdots \cdots$$
$$I_N = y_{Nk}V_k$$

Then, solving for the admittances y_{nk}, we obtain

$$y_{nk} = \frac{I_n}{V_k} \qquad (9\text{-}26)$$

Thus the admittances y_{nk} are given by the ratio of the complex amplitude of the short-circuit current flowing in the nth terminal pair to the complex amplitude of the voltage applied to the kth terminal pair (when all other terminal pairs are short-circuited); whence the name "short-circuit transfer admittance." If

$$y_{nk} = y_{kn} \qquad (9\text{-}27)$$

for all values of n and k, the network is said to be *bilateral*.

It is convenient at this time to introduce the Thévenin equivalent circuit of our network as seen from the output terminals. This equivalent network, shown in Fig. 9-4, consists of a generator whose voltage equals the open-circuit output voltage of the original network, in series with an impedance Z_0, which is the impedance seen looking back into

[†] Cf. Guillemin (I, Vol. I, Chap. 4).

the original network when all of the inputs are short-circuited. From this definition of output impedance, and from the network equations (9-25), it follows that

$$I_0 = y_{00}V_0 = \frac{V_0}{Z_0}$$

Therefore
$$Z_0 = \frac{1}{y_{00}} \tag{9-28}$$

i.e., the output impedance equals the reciprocal of the short-circuit driving-point admittance of the output terminal pair.

Let us now determine the open-circuit output which results from the application of the input voltages $V_1, V_2, \ldots, V_N$. On short-circuiting

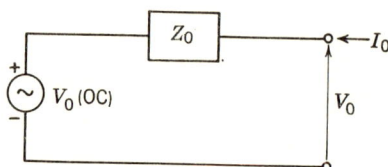

FIG. 9-4. The Thévenin equivalent network.

the output terminals, we get from the Thévenin equivalent network the relations

$$V_0(\text{OC}) = -I_0(\text{SC})Z_0 = -\frac{I_0(\text{SC})}{y_{00}}$$

The short-circuit output current $I_0(\text{SC})$ flowing in response to the given set of input voltages follows from the first of Eqs. (9-25):

$$I_0(\text{SC}) = \sum_{n=1}^{N} y_{0n}V_n$$

Hence the open-circuit output voltage $V_0(\text{OC})$, can be expressed in terms of the input voltages by the equation

$$V_0(\text{OC}) = \sum_{n=1}^{N} \left(-\frac{y_{0n}}{y_{00}}\right) V_n$$

From the definition of the system functions $H_n(j\omega)$ it then follows that

$$V_0(\text{OC}) = \sum_{n=1}^{N} H_n(j\omega) V_n$$

On comparing these last two equations, we see that the system functions can be evaluated in terms of the short-circuit transfer admittances by the relations

$$H_n(j\omega) = -\frac{y_{0n}}{y_{00}} \qquad (9\text{-}29)$$

We shall use these results later in our studies of thermal noise in linear networks.

9-2. Random Inputs

The input–output relations discussed in the preceding section determine the output of a fixed-parameter linear system when the input is a *known* function of time. Suppose now that the input is a sample function of some random process. If we know what function of time the given sample function is, we can apply directly the results of Art. 9-1. More often, however, the only facts known about the input sample function are the properties of its random process; e.g., it might be stated only that the input is a sample function of a gaussian random process whose mean and autocorrelation function are given. A question then arises as to whether the results of the preceding article are applicable in such a case.

Suppose that the input to a stable fixed-parameter linear system is a sample function of a *bounded* random process. Since all sample functions of such a random process are bounded, it follows from the definition of stability that the convolution integral

$$y(t) = \int_{-\infty}^{+\infty} h(\tau)x(t - \tau)\, d\tau \qquad (9\text{-}30)$$

converges for every sample function and hence gives the relation between the input and output sample functions. More generally, it follows from the facts stated in Art. 4-7 that the integral of Eq. (9-30) converges for all sample functions except a set of probability zero if

$$\int_{-\infty}^{\infty} |h(t - \tau)| E[|x(\tau)|]\, d\tau < \infty \qquad (9\text{-}31)$$

Since the filter is assumed to be stable, this condition is satisfied if $E(|x_\tau|)$ is bounded for all τ. In particular, if the input random process is stationary or wide-sense stationary and has an absolute first moment, then $E(|x_\tau|) = \text{constant}$ and the condition above is satisfied. It may be noted in passing that from the Schwarz inequality

$$E(|x|) \le \sqrt{E(|x|^2)} \qquad (9\text{-}32)$$

so that if $x(t)$ is a sample function from any stationary random process with a finite second moment, the integral of Eq. (9-30) converges for all sample functions except a set of probability zero. Furthermore, if the input process is stationary,

$$E(y_t) = \int_{-\infty}^{\infty} h(\tau)E[x(t - \tau)] \, d\tau = m_x \int_{-\infty}^{\infty} h(\tau) \, d\tau \qquad (9\text{-}33)$$

9-3. Output Correlation Functions and Spectra

Suppose that the input to a fixed-parameter stable linear system is a sample function $x(t)$ of a given random process and that $y(t)$ is the corresponding sample function of the output random process. The output autocorrelation function is, by definition,

$$R_y(t_1,t_2) = E(y_{t_1}y_{t_2}^*)$$

Since the system is stable, y_t may be expressed in terms of x_t by the convolution integral in Eq. (9-30). Hence we can write

$$R_y(t_1,t_2) = E\left[\int_{-\infty}^{+\infty} h(\alpha)x_{t_1-\alpha} \, d\alpha \int_{-\infty}^{+\infty} h(\beta)x_{t_2-\beta}^* \, d\beta \right]$$

which, on interchanging the order of averaging and integration, becomes

$$R_y(t_1,t_2) = \int_{-\infty}^{+\infty} h(\alpha) \, d\alpha \int_{-\infty}^{+\infty} h(\beta) \, d\beta \, E(x_{t_1-\alpha}x_{t_2-\beta}^*)$$

or $\qquad R_y(t_1,t_2) = \int_{-\infty}^{+\infty} h(\alpha) \, d\alpha \int_{-\infty}^{+\infty} h(\beta) \, d\beta \, R_x(t_1 - \alpha, t_2 - \beta) \qquad (9\text{-}34)$

The autocorrelation function of the output of a stable linear system is therefore given by the double convolution of the input autocorrelation function with the unit impulse response of the system.

If the input random process is stationary in the wide sense, we have

$$R_x(t_1 - \alpha, t_2 - \beta) = R_x(\tau + \beta - \alpha)$$

where $\tau = t_1 - t_2$. In this case the output autocorrelation function becomes

$$R_y(\tau) = \int_{-\infty}^{+\infty} h(\alpha) \, d\alpha \int_{-\infty}^{+\infty} h(\beta) \, d\beta \, R_x(\tau + \beta - \alpha) \qquad (9\text{-}35)$$

The times t_1 and t_2 enter into Eq. (9-35) only through their difference τ. This result, in conjunction with Eq. (9-33), shows that if the random process input to a stable linear system is stationary in the wide sense, then so is the output random process.

The spectral density $S_y(f)$ of the system output is given by,

$$S_y(f) = \int_{-\infty}^{+\infty} R_y(\tau)e^{-j\omega\tau} \, d\tau$$

where $\omega = 2\pi f$. Hence, from Eq. (9-35),

$$S_y(f) = \int_{-\infty}^{+\infty} h(\alpha) \, d\alpha \int_{-\infty}^{+\infty} h(\beta) \, d\beta \int_{-\infty}^{+\infty} R_x(\tau + \beta - \alpha)e^{-j\omega\tau} \, d\tau$$

On introducing a new variable $\gamma = \tau + \beta - \alpha$, we get

$$S_y(f) = \int_{-\infty}^{+\infty} h(\alpha)e^{-j\omega\alpha}\, d\alpha \int_{-\infty}^{+\infty} h(\beta)e^{+j\omega\beta}\, d\beta \int_{-\infty}^{+\infty} R_x(\gamma)e^{-j\omega\gamma}\, d\gamma$$

The integrals in this equation may be identified in terms of the system function and the input spectral density. Thus

$$S_y(f) = H(j2\pi f)H^*(j2\pi f)S_x(f) = |H(j2\pi f)|^2 S_x(f) \qquad (9\text{-}36)$$

The spectral density of the output of a stable linear system, in response to a wide-sense stationary input random process, is therefore equal to the square of the magnitude of the system function times the spectral density of the system input.

Multiple-input Systems. Let us consider next a fixed-parameter stable linear system having N inputs $x_n(t)$ and a single output $y(t)$. We showed in Art. 9-1 that since the system is linear, the superposition principle may be applied and the system output can be expressed as

$$y(t) = \sum_{n=1}^{N} y_n(t)$$

where $y_n(t)$ equals $y(t)$ when all the inputs are zero except the nth. We also showed there that we may define a system function $H_n(j\omega)$ and an impulse response $h_n(t)$ relating $y_n(t)$ and $x_n(t)$. From the definition of $h_n(t)$, it follows that $y_n(t)$ can be expressed in terms of $x_n(t)$ by the convolution integral

$$y_n(t) = \int_{-\infty}^{+\infty} h_n(\tau)x_n(t - \tau)\, d\tau$$

The total output may therefore be expressed in terms of the various inputs by the equation

$$y(t) = \sum_{n=1}^{N} \int_{-\infty}^{+\infty} h_n(\tau)x_n(t - \tau)\, d\tau$$

When the various inputs are sample functions of random processes, we can determine the output autocorrelation function by paralleling the steps followed in the single-input single-output case above. In this way we obtain

$$R_y(t_1,t_2) = \sum_{n=1}^{N} \sum_{m=1}^{N} \int_{-\infty}^{+\infty} h_n(\alpha)\, d\alpha \int_{-\infty}^{+\infty} h_m(\beta)\, d\beta\; R_{nm}(t_1 - \alpha, t_2 - \beta)$$

$$(9\text{-}37)$$

where R_{nm} is the cross-correlation function of the nth and mth input processes:

$$R_{nm}(t,t') = E(x_{n,t}x^*_{m,t'})$$

If the various inputs are mutually uncorrelated and all have zero means,

$$R_{nm}(t,t') = \begin{cases} R_n(t,t') & \text{when } n = m \\ 0 & \text{when } n \neq m \end{cases}$$

where R_n is the autocorrelation function of the nth input. The output autocorrelation function then becomes

$$R_y(t_1,t_2) = \sum_{n=1}^{N} \int_{-\infty}^{+\infty} h_n(\alpha)\, d\alpha \int_{-\infty}^{+\infty} h_n(\beta)\, d\beta\, R_n(t_1 - \alpha, t_2 - \beta) \quad (9\text{-}38)$$

If the various input random processes are all stationary in the wide sense, then

$$R_{nm}(t,t') = R_{nm}(t - t')$$

On setting $\tau = t_1 - t_2$, the output autocorrelation function therefore becomes

$$R_y(\tau) = \sum_{n=1}^{N} \sum_{m=1}^{N} \int_{-\infty}^{+\infty} h_n(\alpha)\, d\alpha \int_{-\infty}^{+\infty} h_m(\beta)\, d\beta\, R_{nm}(\tau + \beta - \alpha) \quad (9\text{-}39)$$

when the input processes are correlated, and becomes

$$R_y(\tau) = \sum_{n=1}^{N} \int_{-\infty}^{+\infty} h_n(\alpha)\, d\alpha \int_{-\infty}^{+\infty} h_n(\beta)\, d\beta\, R_n(\tau + \beta - \alpha) \quad (9\text{-}40)$$

when the input processes are uncorrelated and have zero means. The corresponding output spectral densities may be obtained by taking the Fourier transforms of these results. Thus we obtain, where S_{nm} is the cross-spectral density of the nth and mth inputs,

$$S_y(f) = \sum_{n=1}^{N} \sum_{m=1}^{N} H_n(j2\pi f) H_m^*(j2\pi f) S_{nm}(f) \quad (9\text{-}41)$$

when the input processes are correlated, and

$$S_y(f) = \sum_{n=1}^{N} |H_n(j2\pi f)|^2 S_n(f) \quad (9\text{-}42)$$

when the input processes are uncorrelated and have zero means (where S_n is the spectral density of the nth-input process).

A particular case of interest is that in which the stable fixed-parameter linear system under study is an $N + 1$ terminal-pair electrical network as shown in Fig. 9-2. We saw in Art. 9-1 that the system function of such a network could be expressed in terms of the various short-circuit transfer admittances by the relations

$$H_n(j2\pi f) = -\frac{y_{0n}}{y_{00}}$$

On substituting these relations into Eqs. (9-41) and (9-42), we get

$$S_0(f) = \sum_{n=1}^{N} \sum_{m=1}^{N} \frac{y_{0n}y_{0m}^*}{|y_{00}|^2} S_{nm}(f) \tag{9-43}$$

when the inputs are correlated, and

$$S_0(f) = \sum_{n=1}^{N} \left| \frac{y_{0n}}{y_{00}} \right|^2 S_n(f) \tag{9-44}$$

when the input processes are uncorrelated and have zero means.

9-4. Thermal Noise†

The randomness of the thermally excited motion of free electrons in a resistor gives rise to a fluctuating voltage which appears across the terminals of the resistor. This fluctuation is known as *thermal noise*. Since the total noise voltage is given by the sum of a very large number of electronic voltage pulses, one might expect from the central limit theorem that the total noise voltage would be a gaussian process. This can indeed be shown to be true. It can also be shown that the individual voltage pulses last for only very short intervals in time, and hence the noise-voltage spectral density can be assumed practically to be flat; i.e., approximately

$$S_{v_n}(f) = S_{v_n}(0) \tag{9-45}$$

as in the study of shot noise.

It is often convenient in analyses of thermal noise in electrical networks to represent a noisy resistor either by a Thévenin equivalent circuit consisting of a noise voltage generator in series with a *noiseless* resistor or by an equivalent circuit consisting of a noise current generator in parallel with a *noiseless* conductance. These equivalent circuits are shown in Fig. 9-5.

The zero-frequency value of the noise-voltage spectral density may be determined by studying the thermal equilibrium behavior of the system composed of a noisy resistor connected in parallel with a lossless inductor. Such a system with its equivalent circuit is shown in Fig. 9-6.

† Cf. Lawson and Uhlenbeck (I, Arts. 4.1 through 4.5).

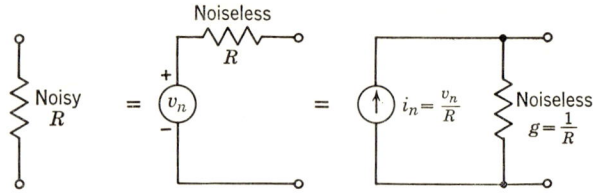

FIG. 9-5. A noisy resistor and its equivalent circuits.

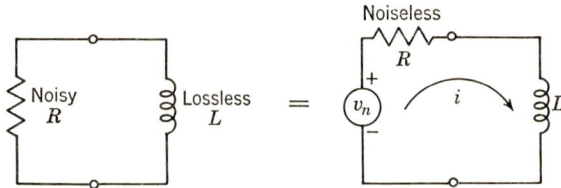

FIG. 9-6. The resistor–inductor system.

From the equipartition theorem of statistical mechanics,† it follows that the average free energy in the current flowing around the loop must equal $kT/2$, where k is Boltzmann's constant and T is the temperature of the system in degrees Kelvin.   Hence

$$E\left(\frac{Li^2}{2}\right) = \frac{L}{2} E(i^2) = \frac{kT}{2} \tag{9-46}$$

The mean-square value of the current can be evaluated from its spectral density:

$$E(i^2) = \int_{-\infty}^{+\infty} S_i(f)\, df$$

Defining $H(j2\pi f)$ to be the system function relating loop current to noise voltage, we have

$$H(j2\pi f) = \frac{1}{R + j2\pi fL}$$

and hence, from Eqs. (9-36) and (9-45),

$$S_i(f) = |H(j2\pi f)|^2 S_{v_n}(f) = \frac{S_{v_n}(0)}{R^2 + (2\pi fL)^2}$$

The mean-square value of the current is, therefore,

$$E(i^2) = S_{v_n}(0) \int_{-\infty}^{+\infty} \frac{df}{R^2 + (2\pi fL)^2} = \frac{S_{v_n}(0)}{2RL}$$

On substituting this result into Eq. (9-46) and solving for $S_{v_n}(0)$, we get

$$S_{v_n}(0) = 2kTR \tag{9-47}$$

† See Valley and Wallman (I, pp. 515–519) or van der Ziel (I, Art. 11.2).

The spectral density of the equivalent noise current generator, shown in Fig. 9-5, is then[†]

$$S_{i_n}(f) = 2kTg \qquad (9\text{-}48)$$

The frequency above which it is unreasonable to assume a flat spectral density for thermal noise is subject to some question.[‡] However, it is undoubtedly above the frequency range in which the concept of a physical resistor as a lumped-constant circuit element breaks down.

Thermal Noise in Linear Networks. Let us now determine the spectral density of the noise voltage appearing across any two terminals of a lossy linear electrical network. Suppose, for the moment, that the loss

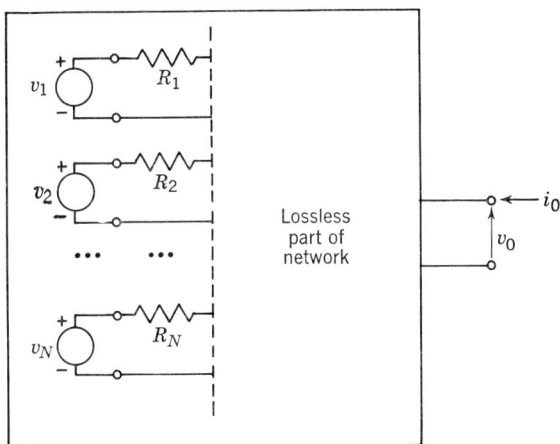

FIG. 9-7. An N-resistor lossy network.

in the network under consideration is due entirely to the presence of N resistors contained within the network. Let the resistance and temperature of the nth resistor be R_n and T_n respectively. We pointed out above that a noisy resistor could be represented by a noiseless resistor in series with a noise-voltage source. The network under study may be considered therefore to be an $N + 1$ terminal-pair network having as inputs the noise-voltage sources of the N resistors. Aside from the N resistors, the network is lossless; the system can be represented schematically as in Fig. 9-7.

The noise voltages generated by the different resistors are statistically independent; hence, from Eq. (9-44), the spectral density of the output noise voltage is

$$S_0(f) = \sum_{n=1}^{N} \left| \frac{y_{0n}}{y_{00}} \right|^2 S_n(f)$$

[†] These results were first derived by Nyquist. See Nyquist (I).
[‡] See Lawson and Uhlenbeck (I, Art. 4.5).

where the short-circuit transfer admittance y_{0n} relates the output terminals to the terminals of the nth noise voltage source, and where $S_n(f)$ is the spectral density of the nth noise voltage source. Since, from Eq. (9-47),

$$S_n(f) = 2kT_nR_n$$

it follows that the spectral density of the output noise voltage of the lossy network can be expressed as

$$S_0(f) = 2k \sum_{n=1}^{N} \left| \frac{y_{0n}}{y_{00}} \right|^2 T_n R_n \qquad (9\text{-}49)$$

This result is sometimes known as Williams's theorem.† When all resistors in the network are at the same temperature T, Eq. (9-4) simplifies to

$$S_0(f) = 2kT \sum_{n=1}^{N} \left| \frac{y_{0n}}{y_{00}} \right|^2 R_n \qquad (9\text{-}50)$$

If, in addition, the requirement that the lossy network be bilateral is added, at least in the sense that

$$y_{0n} = y_{n0}$$

for all n, the spectral density of the output noise voltage becomes

$$S_0(f) = 2kTR_0(f) \qquad (9\text{-}51)$$

where $R_0(f)$ is the real part of the impedance seen looking back into the output terminals of the lossy network. This result is commonly known as the *generalized Nyquist theorem;* by suitably defining "resistance," it can be extended to apply to linear dissipative systems in general,‡ e.g., to Brownian motion, gas-pressure fluctuations, etc.

In order to prove the generalized Nyquist theorem, let us suppose that we can short-circuit all the idealized noise voltage sources v_n and apply a sinusoidal voltage across the output terminals of the lossy network. The complex amplitude I_n of the current flowing in the resistor R_n is related to the complex amplitude V_0 of the applied voltage by

$$I_n = y_{n0}V_0$$

The average power $\bar{P}_n$ dissipated in R_n under these conditions is

$$\bar{P}_n = \tfrac{1}{2}|I_n|^2 R_n = \tfrac{1}{2}|y_{n0}|^2 R_n |V_0|^2$$

† Williams (I).
‡ Callen and Welton (I).

The total average power $\bar{P}_0$ dissipated in the lossy network is the sum of the average powers dissipated in the N resistors:

$$\bar{P}_0 = \frac{|V_0|^2}{2} \sum_{n=1}^{N} |y_{n0}|^2 R_n$$

This total average power can also be expressed in terms of the conditions at the output terminals; thus

$$P_0 = \frac{1}{2} \Re[V_0 I_0^*] = \frac{|V_0|^2}{2} \Re\left[\frac{1}{z_0^*}\right] = \frac{|V_0|^2}{2} \frac{R_0}{|z_0|^2}$$

where $\Re$ denotes real part and R_0 is the real part of the output impedance z_0 of the lossy network. Comparing the last two results, we find that

$$R_0 = |z_0|^2 \sum_{n=1}^{N} |y_{n0}|^2 R_n = \sum_{n=1}^{N} \left|\frac{y_{n0}}{y_{00}}\right|^2 R_n \tag{9-52}$$

If $y_{n0} = y_{0n}$ for all n, then

$$R_0 = \sum_{n=1}^{N} \left|\frac{y_{0n}}{y_{00}}\right|^2 R_n \tag{9-53}$$

If this result is substituted into Eq. (9-50), the generalized Nyquist theorem, Eq. (9-51), is obtained.

9-5. Output Probability Distributions

Our concern in this chapter is to characterize the random-process output of a linear system which is excited by a random-process input. Up to this point we have concerned ourselves chiefly with the problem of finding the autocorrelation function of the output when the autocorrelation function of the input and the impulse response of the system are known; we have also considered the problem, which is equivalent for a fixed-parameter system excited by a stationary process, of finding the spectrum of the output when the spectrum of the input and the system function are known. We have seen that these problems can always be solved, at least in principle, if the system can be characterized by an integral operator as in Eq. (9-18). If the input random process is gaussian, the output random process is also gaussian, as discussed in Art. 8-4; hence, from a knowledge of the mean and autocorrelation functions for the output, any n-variate probability density function for the output process can be written explicitly.[†] If the input random process is non-

[†] Cf. Wang and Uhlenbeck (I, Arts. 10 and 11).

gaussian, the probability distributions characterizing the output process are not, in general, determined by the autocorrelation and mean of the output.

This being so, it is natural to ask what can be done toward obtaining output probability distributions for a linear system driven by a nongaussian random process. The answer is, usually very little. There seems to be no general method for finding even a single-variate probability distribution of the output, beyond the brute-force method of calculating all its moments. This is unsatisfactory, of course, first because not all distributions are determined by their moments, and second because it is usually a practical impossibility to get all the moments in an easily calculable form. However, a few examples of this type of problem have been worked out by one device or another; in most of these examples the input random process, although not gaussian, is some particular nonlinear functional or class of functionals of a gaussian process,[†] such as the envelope, the absolute value, or the square.

The Square-law Detector and Video Amplifier. An interesting and important calculation is that for the probability distribution at time t of the output of a system consisting of an IF filter, square-law detector, and video filter, and for which the system input is either white gaussian noise or signal plus white gaussian noise. This was first done by Kac and Siegert,[‡] and the results have been extended by others.[§] We shall follow essentially Emerson's treatment.[§]

First we note that this example is indeed one in which a linear system is driven by a nongaussian random process; for although the input to the IF filter is gaussian and hence the output of that filter is gaussian, the video filter input is the output of the detector, which is the square of a gaussian process. The essential difficulties of the problem arise because of the inclusion of the video filter; the probability distribution of the output of the square-law detector at time t can be calculated directly (e.g., see Arts. 3-6 and 12-2). The IF filter is included in the problem setup for two reasons: engineering verisimilitude and mathematical expediency. The problem can be treated by starting with a stationary gaussian process into the square-law detector, but to give a rigorous justification of some of the steps in the analysis (which we do not do here) it is necessary to impose some conditions on the nature of the input process; a sufficient condition is to suppose it has a spectrum which is the square of the system function of a realizable stable filter (e.g., such a spectrum as would be obtained by passing white gaussian noise through an IF filter).

† Cf. Siegert (I, pp. 12–25).
‡ Kac and Siegert (I).
§ See Emerson (I) and Meyer and Middleton(I).

Let $H_1(j\omega)$ and $H_2(j\omega)$ be the system functions of the IF and video filters, respectively, and let $h_1(t)$ and $h_2(t)$ be their unit impulse responses (Fig. 9-8). Let $v_o(t)$ be the IF filter input, $v_1(t)$ the IF filter output, and $v_2(t)$ the output from the video filter. Assuming both filters to be stable, we can write

$$v_1(t) = \int_{-\infty}^{+\infty} h_1(t - s)v_o(s)\, ds \tag{9-54}$$

$$v_1{}^2(t) = \int_{-\infty}^{+\infty} \int_{-\infty}^{+\infty} h_1(t - s)h_1(t - u)v_o(s)v_o(u)\, ds\, du \tag{9-55}$$

and

$$v_2(t) = \int_{-\infty}^{+\infty} h_2(t - \tau)v_1{}^2(\tau)\, d\tau \tag{9-56}$$

Hence

$$v_2(t) = \int_{-\infty}^{+\infty} \int_{-\infty}^{+\infty} \int_{-\infty}^{+\infty} h_2(t - \tau)h_1(\tau - s)h_1(\tau - u)v_o(s)v_o(u)\, ds\, du\, d\tau \tag{9-57}$$

Upon letting $\sigma = t - \tau$, $\nu = t - u$, and $\mu = t - s$, Eq. (9-57) becomes

$$v_2(t) = \int_{-\infty}^{+\infty} \int_{-\infty}^{+\infty} \Lambda\,(\mu,\nu)v_o(t - \mu)v_o(t-\nu)\, d\mu\, d\nu \tag{9-58}$$

where

$$\Lambda(\mu,\nu) = \int_{-\infty}^{+\infty} h_1(\mu - \sigma)h_1(\nu - \sigma)h_2(\sigma)\, d\sigma \tag{9-59}$$

The central idea in the solution is to expand the input signal and noise in a series of orthogonal functions chosen so that the output at any time

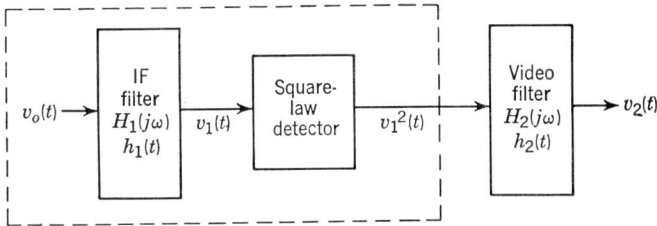

FIG. 9-8. A linear system with a nongaussian input.

t can be written as a series of independent random variables. It turns out, as we shall show below, that this can be done by choosing for the orthogonal functions an orthonormal set of characteristic functions of the integral equation

$$\int_{-\infty}^{+\infty} \Lambda(\mu,\nu)\phi(\nu)\, d\nu = \lambda\phi(\mu) \tag{9-60}$$

From Eq. (9-59) we see that $\Lambda(\mu,\nu)$ is a symmetric function of μ and ν. Also, if $h_1(t)$ is square integrable and $h_2(t)$ is absolutely integrable, it can be shown that

$$\int_{-\infty}^{+\infty} \int_{-\infty}^{+\infty} \Lambda^2(\mu,\nu)\, d\mu\, d\nu < +\infty \tag{9-61}$$

Hence Eq. (9-60) has a discrete set of characteristic values and functions under these conditions.†   Furthermore, $\Lambda(\mu,\nu)$ is non-negative definite if in addition $h_2(t) \geq 0$ for all t; the condition that $\Lambda(\mu,\nu)$ be non-negative definite‡ is equivalent to the condition that the right side of Eq. (9-58) be nonnegative, i.e., that $v_2(t)$ be nonnegative. But since $v_1{}^2(t) \geq 0$, it follows from Eq. (9-56) that $v_2(t) \geq 0$ if $h_2(t) \geq 0.$§   Hence it follows from Mercer's theorem† that

$$\Lambda(\mu,\nu) = \sum_k \lambda_k \phi_k(\mu) \phi_k(\nu) \tag{9-62}$$

where the λ_k and ϕ_k are the characteristic values and orthonormal characteristic functions of Eq. (9-60).

On using Eq. (9-62) in Eq. (9-58) we obtain

$$v_2(t) = \sum_k \lambda_k \left[\int_{-\infty}^{+\infty} v_o(t - \mu) \phi_k(\mu) \, d\mu \right]^2. \tag{9-63}$$

Suppose now that the system input is composed of an input signal function $s(t)$ and a sample function $n(t)$ of a white gaussian random process:

$$v_o(t) = s(t) + n(t) \tag{9-64}$$

If then we define

$$s_k(t) = \int_{-\infty}^{+\infty} s(t - \mu) \phi_k(\mu) \, d\mu \tag{9-65}$$

and

$$n_k(t) = \int_{-\infty}^{+\infty} n(t - \mu) \phi_k(\mu) \, d\mu \tag{9-66}$$

the output of the video filter can be expressed as

$$v_2(t) = \sum_k \lambda_k [s_k(t) + n_k(t)]^2 \tag{9-67}$$

Let v_2 be the random variable defined by $v_2(t)$ at time t, n_k the random variable defined by $n_k(t)$, and s_k the value of $s_k(t)$.   Each n_k is a gaussian random variable, since it is the integral of a gaussian process.   Hence, each $(s_k + n_k)$ is a gaussian random variable with mean value s_k.   Furthermore, the $(s_k + n_k)$ are mutually independent and have variance S_n, since

$$E(n_j n_k) = \int_{-\infty}^{+\infty} \int_{-\infty}^{+\infty} E[n(t - u)n(t - v)] \phi_j(u) \phi_k(v) \, du \, dv$$

$$= \int_{-\infty}^{+\infty} \int_{-\infty}^{+\infty} S_n \delta(u - v) \phi_j(u) \phi_k(v) \, du \, dv$$

$$= \begin{cases} S_n & \text{if } k = j \\ 0 & \text{if } k \neq j \end{cases} \tag{9-68}$$

† See Appendix 2, Art. A2-2.
‡ See Appendix 2, Art. A2-1.
§ Note that we have shown only that $h_2(t) \geq 0$ is a sufficient condition for the non-negativeness of $\Lambda(\mu,\nu)$ and not that it is a necessary condition.

where S_n is the spectral density of the input noise, and $\delta(u - v)$ is the impulse function.

The key part of the problem of finding the single-variate probability distribution of the output is now done. Equation (9-67) gives v_2 as a sum of squares of independent gaussian random variables, so the determination of the characteristic function of v_2 is straightforward. However, after getting the characteristic function, there will still remain the difficulty of obtaining the probability density function of v_2, i.e., of finding the Fourier transform of the characteristic function.

The characteristic function of v_2 is

$$M_{v_2}(jz) = E\left\{\exp\left[jz\sum_k \lambda_k(s_k + n_k)^2\right]\right\}$$

$$= \prod_k E\{\exp[jz\lambda_k(s_k + n_k)^2]\}$$

Since the averaging is over the probability distribution of the gaussian random variable n_k,

$$M_{v_2}(jz) = \prod_k \int_{-\infty}^{+\infty} \exp[jz\lambda_k(s_k + n_k)^2]\frac{\exp(-n_k^2/2S_n)}{\sqrt{2\pi S_n}}\,dn_k \quad (9\text{-}69)$$

The integrals may be evaluated by completing the square to give

$$M_{v_n}(jz) = \prod_k \frac{\exp[(s_k^2/2S_n)2jz\lambda_k S_n/(1 - 2jz\lambda_k S_n)]}{(1 - 2jz\lambda_k S_n)^{\frac{1}{2}}} \quad (9\text{-}70)$$

If the input to the system is noise alone, $s_k = 0$ and Eq. (9-70) becomes

$$M_{v_2}(jz) = \prod_k \frac{1}{(1 - 2jz\lambda_k S_n)^{\frac{1}{2}}} \quad (9\text{-}71)$$

Square-law Envelope Detector. Before going further, let us backtrack and consider a problem, only slightly different from the one just discussed, in which instead of having a square-law detector after the IF filter we have a square-law envelope detector.[†] That is, supposing that the detector input $v_1(t)$ may be written in the form of a narrow-band wave about a center angular frequency ω_o:

$$v_1(t) = V_x(t) \cos \omega_o t - V_y(t) \sin \omega_o t \quad (9\text{-}72)$$

then the action of the detector is to transform this input into $V_x{}^2(t) + V_y{}^2(t)$. This detector action can be interpreted in two ways: first, as

† This is the problem originally treated by Kac and Siegert.

taking the envelope of $v_1(t)$ and squaring the result; second, since

$$v_1{}^2(t) = \tfrac{1}{2}[V_x{}^2(t) + V_y{}^2(t)] + \tfrac{1}{2}[V_x{}^2(t) \cos 2\omega_o t - V_y{}^2(t) \cos 2\omega_o t]$$
$$+ V_x(t)V_y(t) \sin 2\omega_o t$$

as squaring $v_1(t)$ and then filtering out completely the high-frequency components (assuming, of course, that $V_x(t)$ and $V_y(t)$ have no high-frequency components). Introducing the square-law envelope detector instead of the square-law detector implies, strictly speaking, the existence of a nonrealizable filter. For the usual case in which the video pass band cuts off well below the IF, however, the filtering is practically realizable and the assumption of square-law envelope detection changes the problem very little from the problem with square-law detection.

Since we assume in this section that the detector input is a narrow-band wave, we may as well stipulate that the IF pass band and the bandwidth of the signal into the IF amplifier are both narrow relative to ω_o. Further, since only that part of the white-noise spectrum which coincides with the IF pass band effects the IF output $v_1(t)$, we can assume that the input noise, instead of being white, is band-limited with a flat spectrum in some band about ω_o. With these assumptions it is possible to describe approximately the action of the IF amplifier, which is described accurately by Eq. (9-54), in terms of a low-pass equivalent. This approximation is useful here, and indeed in many other circumstances where narrow-band waves are fed into narrow-band filters.

In order to get this low-pass equivalent, we can write the input signal in the form of a modulated sine wave

$$s(t) = a(t) \cos \omega_o t - b(t) \sin \omega_o t \tag{9-73}$$

where $a(t)$ and $b(t)$ have frequency components only in a narrow band (relative to ω_o) about zero frequency. We represent the noise input by

$$n(t) = x(t) \cos \omega_o t - y(t) \sin \omega_o t \tag{9-74}$$

where $x(t)$ and $y(t)$ are sample functions of independent stationary gaussian processes, each with constant spectral density $S_n/2$ in a band of width less than ω_o about zero and spectral density zero outside that band. It may be immediately verified that $n(t)$ is stationary and has spectral density $S_n/2$ in bands around $-\omega_o$ and $+\omega_o$. On introducing the complex amplitudes

$$s_e(t) = a(t) + jb(t)$$
and
$$n_e(t) = x(t) + jy(t) \tag{9-75}$$

we can express the input signal and noise as

$$s(t) = \Re[s_e(t)e^{j\omega_o t}]$$
and
$$n(t) = \Re[n_e(t)e^{j\omega_o t}]$$

The expression for the detector input, Eq. (9-54), then becomes

$$v_1(t) = \int_{-\infty}^{+\infty} h_1(t - u)[s(u) + n(u)]\, du$$

$$= \Re\left\{ \int_{-\infty}^{+\infty} h_1(t - u)e^{j\omega_o u}[s_e(u) + n_e(u)]\, du \right\}$$

$$= \Re\left\{ e^{j\omega_o t} \int_{-\infty}^{+\infty} h_1(t - u)e^{j\omega_o(u-t)}[s_e(u) + n_e(u)]\, du \right\} \qquad (9\text{-}76)$$

Now since

$$h_1(t) = \frac{1}{2\pi} \int_{-\infty}^{+\infty} H_1(j\omega)e^{j\omega t}\, d\omega$$

we have

$$h_1(t - u)e^{-j\omega_o(t-u)} = \frac{1}{2\pi} \int_{-\infty}^{+\infty} H_1[j(\omega + \omega_o)]e^{j\omega(t-u)}\, d\omega$$

Substituting this in Eq. (9-76) yields

$$v_1(t) = \Re[V_e(t)e^{j\omega_o t}]$$

where

$$V_e(t) = \frac{1}{2\pi} \int_{-\infty}^{+\infty} [s_e(u) + n_e(u)]\, du \int_{-\infty}^{+\infty} H_1[j(\omega + \omega_o)]e^{j\omega(t-u)}\, d\omega$$

The narrow-band functions $V_x(t)$ and $V_y(t)$ in Eq. (9-72) can therefore be expressed as the real and imaginary parts, respectively, of $V_e(t)$:

$$V_x(t) = \Re[V_e(t)] \qquad \text{and} \qquad V_y(t) = \Im[V_e(t)]$$

To get $V_x(t)$ and $V_y(t)$ we then need expressions for the real and imaginary parts of

$$\int_{-\infty}^{\infty} H_1[j(\omega + \omega_o)]e^{j\omega(t-u)}\, d\omega$$

Since $H_1^*(j\omega) = H_1(-j\omega)$,

$$\Re\left\{ \int_{-\infty}^{\infty} H_1[j(\omega + \omega_o)]e^{j\omega(t-u)}\, d\omega \right\}$$

$$= \tfrac{1}{2} \int_{-\infty}^{\infty} \{H_1[j(\omega + \omega_o)] + H_1[j(\omega - \omega_o)]\}e^{j\omega(t-u)}\, d\omega \qquad (9\text{-}77a)$$

and

$$\Im\left\{ \int_{-\infty}^{\infty} H_1[j(\omega + \omega_o)]e^{j\omega(t-u)}\, d\omega \right\}$$

$$= \frac{1}{2j} \int_{-\infty}^{\infty} \{H_1[j(\omega + \omega_o)] - H_1[j(\omega - \omega_o)]\}e^{j\omega(t-u)}\, d\omega \qquad (9\text{-}77b)$$

Now $H_1(j\omega)$ is the system function of a narrow-band filter with its pass band centered at ω_o. If it is nearly symmetric in the band about ω_o and if it introduces negligible phase shift, then the two terms in the integrand of Eq. (9-77b) nearly cancel in the frequency zone about $\omega = 0$. Therefore, in the right-hand side of Eq. (9-76) we may neglect the imaginary part of $h_1(t - u) \exp[-j\omega_o(t - u)]$, because, as we have just seen, its frequency components in the zone around zero frequency are negligible and because its higher-frequency components (in the zones around $\pm 2\omega_o$) are erased when it is convolved with $s_e(t) + n_e(t)$, which has only low-frequency components.

Thus, introducing the notations

$$H_{1e}(j\omega) = \tfrac{1}{2}\{H_1[j(\omega + \omega_o)] + H_1[j(\omega - \omega_o)]\} \qquad (9\text{-}78)$$

$$h_{1e}(t) = \frac{1}{2\pi} \int_{-\infty}^{\infty} H_{1e}(j\omega)e^{j\omega t}\, d\omega = h_1(t)\cos\omega_o t$$

Eq. (9-76) becomes approximately

$$v_1(t) = \Re\left\{e^{j\omega_o t} \int_{-\infty}^{+\infty} h_{1e}(t - u)[s_e(u) + n_e(u)]\, du\right\}$$

$$= \left\{\int_{-\infty}^{+\infty} h_{1e}(t - u)[a(u) + x(u)]\, du\right\}\cos\omega_o t$$

$$- \left\{\int_{-\infty}^{+\infty} h_{1e}(t - u)[b(u) + y(u)]\, du\right\}\sin\omega_o t \qquad (9\text{-}79)$$

This result is in the form of Eq. (9-72), with

$$V_x(t) = \int_{-\infty}^{+\infty} h_{1e}(t - u)[a(u) + x(u)]\, du$$

and $$V_y(t) = \int_{-\infty}^{+\infty} h_{1e}(t - u)[b(u) + y(u)]\, du \qquad (9\text{-}80)$$

Note now that both $V_x{}^2(t)$ and $V_y{}^2(t)$ are of the same form as the input signal to the video filter when the detector is a square-law detector. Since the video filter is linear, the output signal $v_2(t)$ is now of the form of the sum of two output signals as given by Eq. (9-67). Furthermore, since $x(u)$ and $y(u)$ are statistically independent processes, $V_x{}^2(t)$ and $V_y{}^2(t)$ are statistically independent processes and the terms in $v_2(t)$ due to $V_x{}^2(t)$ are independent of those due to $V_y{}^2(t)$. From these comments it follows that for the case of the square-law envelope detector, the video-filter output is given by

$$v_2(t) = \sum_k \mu_k\left\{[a_k(t) + x_k(t)]^2 + [b_k(t) + y_k(t)]^2\right\} \qquad (9\text{-}81)$$

where the μ_k are the characteristic values of the integral equation

$$\int_{-\infty}^{+\infty} M(u,v)\psi(v)\, dv = \mu\psi(u) \qquad (9\text{-}82)$$

with $$M(u,v) = \int_{-\infty}^{+\infty} h_{1e}(u - \sigma)h_{1e}(v - \sigma)h_2(\sigma)\, d\sigma \qquad (9\text{-}83)$$

and where, if the $\psi_k(u)$ are the orthonormal characteristic functions of the same integral equation,

$$a_k(t) = \int_{-\infty}^{+\infty} a(t - u)\psi_k(u)\, du \qquad b_k(t) = \int_{-\infty}^{+\infty} b(t - u)\psi_k(u)\, du$$

and

$$x_k(t) = \int_{-\infty}^{+\infty} x(t - u)\psi_k(u)\, du \qquad y_k(t) = \int_{-\infty}^{+\infty} y(t - u)\psi_k(u)\, du \quad (9\text{-}84)$$

From our assumptions about $x(t)$ and $y(t)$, x_k and y_k are, for all k, gaussian random variables which satisfy the relations

$$E(x_k) = E(y_k) = 0$$

$$E(y_k y_j) = E(x_k x_j) = \begin{cases} \dfrac{S_n}{2} & \text{for } k = j \\ 0 & \text{for } k \neq j \end{cases}$$

and
$$E(x_k y_j) = 0 \qquad \text{for all } k \text{ and } j \qquad (9\text{-}85)$$

Since v_2, as given by Eq. (9-81), is exactly the sum of two independent random variables as given by Eq. (9-67), the characteristic function of v_2 may be written immediately from Eq. (9-70):

$$M_{v_2}(jz) = \prod_k \frac{\exp[jz\mu_k(a_k^2 + b_k^2)/(1 - jz\mu_k S_n)]}{(1 - jz\mu_k S_n)} \qquad (9\text{-}86)$$

If the input to the system is noise alone, this becomes

$$M_{v_2}(jz) = \prod_k \frac{1}{(1 - jz\mu_k S_n)} \qquad (9\text{-}87)$$

The Output Distributions. We now have expressions for the characteristic function of the output random variable v_2 for both kinds of detector. Getting the Fourier transforms of these expressions, i.e., the output probability density functions, in usable form is difficult. One of them, the transform of Eq. (9-87), can be integrated by the calculus of residues if the characteristic values are of multiplicity one. We have

$$p(v_2) = \frac{1}{2\pi} \int_{-\infty}^{+\infty} e^{-jzv_2} \prod_k \frac{1}{(1 - jz\mu_k S_n)}\, dz \qquad (9\text{-}88)$$

If we assume as we have previously that $h_2(t) \geq 0$, so that the output voltage is nonnegative, the μ_k are all positive and the poles of the integrand of Eq. (9-88) all occur in the lower half of the z plane at $z_k = -j/\mu_k S_n$. The integral of Eq. (9-88) can then be evaluated by closing a contour in the lower half plane,† and $p(v_2)$ is given by j times

† E.g., see Kac and Siegert (I).

the sum of the residues at the poles, which are simple when the characteristic values μ_k are all distinct; thus

$$p(v_2) = \sum_l \frac{\exp(-v_2/\mu_l S_n)}{\mu_l S_n \prod_{k \neq l} (1 - \mu_k/\mu_l)} \qquad \text{when } v_2 \geq 0 \qquad (9\text{-}89)$$

and is zero otherwise. Usually there are infinitely many characteristic values μ_k and hence infinitely many poles, so this evaluation as it stands is only formal.

Emerson† has calculated the output probability distribution approximately for a number of examples using the characteristic function as given by Eq. (9-70). His method is to expand log $[M_{v_2}(jz)]$ in power

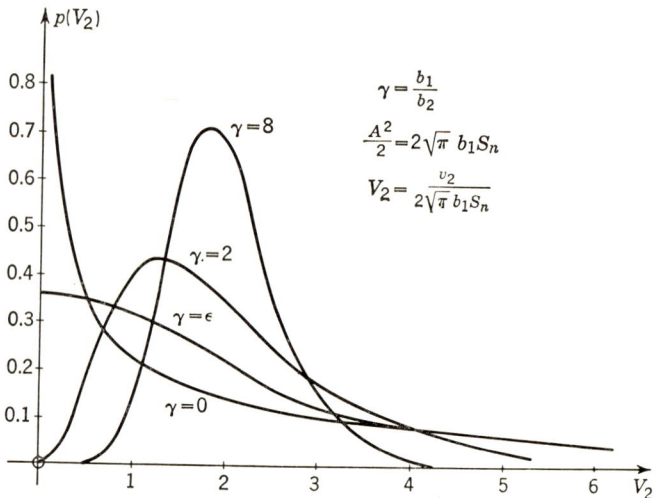

FIG. 9-9. Probability density of the normalized video-filter output for a sine wave plus white gaussian noise input with unity signal-to-noise ratio at the detector input. (*From Fig. 4 of Emerson (I). By permission of the American Institute of Physics.*)

series and use the coefficients in this series to get an asymptotic expansion for the probability density. The reader is referred to that paper for details and also for further references to the mathematical background. Some of his results are shown in Fig. 9-9 for the case of a sine wave plus white gaussian noise input, i.e., for

$$v_o(t) = A \cos \omega_o t + n(t)$$

Emerson's calculations are for cases in which the absolute value of the

† Emerson (I).

system functions of both the IF and video filters are gaussian.† Then the impulse response functions are gaussian, specifically

$$h_1(t) = 2(2\pi b_1)^{1/2} \exp\left[-(2\pi b_1)^2 \frac{t^2}{2}\right] \cos \omega_o t$$

and
$$h_2(t) = (2\pi b_2)^{1/2} \exp\left[-(2\pi b_2)^2 \frac{t^2}{2}\right] \tag{9-90}$$

where b_1 and b_2 are parameters determining the bandwidth and where ω_o is the band-center angular frequency of the IF filter. The parameter γ in Fig. 9-9 is the bandwidth ratio b_1/b_2; hence large values of γ correspond to wide-band IF and narrow-band video filters. These calculations neglect the high-frequency terms at the output of the detector for all curves except the one labeled $\gamma = 0$. Thus the $\gamma = \epsilon$ curve is for a square-law *envelope* detector with no additional filtering, and the $\gamma = 0$ curve is for a square-law detector with no video filter.

The reader is also referred to the paper by Kac and Siegert for the solution of the problem with a single-tuned IF filter and simple low-pass video filter and for a discussion of the limiting cases $(b_1/b_2) \to 0$ and $(b_1/b_2) \to \infty$. It is pointed out there that in the latter case the output probability density tends toward the gaussian.

Example 9-5.1. As a particular example of the foregoing theory, consider the following: A white gaussian noise is passed first through a single-tuned resonant circuit of small fractional bandwidth and then through a square-law envelope detector; it is then integrated. What is the probability distribution of the output of the integrator after a time T?

First, we observe that the integral equation (9-82) may be transformed so that we can work with the autocorrelation function of $V_x(t)$ and $V_y(t)$ (see Eq. (9-72)) directly instead of with $h_{1e}(t)$. This can always be done when the input is noise alone. In fact, if we let

$$g(\sigma) = \int_{-\infty}^{\infty} h_{1e}(v - \sigma)\psi(v) \, dv \tag{9-91}$$

Eq. (9-82) becomes

$$\int_{-\infty}^{\infty} h_{1e}(u - \sigma)h_2(\sigma)g(\sigma) \, d\sigma = \mu\psi(\mu) \tag{9-92}$$

Multiplying by $h_{1e}(u - s)$ and integrating with respect to u gives

$$\int_{-\infty}^{\infty} h_2(\sigma) \left[\int_{-\infty}^{\infty} h_{1e}(u - \sigma)h_{1e}(u - s) \, du\right] g(\sigma) \, d\sigma = \mu g(s)$$

The expression in brackets is the autocorrelation function of $V_x(t)$ and $V_y(t)$ (they are the same) for a white-noise input to the IF. Hence we can write

$$\int_{-\infty}^{\infty} h_2(\sigma) R_{1e}(\sigma - s)g(\sigma) \, d\sigma = g(s) \tag{9-93}$$

† Strictly speaking, such system functions are nonrealizable (cf. Wallman (I)). However, they do approximate rather well a multiple stage cascade of identically tuned single-pole filters.

The characteristic functions of Eq. (9-93) are related to those of Eq. (9-82) by Eq. (9-91); the characteristic values are the same.

Returning to our particular example, we may consider the single-tuned IF to have approximately the same response as that of an R-C filter shifted up in frequency. Hence we have approximately

$$R_{1e}(\tau) = e^{-\alpha|\tau|} \tag{9-94}$$

The weighting function for the second filter $h_2(t)$ is given by

$$h_2(t) = \begin{cases} 1 & \text{for } 0 \le t \le T \\ 0 & \text{otherwise} \end{cases} \tag{9-95}$$

Hence Eq. (9-94) becomes

$$\int_0^T \exp(-\alpha|\sigma - s|)g(\sigma)\, d\sigma = \mu g(s) \tag{9-96}$$

The characteristic values of this equation can be got from Example 6-4.1 by making a change of variable in the integral equation discussed there. The characteristic values are

$$\mu_k = \frac{2}{\alpha}\left(\frac{1}{1 + b_k^2}\right) \tag{9-97}$$

where b_k is a solution of either

$$b \tan\left(\frac{b\alpha T}{2}\right) = 1$$

or

$$b \cot\left(\frac{b\alpha T}{2}\right) = 1 \tag{9-98}$$

These values of b_k substituted into Eqs. (9-87) and (9-89) give the characteristic function and probability density of the output of the integrator. From Eq. (9-91) it follows that $S_n = 1$.

9-6. Problems

1. Let $x(t)$ and $y(t)$ be the input and output functions respectively of a fixed-parameter linear system. Show that if the system is stable and the input has finite energy, i.e., if

$$\int_{-\infty}^{+\infty} |h(\tau)|^2\, d\tau < +\infty \qquad \text{and} \qquad \int_{-\infty}^{+\infty} |x(t)|^2\, dt < +\infty$$

then the output also has finite energy.

2. Let $x(t)$ and $y(t)$ be sample functions of the input and output random processes respectively of a stable fixed-parameter linear system. Suppose that the input process is stationary. Show that

$$R_{xy}(\tau) = \int_{-\infty}^{+\infty} h(\alpha)R_x(\tau - \alpha)\, d\alpha \tag{9-99}$$

Suppose that the input spectral density is flat; i.e., suppose

$$S_x(f) = S_x(0)$$

for all f. Show that, in this case,

$$R_{xy}(\tau) = S_x(0)h(\tau) \tag{9-100}$$

3. Consider the fixed-parameter linear system shown in Fig. 9-10.

a. Determine its unit impulse response.
b. Determine its system function.
c. Determine whether or not the system is stable.

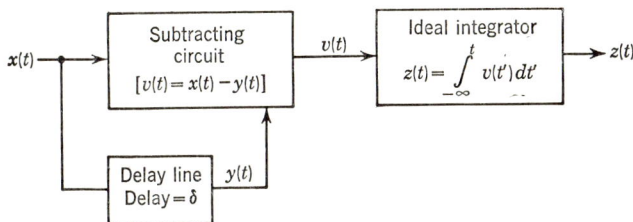

FIG. 9-10. A linear system.

4. The *effective noise bandwidth* B_N of a stable fixed-parameter linear system is defined by

$$B_N = \frac{\int_{-\infty}^{+\infty} |H(j2\pi f)|^2 \, df}{|H_{max}|^2} \tag{9-101}$$

where $|H_{max}|$ is the maximum value of $|H(j2\pi f)|$. Determine B_N for the system in Prob. 3.

5. Determine the effective noise bandwidth of the network shown in Fig. 9-11.

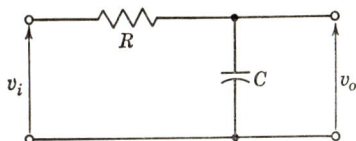

FIG. 9-11. The R-C low-pass filter.   FIG. 9-12. The lossless coupling network.

6. Suppose that the input to the network shown in Fig. 9-11 consists of a sample function of a stationary random process with a flat spectral density of height S_o, plus a sequence of constant-amplitude rectangular pulses. The pulse duration is δ, and the minimum interval between pulses is T, where $\delta << T$.

A signal-to-noise ratio at the system output is defined here to be the ratio of the maximum amplitude of the output signal to rms value of the output noise.

a. Derive an expression relating the output signal-to-noise ratio as defined above to the input pulse duration and the effective noise bandwidth of the network.
b. Determine what relation should exist between the input pulse duration and the effective noise bandwidth of the network to obtain the maximum output signal-to-noise ratio.

7. Let a resistor R at a temperture T be shunted by a condenser C.

a. Determine the spectral density of the noise voltage developed across C.
b. Determine the mean-square value of the noise voltage developed across C.

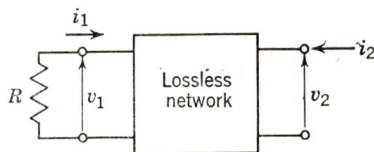

8. Consider a lossless coupling network with a resistor R across its input terminal pair, as shown in Fig. 9-12.

  a. Determine the system function relating v_2 to v_1.

  b. Determine the real part of the impedance seen looking into the output terminals.

  c. Suppose that the temperature of R is T. Show by direct calculation for this specific case that the spectral density of the output voltage v_2 may be obtained either via the system function or via the real part of the output impedance.

9.† Suppose that a sample function $x(t)$ of a wide-sense stationary random process is applied at $t = 0$ to the input of a stable low-pass filter. Let $R_x(\tau)$ be the autocorrelation function of the input process and $h(t)$ be the unit impulse response of the low-pass filter. Suppose that the output $M(t)$ of the filter is observed at $t = T$. Then

$$M(T) = \int_0^T h(t')x(T - t')\, dt' \qquad (9\text{-}102)$$

It is convenient to define a new unit impulse response $h(t,T)$ for the filter by the relations

$$h(t,T) = \begin{cases} h(t) & \text{for } 0 \le t \le T \\ 0 & \text{otherwise} \end{cases} \qquad (9\text{-}103)$$

The new impulse response thus combines the effects of the filter weighting and the observation time.

  a. Let M_T be the random variable referring to the possible values of $M(T)$. Show that

$$E(M_T) = m_x \int_{-\infty}^{+\infty} h(t,T)\, dt \qquad (9\text{-}104)$$

  and hence that

$$E(M_T) = m_x H(0,T) \qquad (9\text{-}105)$$

  where $m_x = E(x)$ and $H(j2\pi f,T)$ is the system function corresponding to $h(t,T)$.

  b. Let

$$R_h(\tau) = \int_{-\infty}^{+\infty} h(t,T)h(t + \tau,T)\, dt \qquad (9\text{-}106)$$

  Show that

$$R_h(\tau) = \int_{-\infty}^{+\infty} |H(j2\pi f,T)|^2 e^{-j2\pi f\tau}\, df \qquad (9\text{-}107)$$

  and hence that $R_h(\tau)$ may be thought of as the autocorrelation function of the observed output of the low-pass filter when the input process has a flat spectral density of unit height.

  c. Show that

$$\sigma^2(M_T) = \int_{-\infty}^{+\infty} R_h(\tau)R_\xi(\tau)\, d\tau \qquad (9\text{-}108)$$

  and hence that

$$\sigma^2(M_T) = \int_{-\infty}^{+\infty} |H(j2\pi f,T)|^2 S_\xi(f)\, df \qquad (9\text{-}109)$$

  where $\xi(t) = x(t) - m_x$ and $S_\xi(f)$ is the spectral density of $\xi(t)$.

† Cf. Davenport, Johnson, and Middleton (I).

10. Referring to Prob. 9,

a. Show that as $T \to 0$, approximately,

$$\frac{\sigma(M_T)}{E(M_T)} = \frac{\sigma_x}{m_x} \tag{9-110}$$

b. Show that as $T \to \infty$, approximately,

$$\frac{\sigma(M_T)}{E(M_T)} = \frac{[S_\xi(0)B_N]^{1/2}}{m_x} \tag{9-111}$$

where B_N is the effective noise bandwidth of the low-pass filter (for $T \to \infty$).

11. Carry out the indicated integration in Eq. (9-69) to obtain Eq. (9-70).

12. The probability density of the output of a square-law detector can be determined as a limiting case of the results of Art. 9-5, when $h_2(t) = \delta(t)$, although it may more easily be done directly.

a. Taking $h_2(t) = \delta(t)$ and noting the uniqueness of the expansion for $\Lambda(\mu,\nu)$ given by Mercer's theorem, show that Eq. (9-60) has only one characteristic value λ, and then find the characteristic function of the output voltage in terms of λ.

b. If the input is noise alone, show that the probability density function for the output is

$$p(v_2) = \frac{\exp(-v_2/2S_n\lambda)}{(2\pi S_n\lambda v_2)^{1/2}} \tag{9-112}$$

13. Repeat Prob. 12 for a square-law envelope detector. Compare the probability density function for the output with that of Prob. 12.

14. a. Show that the logarithm of the characteristic function of v_2, $\log M_{v_2}(jz)$, given by Eq. (9-70), can be expanded in a power series in z which converges for $|z| < (\frac{1}{2}\lambda_1 S_n)$ where λ_1 is the largest characteristic value and in which the mth coefficient contains the λ_k and s_k only in the forms

$$\sum_k \lambda_k^m$$

and

$$\sum_k \lambda_k^m s_k^2 \qquad m = 1, 2, \ldots$$

b. Show that the same result is true if $M_{v_2}(jz)$ is given by Eq. (9-86), except that $\sum_k y_k^m s_k^2$ is replaced by

$$\sum_k \lambda_k^m (a_k^2 + b_k^2) \qquad m = 1, 2, \ldots$$

15. From the results of Prob. 14 and Appendix 2 show that the mth coefficient in the power series for $\log M_{v_2}(jz)$ can be expressed in terms

of

$$\int_{-\infty}^{+\infty} \Lambda^{(m)}(u,u) \, du$$

and

$$\int_{-\infty}^{+\infty} \int_{-\infty}^{+\infty} \Lambda^{(m)}(u,v) s(t-u) s(t-v) \, du \, dv$$

Thus an expression for $M_{v_2}(jz)$ can be written for which the characteristic values and functions do not have to be calculated explicitly.

CHAPTER 10

NOISE FIGURE

Sooner or later it becomes necessary in most communications and data-processing systems to raise the power level of a signal by passing it through a linear amplifier. For various reasons, the output of such an amplifier will generally be an imperfect replica of its input; e.g., a determinate waveform distortion will occur because of the filter characteristics of the amplifier, and a random perturbation of the output will occur whenever there is noise generated within the amplifier itself. The basic tools required for a study of such phenomena were introduced in the preceding chapter.

In certain cases of practical interest, it is not feasible to require from the amplifier any operation upon its input (which may consist of a signal plus a noise) other than that of amplification with as little waveform distortion as possible. A question may well arise in such cases as to how one can *simply* characterize the goodness of performance of an amplifier with respect to its internally generated noise. It has been found in practice that a useful criterion of goodness is the so-called *noise figure* of the amplifier.

In this chapter we shall define the noise figure of an amplifier† and study some of its properties. Our purpose is to introduce the concept and stress a few of the implications involved in its use. The reader is referred to other sources‡ for a comprehensive treatment of the subject.

10-1. Definitions

Our problem is that of studying the effects of noise generated within an amplifier on the system which consists of the amplifier, the source driving the amplifier, and the load presented to the amplifier. The essential elements of the system are shown in Fig. 10-1. In that figure e_s and Z_s represent the Thévenin-equivalent open-circuit voltage and the internal impedance of the source, Z_i the input impedance of the amplifier

---

† Although we shall use the term *amplifier* throughout our discussion, the various concepts and results will apply to any linear system.

‡ E.g., Valley and Wallman (I, Chaps. 12–14) or van der Ziel (I, Chaps. 3, 7, 9, and 10).

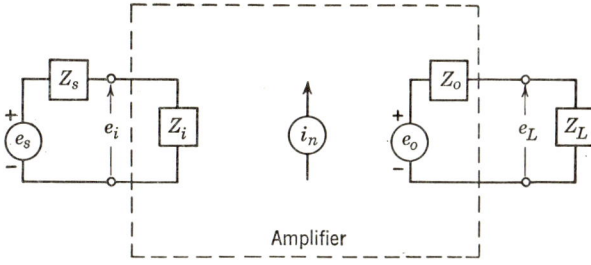

Fig. 10-1. The amplifier and its environment.

(when the amplifier output terminals are open-circuited), e_o and Z_o the Thévenin-equivalent voltage and the internal impedance of the amplifier as seen from its output terminals, and Z_L the load presented to the amplifier. The various sources of noise within the amplifier are represented by the set of noise current generators i_n; all other elements of the amplifier are assumed to be noiseless. The source and the source-amplifier combination are assumed to be *stable linear systems*.

Available Power. The *available power* P_{as} of a source is defined to be the maximum value of the average power P_s obtainable from that source:

$$P_{as} = (P_s)_{max} \tag{10-1}$$

Generally, both the spectral density of the source voltage (or current) and the internal impedance of the source are functions of frequency. We shall therefore commonly refer to the available power from a source within an infinitesimal frequency band of width df which is centered on a frequency f. Such an available power will be called an *incremental available power*.†

Suppose that the source voltage is a sinusoidal function of time:

$$e_s(t) = A \sin \omega t$$

It is a well-known result of circuit analysis that maximum power is delivered by the source when the load impedance connected to it has a value equal to the complex conjugate of the source internal impedance, i.e., when

$$Z_L = Z_s^* = R_s - jX_s$$

where R_s and $+jX_s$ are the real and imaginary parts, respectively, of the source internal impedance. In this case

$$P_{as} = \frac{(A^2/2)}{4R_s} = \frac{<e_s^2(t)>}{4R_s} \tag{10-2}$$

† Incremental power, rather than spectral density, is used in this chapter so as to conform with the available literature on noise figure.

When $e_s(t)$ is a sample function of a real stationary random process, the incremental available power from the source is accordingly defined to be

$$P_{as} = \frac{\overline{(e_s{}^2)_{df}}}{4R_s} \qquad (10\text{-}3)$$

where
$$\overline{(e_s{}^2)_{df}} = 2S_{e_s}(f)\, df \qquad (10\text{-}4)$$

is the mean-square value of the source voltage in the frequency band of width df centered on f.

A particular case of interest is that in which the source is a resistor of value R_s at a temperature T_s. In this case, e_s is the thermal noise voltage generated by R_s; hence, from Eq. (9-47),

$$S_{e_s}(f) = 2kT_sR_s$$

The incremental available thermal noise power from a resistor is therefore

$$P_{as} = kT_s\, df \qquad (10\text{-}5)$$

which depends on the temperature of the resistor but not on the value of its resistance.

Noise Temperature. In view of the simplicity of Eq. (10-5), it is often desirable to ascribe all the noise generated by a source to the real part of its internal impedance and to characterize the source by an *effective noise temperature* T_{es} defined by

$$T_{es} = \frac{N_{as}}{k\, df} \qquad (10\text{-}6)$$

where N_{as} is the incremental available noise power from the source. Since N_{as} may vary with frequency, T_{es} may also. The *relative noise temperature* t_s of a source is defined to be the ratio of the effective noise temperature of the source to a *standard noise temperature* T_o commonly chosen to be 290° Kelvin:†

$$t_s = \frac{T_{es}}{T_o} = \frac{N_{as}}{k(290)\, df} \qquad (10\text{-}7)$$

The relative noise temperature of a source is hence equal to the actual incremental available noise power from the source divided by the incremental available noise power from a resistor whose temperature is standard.

Available Power Gain. The *available power gain* G_a of a stable amplifier is defined to be the ratio of the incremental available power P_{ao} of the amplifier output, in response to a specific source, to the incremental

† I.e., 62.6° Fahrenheit, or a little below normal room temperature. Note that $kT_o/e = 0.25$ v where e is the charge of an electron. Cf. IRE Standards (I).

available power P_{as} of that source:

$$G_a = \frac{P_{ao}}{P_{as}} \tag{10-8}$$

Consider the amplifier shown in Fig. 10-1. From this figure, it follows that the incremental available powers of the source and the amplifier output are

$$P_{as} = \frac{\overline{(e_s{}^2)_{df}}}{4R_s} \quad \text{and} \quad P_{ao} = \frac{\overline{(e_o{}^2)_{df}}}{4R_o}$$

respectively. The available power gain of the amplifier is therefore:

$$G_a = \left(\frac{R_s}{R_o}\right) \frac{\overline{(e_o{}^2)_{df}}}{\overline{(e_s{}^2)_{df}}} = \frac{R_s S_o(f)}{R_o S_s(f)} \tag{10-9}$$

where $S_o(f)$ is the spectral density of e_o and $S_s(f)$ is the spectral density of e_s. The output spectral density can be evaluated by applying the techniques developed in Chap. 9; it is

$$S_o(f) = \left| \frac{H(j2\pi f)Z_i}{Z_i + Z_s} \right|^2 S_s(f)$$

where $H(j2\pi f)$ is the system function of the amplifier relating e_o to e_i. Hence

$$G_a = \frac{R_s}{R_o} \left| \frac{H(j2\pi f)Z_i}{Z_i + Z_s} \right|^2 \tag{10-10}$$

It should be noted that the available power gain of the amplifier depends upon the relation between the amplifier input impedance and the source output impedance.

10-2. Noise Figure

We are now in a position to define the noise figure of an amplifier: The *operating noise figure*† (or *operating noise factor*) F_o of an amplifier is the ratio of the incremental available noise power output N_{ao} from the amplifier to that part N_{ao_s} of the incremental available noise power output which is due solely to the noise generated by the source which drives the amplifier:

$$F_o = \frac{N_{ao}}{N_{ao_s}} \tag{10-11}$$

The noise figure of an amplifier is thus a measure of the noisiness of the amplifier *relative to the noisiness of the source*.

An alternative definition of noise figure may be obtained as follows: The amplifier available noise power output due solely to the source is

† Also called *operating spot noise figure*.

equal to the available noise power of the source N_{as} times the available power gain of the amplifier G_a. If we substitute this result into Eq. (10-11) and then multiply and divide by the available signal power from the source S_{as}, we get

$$F_o = \frac{N_{ao}}{G_a S_{as}} \frac{S_{as}}{N_{as}}$$

Now $G_a S_{as}$ is the available signal power output S_{ao} of the amplifier. Hence we can write

$$F_o = \frac{(S/N)_{as}}{(S/N)_{ao}} \tag{10-12}$$

The noise figure of an amplifier therefore equals the incremental available signal-to-noise power ratio of the source $(S/N)_{as}$ divided by the incremental available signal-to-noise power ratio of the amplifier output $(S/N)_{ao}$. This relation is essentially that originally introduced by Friis[†] as a definition of noise figure.

Let us now determine some of the properties of noise figure. First of all, we note that the noise figure of an amplifier is a function of frequency, since the available powers used in its definition refer to incremental frequency bandwidths.

Next, we note that the amplifier output noise is due to two independent causes: the noise generated by the source and the noise generated within the amplifier itself. The incremental available noise power output is hence the sum of the incremental available noise power output due to the source N_{ao_s} and the incremental available noise power output due to internal noises N_{ao_i}. On using this fact in Eq. (10-11), we get

$$F_o = 1 + \frac{N_{ao_i}}{N_{ao_s}} \tag{10-13}$$

Since both components of the output noise power must be nonnegative, the noise figure of an amplifier must always be greater than or equal to unity:

$$F_o \geq 1 \tag{10-14}$$

The incremental available noise power output due to the source may be expressed in terms of the available power gain of the amplifier and the effective noise temperature of the source. Thus Eq. (10-13) can be rewritten as

$$F_o = 1 + \frac{N_{ao_i}}{G_a k T_{es} \, df} \tag{10-15}$$

The available noise power output due to internal noise sources is generally independent of the effective noise temperature of the source. It

† Friis (I).

therefore follows from Eq. (10-15) that the operating noise figure of an amplifier depends upon the effective noise temperature of the source; the lower is T_{es}, the higher is F_o.

The *standard noise figure* F of an amplifier is defined to be the value of its operating noise figure when the effective noise temperature of the source is standard:

$$F = 1 + \frac{N_{ao_i}}{G_a k T_o \, df} \qquad (10\text{-}16)$$

It then follows that the operating noise figure can be expressed in terms of the standard noise figure by the relation

$$F_o = 1 + \frac{T_o}{T_{es}} (F - 1) \qquad (10\text{-}17)$$

Since in many applications the effective noise temperature of a source differs considerably from the standard value of 290° K (e.g., when the source is an antenna†), care must be taken when using standard noise figure.

From the viewpoint of the amplifier output terminals, the combination of source and amplifier as shown in Fig. 10-1 is simply a two-terminal device. The relative noisiness of the combination could therefore be specified as well by a relative noise temperature as by a noise figure. From Eq. (10-7), it follows that the relative noise temperature t_o of the combination is

$$t_o = \frac{N_{ao}}{k T_o \, df}$$

From the definition of operating noise figure, Eq. (10-11),

$$N_{ao} = F_o N_{ao_s} = F_o G_a k T_{es} \, df \qquad (10\text{-}18)$$

hence the relative noise temperature of the source and amplifier combination is

$$t_o = \frac{T_{es}}{T_o} F_o G_a = t_s F_o G_a \qquad (10\text{-}19)$$

where t_s is the relative noise temperature of the source alone. When the source noise temperature has the standard value we get

$$t_o = F G_a \qquad (10\text{-}20)$$

since $F_o = F$ when $T_{es} = T_o$.

Average Noise Figure. The *average operating noise figure* $\overline{F}_o$ of an amplifier is defined to be the ratio of the *total* available noise power output from the amplifier to that part of the *total* available noise power output which is due solely to the noise generated by the source. These total

† E.g., Lawson and Uhlenbeck (I, Art. 5.2).

available powers can be obtained from the corresponding incremental available powers by integration. Thus we obtain

$$\overline{F}_o = \frac{\int_0^\infty F_o G_a T_{es}\, df}{\int_0^\infty G_a T_{es}\, df} \tag{10-21a}$$

which becomes

$$\overline{F}_o = \frac{\int_0^\infty F_o G_a\, df}{\int_0^\infty G_a\, df} \tag{10-21b}$$

when the effective noise temperature of the source does not change with frequency.

The *average standard noise figure* $\bar{F}$ is the value of $\overline{F}_o$ obtained when the effective noise temperature of the source has the value of 290° K at all frequencies. It then follows from Eq. (10-21b) that

$$\bar{F} = \frac{\int_0^\infty F G_a\, df}{\int_0^\infty G_a\, df} \tag{10-22}$$

10-3. Cascaded Stages

Let us next consider a cascade of amplifier stages. The preceding definitions apply equally well to the over-all cascade as to the individual

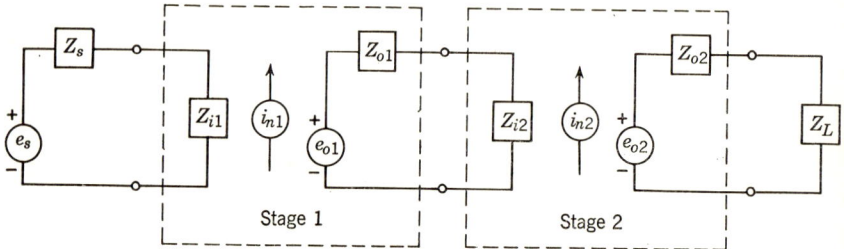

Fig. 10-2. The two-stage cascade.

stages. The problem arises of determining the relation between the available power gain and noise figure of the over-all system and the respective quantities for the individual stages. We shall assume that all stages are stable.

Available Power Gain. For convenience, consider first a two-stage amplifier cascade such as that shown in Fig. 10-2. The available power gain $G_{a_{1,2}}$ of the cascade is

$$G_{a_{1,2}} = \frac{P_{ao_2}}{P_{as}}$$

where P_{as} is the incremental available power from the source which drives the cascade and P_{ao_2} is the incremental available power from the second stage when driven by the combination of the source and the first stage. Let G_{a_2} be the available power gain of the second stage when driven by a source whose internal impedance equals the output impedance of the source-and-first-stage combination. We may then write

$$P_{ao_2} = G_{a_2} P_{ao_1}$$

where P_{ao_1} is the incremental available power output from the first stage in response to the given source. If G_{a_1} is the available power gain of the first stage when driven by the given source, we can also write

$$P_{ao_1} = G_{a_1} P_{as}$$

On combining these equations, we obtain

$$G_{a_{1,2}} = G_{a_1} G_{a_2} \tag{10-23}$$

Thus we see that the over-all available power gain of a two-stage amplifier cascade equals the product of the individual-stage available power gains *when the latter gains are defined for the same input–output impedance relations as are obtained in the cascade.*

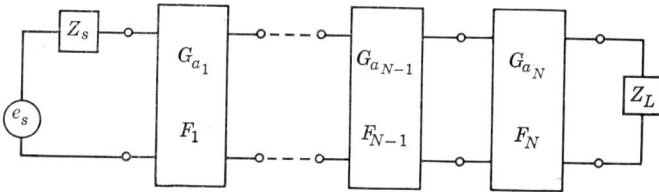

FIG. 10-3. The N-stage cascade.

Consider next a cascade of N stages, as shown in Fig. 10-3. Such a cascade may be split into two parts, one consisting of the first $N - 1$ stages and the other consisting of the last stage. It then follows from Eq. (10-23) that

$$G_{a_{1,N}} = G_{a_{1,(N-1)}} G_{a_N}$$

where $G_{a_{1,N}}$ is the available power gain of the cascade, $G_{a_{1,(N-1)}}$ is the available power gain of the first $N - 1$ stages, and G_{a_N} is the available power gain of the last stage. Successive applications of this result give

$$G_{a_{1,N}} = \prod_{n=1}^{N} G_{a_n} \tag{10-24}$$

Thus the available power gain of an N-stage amplifier cascade equals the product of the available power gains of the individual stages, so long as the latter gains are defined for the same impedance relations as are obtained in the cascade.

Noise Figure. The over-all operating noise figure $F_{o_{1,2}}$ of the two-stage amplifier shown in Fig. 10-2 is, from Eq. (10-15),

$$F_{o_{1,2}} = 1 + \frac{N_{ao_{i1,2}}}{G_{a_{1,2}} k T_{es} \, df}$$

where $N_{ao_{i1,2}}$ is the incremental available noise power output from the cascade due to noise sources internal to both stages, $G_{a_{1,2}}$ is the available power gain of the cascade, and T_{es} is the effective noise temperature of the source which drives the cascade. $N_{ao_{i1,2}}$ equals the sum of the incremental available noise power output due to second-stage internal noises $N_{ao_{i2}}$ and the incremental available noise power output from the first stage due to its internal noise $N_{ao_{i1}}$ multiplied by the available power gain of the second stage. Since

$$F_{o_1} = 1 + \frac{N_{ao_{i1}}}{G_{a_1} k T_{es} \, df} \qquad \text{and} \qquad F_{o_2} = 1 + \frac{N_{ao_{i2}}}{G_{a_2} k T_{e_1} \, df}$$

where T_{e1} is the effective noise temperature of the source and first-stage combination, it follows that we can express the two-stage noise figure in terms of the single-stage noise figures by the equation

$$F_{o_{1,2}} = F_{o_1} + \frac{T_{e1}}{T_{es}} \left(\frac{F_{o_2} - 1}{G_{a_1}} \right) \tag{10-25a}$$

This becomes

$$F_{o_{1,2}} = F_{o_1} + \left(\frac{F_{o_2} - 1}{G_{a_1}} \right) \tag{10-25b}$$

when $T_{e_1} = T_{es}$. In particular when both effective noise temperatures are standard, we get

$$F_{1,2} = F_1 + \frac{F_2 - 1}{G_{a_1}} \tag{10-26}$$

This is the cascading relation for standard noise figures.

Consider now a cascade of N amplifier stages, as shown in Fig. 10-3. Again we can split up the N-stage cascade into two parts. If we let $F_{o_{1,N}}$ be the over-all operating noise figure of the cascade, $F_{o_{1,(N-1)}}$ the over-all operating noise figure of the first $N - 1$ stages of the cascade $T_{e(N-1)}$ the effective noise temperature of the combination of the source and the first $N - 1$ stages, and F_{o_N} the noise figure of the last stage when driven by the first $N - 1$ stages, we obtain, from Eqs. (10-24) and (10-25),

$$F_{o_{1,N}} = F_{o_{1,(N-1)}} + \frac{T_{e(N-1)}}{T_{es}} (F_{o_N} - 1) \Big/ \prod_{m=1}^{N-1} G_{a_m}$$

Successive applications of this result enable us to express $F_{o_{1,N}}$ in terms of the noise figures of the individual stages as

$$F_{o_{1,N}} = F_{o_1} + \sum_{n=2}^{N} \frac{T_{e(n-1)}}{T_{es}} (F_{o_n} - 1) \bigg/ \prod_{m=1}^{n-1} G_{a_m} \qquad (10\text{-}27a)$$

which becomes

$$F_{o_{1,N}} = F_{o_1} + \sum_{n=2}^{N} (F_{o_n} - 1) \bigg/ \prod_{m=1}^{n-1} G_{a_m} \qquad (10\text{-}27b)$$

when all $T_{en} = T_{es}$. It therefore follows that when all the effective noise temperatures are standard, we obtain

$$F_{1,N} = F_1 + \sum_{n=2}^{N} (F_n - 1) \bigg/ \prod_{m=1}^{n-1} G_{a_m} \qquad (10\text{-}28)$$

This is the cascading relation for the standard noise figures. The *operating* noise figure of the cascade may then be obtained either by substitution of the individual-stage *operating* noise figures into Eqs. (10-27) or by substitution of the individual-stage *standard* noise figures into Eq. (10-28), substituting the result into Eq. (10-17). The latter procedure may well be more practical.

Equation (10-28) shows that when the available power gains of all stages after the kth are much larger than unity, the summation converges rapidly after the $(n = k + 1)$th term. Only the noise figures of the first $k + 1$ stages are then particularly significant. For example, if every stage has a large available power gain, the noise figure of the cascade is essentially that of the first stage.

10-4. Example†

As an illustration of the application of the results above, let us determine the noise figure of a single-stage triode amplifier such as that shown in Fig. 10-4. We shall assume that any reactive coupling between grid and plate has been neutralized over the amplifier pass band and that the frequency of operation is high enough that we must consider the effect of both transit-time loading of the grid circuit and induced grid noise.‡ Under these conditions the equivalent circuit of the system is that shown in Fig. 10-5.

All the passive elements in Fig. 10-5 are assumed to be noiseless (except R_{eq}). The source noise current i_s is assumed to result from the source conductance g_s with an effective temperature T_s. The amplifier input admittance is split into its real and imaginary parts g_i and B_i, respectively; the noise generated by the real part is represented by i_i. The induced grid noise is represented by the noise current source i_r and

† Cf. Valley and Wallman (I, Arts. 13.5 through 13.7).
‡ See, for example, van der Ziel (I, Art. 6.3) or Lawson and Uhlenbeck (I, Art. 4.10).

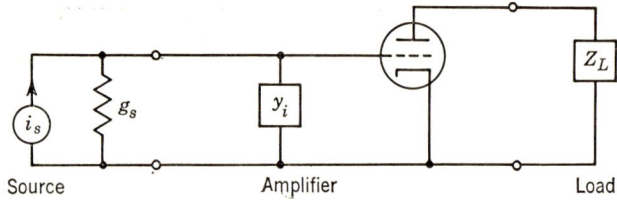

FIG. 10-4. The triode amplifier.

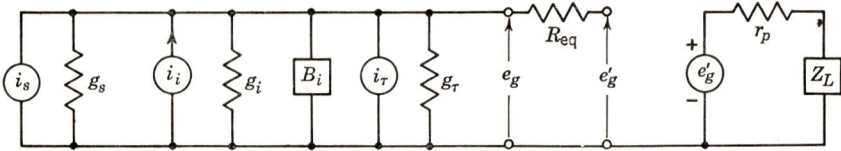

FIG. 10-5. Noise equivalent circuit of the triode amplifier.

is assumed to result from the grid loading conductance g_τ with an effective temperature T_τ. The plate-current noise of the triode is accounted for by the equivalent noise resistor† R_{eq}. For convenience, we have lumped the tube input susceptance with B_i and the tube output reactance with Z_L.

The incremental available noise power from the source is

$$P_{as} = \frac{\overline{(i_s{}^2)}\,df}{4g_s} = kT_s\,df$$

The incremental available noise power output of the amplifier due to the input source noise is

$$P_{ao_s} = \frac{\mu^2}{4r_p}\left[\frac{\overline{(i_s{}^2)}\,df}{(g_s + g_i + g_\tau)^2 + B_i{}^2}\right]$$

The available power gain of the amplifier is therefore

$$G_a = \frac{\mu^2 g_s}{r_p[(g_s + g_i + g_\tau)^2 + B_i{}^2]} \qquad (10\text{-}30)$$

Here the most convenient of our various expressions for noise figure is that given by Eq. (10-15). In order to apply that expression, we must determine the incremental

† It is often convenient to represent the plate-current noise of a tube as being due to a noisy resistor R_{eq}, with a standard effective temperature T_o, in the grid circuit of the tube (as in Fig. 10-5). In this case the spectral density of the plate-current noise may be written as

$$S_i(f) = 2kT_oR_{eq}g_m{}^2$$

It then follows from Eq. (7-91) that, for a triode,

$$R_{eq} = \left(\frac{0.644}{\sigma}\right)\left(\frac{T_c}{T_o}\right)\frac{1}{g_m} \qquad (10\text{-}29)$$

It should be emphasized that no current flows through R_{eq} at any time; only its noise voltage is significant.

available noise power output due to internal noises N_{ao_i}. Assuming that the internal noises are uncorrelated, it follows that

$$N_{ao_i} = \frac{\mu^2}{4r_p} \left[4kT_o R_{eq}\, df + \frac{\overline{(i_i)_{df}^2} + \overline{(i_\tau)_{df}^2}}{(g_s + g_i + g_\tau)^2 + B_i^2} \right]$$

and hence (assuming that the temperature of g_i is T_i) that

$$N_{ao_i} = \frac{\mu^2 kT_o\, df}{r_p} \left[R_{eq} + \frac{t_i g_i + t_\tau g_\tau}{(g_s + g_i + g_\tau)^2 + B_i^2} \right]$$

where t_i and t_τ are the relative noise temperatures of the grid input circuit and loading conductances:

$$t_i = \frac{T_i}{T_o} \quad \text{and} \quad t_\tau = \frac{T_\tau}{T_o} \tag{10-31}$$

Upon substituting these results into Eq. (10-15) we obtain

$$F_o = 1 + \frac{T_o}{T_s} \left\{ \frac{t_i g_i + t_\tau g_\tau}{g_s} + \frac{R_{eq}}{g_s} [(g_s + g_i + g_\tau)^2 + B_i^2] \right\} \tag{10-32}$$

for the operating noise figure of the single-stage triode amplifier.

Optimization of F_o. It is interesting to determine what can be done to minimize the noise figure of the given amplifier. First, assuming fixed circuit parameters, we see from Eq. (10-32) that F_o is a minimum when the frequency of operation is such that

$$B_i = 0 \tag{10-33}$$

Although this condition is satisfied only at the resonant frequency of the input circuit, it may be approximated over the bandwidth of the input signal when that bandwidth is much narrower than the bandwidth of the amplifier input circuit.

Of the various other amplifier parameters, certain ones (t_τ, g_τ, and R_{eq}) are determined by the type of tube used. The best one can do with these is to select a type for which they are as small as possible. The minimum value of the input-circuit conductance g_i is determined by the maximum Q's which are physically obtainable for the circuit elements and by the required bandwidth of the input circuit. The relative noise temperature of the grid input circuit t_i might well be reduced by cooling the input circuit.

Since the effective noise temperature of the source is generally not under control, the only remaining possibility is to adjust the source conductance g_s so as to minimize F_o. That there is an optimum value of g_s follows from Eq. (10-32); $F_o \to \infty$ when either $g_s \to 0$ or $g_s \to \infty$. The optimum value can be found by equating to zero the partial derivative of F_o with respect to g_s. In this way we obtain, assuming $B_i = 0$,

$$g_s \bigg]_{\text{opt}} = \left[(g_i + g_\tau)^2 + \frac{t_i g_i + t_\tau g_\tau}{R_{eq}} \right]^{\frac12} \tag{10-34}$$

Substituting this result into Eq. (10-32) and setting $B_i = 0$, we get

$$F_{o_{\min}} = 1 + 2R_{eq} \left(\frac{T_o}{T_s} \right) \left\{ g_i + g_\tau + \left[(g_i + g_\tau)^2 + \frac{t_i g_i + t_\tau g_\tau}{R_{eq}} \right]^{\frac12} \right\} \tag{10-35}$$

In certain practical amplifiers, the condition†

$$\frac{t_i g_i + t_\tau g_\tau}{R_{eq}} >> (g_i + g_\tau)^2 \tag{10-36}$$

is satisfied. In these cases, the optimum value of the source conductance becomes, approximately,

$$g_s \bigg]_{opt} = \left[\frac{t_i g_i + t_\tau g_\tau}{R_{eq}} \right]^{\frac{1}{2}} \tag{10-37}$$

and the minimum value of the operating noise figure becomes

$$F_{o_{min}} = 1 + 2 \left(\frac{T_o}{T_s} \right) [R_{eq}(t_i g_i + t_\tau g_\tau)]^{\frac{1}{2}} \tag{10-38}$$

Thus we see that matching the source impedance to the amplifier input impedance in the usual sense (i.e., for maximum average power transfer) does not necessarily lead to the lowest possible value of the amplifier noise figure.

The purpose of the above example is mainly to illustrate the application of our previous noise-figure results. The conclusions reached are only as valid as the noise equivalent circuit shown in Fig. 10-5.

10-5. Problems

1. Consider the system shown in Fig. 10-1. Let the internal noise current i_n be due to a conductance g_n at an effective temperature T_n, and let Z_{on} be the transfer impedance relating the amplifier Thévenin equivalent output voltage e_o to the current i_n.

Derive an expression for the operating noise figure of the amplifier in terms of the various system parameters.

2. Consider the system shown in Fig. 10-6. Suppose that $a << 1$ and $b >> 1$ and that all resistors are at the standard temperature.

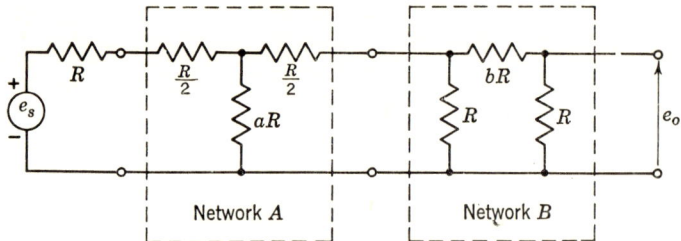

FIG. 10-6. An attenuator cascade.

a. Determine the available power gains of the networks A and B in terms of R, a, and b.

b. Determine the standard noise figures of networks A and B.

3. Referring to Prob. 2 above and Fig. 10-6,

a. Determine the available power gain of the cascade of networks A and B.

b. Determine the standard noise figure of the cascade.

c. Determine the relative noise temperature of the cascade.

† Cf. Valley and Wallman (I, Art. 13.6).

4. Suppose that the effective noise temperature T_{es} of the source driving a given amplifier does not vary with frequency. Show that the average operating noise figure $\bar{F}_o$ of the amplifier is related to the average standard noise figure $\bar{F}$ of the amplifier by the equation

$$\bar{F}_o = 1 + \frac{T_o}{T_{es}}(\bar{F} - 1) \tag{10-39}$$

5. Consider the system shown in Fig. 10-7. Suppose that R_s, R_i, and R_L are at the standard temperature.

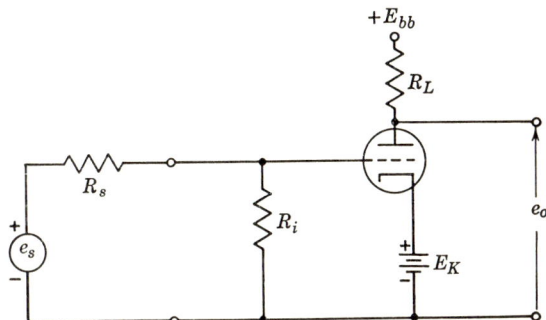

Fig. 10-7. An amplifier circuit.

a. Determine the available power gain of the amplifier.
b. Determine the standard noise figure of the amplifier.
c. Assuming $R_i \gg R_s$ and a high g_m tube, obtain approximations to the results in a and b.

6. Suppose the circuit elements in Fig. 10-7 have the values

$$R_s = 600 \text{ ohms}, \ R_i = 0.5 \text{ megohm, and } R_L = 100 \text{ K ohms}$$

and that the tube parameters are

$$g_m = 1600 \text{ mhos}, \ r_p = 44{,}000 \text{ ohms, and } R_{eq} = 1560 \text{ ohms}$$

For these values

a. Evaluate the available power gain of the amplifier.
b. Evaluate the standard noise figure.

7. For the system of Prob. 5,

a. Determine the optimum value of the source resistance R_s.
b. Using the values stated in Prob. 6, evaluate a.
c. Using b, determine the available power gain and noise figure of the amplifier when the source resistance has its optimum value.

8. Consider the system shown in Fig. 10-8. The *power meter* measures the power output from the amplifier in a frequency band of width df centered on f. The source resistor is at the standard temperature.

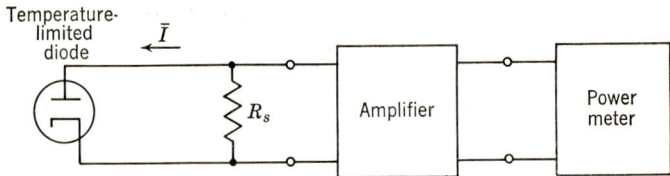

FIG. 10-8. Noise-figure measuring system.

a. Show that the ratio of the power-meter reading P_d with the diode connected across R_s to the reading P_o when the diode is disconnected is

$$\frac{P_d}{P_o} = 1 + \frac{e\bar{I}R_s}{2kT_oF} \tag{10-40}$$

where $\bar{I}$ is the average current flowing through the temperature-limited diode.

b. Show that when $P_d/P_o = 2$, the standard noise figure of the amplifier is given by

$$F = 20\bar{I}R_s \tag{10-41}$$

9.† Consider the system shown in Fig. 10-9. The inputs of N identical high-gain amplifiers of the type shown in Figs. 10-4 and 10-5 are connected to the source through a linear passive coupling network, and their outputs are added to drive a load.

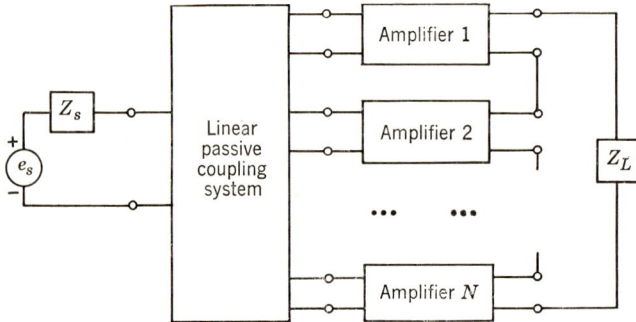

FIG. 10-9. An N-amplifier combination.

Show that the standard noise figure of the system is greater than, or equal to, the optimum standard noise figure of one of the amplifiers.

10. It has been suggested‡ that the noise performance of an amplifier is better characterized by the *noise measure* M of the amplifier, where

$$M = \frac{F - 1}{1 - 1/G_a} \tag{10-42}$$

than by the noise figure alone.

Consider a cascade of two identical amplifiers. Show that the noise measure of the cascade is equal to the noise measure of a single stage, whereas the noise figure of the cascade exceeds that of a single stage.

† Cf. Bose and Pezaris (I).
‡ Haus and Adler (I).

CHAPTER 11

OPTIMUM LINEAR SYSTEMS

11-1. Introduction

One application of the theory developed in Chaps. 6 and 9 is to the design of linear systems to perform operations in an optimum way when some or all of the inputs are sample functions from random processes. Thus, for example, a so-called *smoothing* filter is designed to extract as well as possible a wanted signal from a mixture of signal and noise; a *predicting* filter is designed to yield a future value of a signal, where the signal may again be mixed with noise. In this chapter we shall consider how to find certain optimum linear systems. For convenience, we shall suppose throughout that signal and noise terms are real-valued.

Before treating particular problems, let us consider briefly four conditions which largely determine any problem in system optimization. These are the purpose of the system, the nature of the inputs, the criterion of goodness of performance to be used, and the freedom of choice to be allowed in the design. Whenever these four conditions are specified, some kind of optimization problem is defined, although the specifications may, of course, be such that the problem has no solution at all, or no best solution, or no unique best solution. In practical problems another consideration is usually added, the cost of the system (perhaps in some generalized sense of the word cost). For our present purpose, however, we assume that cost is not to be a factor.

Rather than discuss these four conditions completely abstractly, let us observe, as an illustration, how they apply in a particular situation, e.g., the design of an optimum smoothing filter. We suppose that there is available a corrupted signal $y(t)$ which is the sum of a wanted signal $s(t)$ and unwanted noise $n(t)$,

$$y(t) = s(t) + n(t)$$

The first condition on the problem is the purpose of the system. Here we are assuming that it is to recover the wanted signal $s(t)$ from the corrupted signal $y(t)$.

Next we have to specify the nature of both inputs, $s(t)$ and $n(t)$. There are various possibilities, including some uninteresting ones. For example, if $s(t)$ is known exactly, there is no problem, at least in principle; if $n(t)$ is known exactly, it is trivial to obtain $s(t) = y(t) - n(t)$, and there is again no problem. At the other extreme, if there is no a priori information at all about $s(t)$ and $n(t)$, there is no hope for extracting $s(t)$ even approximately. Obviously the interesting cases are those in which there is some uncertainty about the inputs but not too much. It is often reasonable to assume that the noise input $n(t)$ is caused by some random physical phenomenon whose statistics are known, as, for example, when $n(t)$ is thermal noise or shot noise. If this is true, we may suppose $n(t)$ to be a sample function of a random process. Depending on its origin, the signal $s(t)$ might be represented, to mention only some of the possibilities, by a polynomial of degree m with unknown coefficients, by a finite trigonometric series with unknown coefficients, by a sample function of a random process, or by a combination of these. One common specification of the nature of the signal and noise inputs is that both signal and noise are sample functions of stationary random processes and that both the autocorrelation functions and the cross-correlation function are known. These are the assumptions made by Wiener in his analysis of the linear smoothing filter, which we shall discuss below.

There are as many possibilities as one desires for a criterion of goodness of performance, that is, for a measure of how well the system performs its intended task. The desired output of the system is $s(t)$; if the actual output is $z(t)$, then any functional of $s(t)$ and $z(t)$ is some kind of measure of how well the system operates. Usually, however, one thinks of a measure of system performance as being some quantity which depends on the error $z(t) - s(t)$, which is a minimum (maximum) when the error is zero and becomes larger (smaller) when the error is increased. Since we are discussing systems which have random inputs and hence random outputs, it is appropriate to use probabilities and statistical averages. Some reasonable choices for a measure of how good the performance is, or how small the error is, are

1. $p(z_t = s_t | y(\tau), \tau \leq t)$
2. $P(|z_t - s_t| > \epsilon)$
3. $E(|z_t - s_t|)$
4. $E(|z_t - s_t|^2)$

The reader can readily add to this list. If we use (1), we ask for that system whose output has the largest conditional probability, using all the past history of the signal, of being the right value. If the conditional probability densities are continuous and our only interest is that the output have a small error as much of the time as possible, all errors larger than a certain value being roughly equally bad, such a criterion is attractive. It has the disadvantage, however, that it requires a complete statistical knowledge of the inputs, and this is often not available. Choice (2) is a measure of performance for which all errors greater than some threshold are considered exactly equally bad, while small errors are tolerated. In this case, of course, one asks for a system which minimizes the specified probability. In the criteria given by (3) and (4), errors are weighted according to their magnitude; in (4), large errors are weighted quite heavily. Choice (4) provides no better criterion for many applications than various other choices, and often it gives a worse one. However, it has the advantage that it leads generally to a workable analysis. The calculation of $E(|z_t - s_t|^2)$ can be made straightforwardly, given the input correlation functions and the system function of the linear system. This criterion, of least mean-square error, is used in the Wiener theory of linear smoothing and in most of the extensions of that theory.

Finally, we must consider the freedom of choice to be allowed in the design. We have assumed from the beginning of this chapter that the systems were to be linear. This restriction is imposed not because linear systems are necessarily best but simply because it is too hard mathematically in most instances to allow a wider class of possibilities. In addition, we shall usually require that the systems be fixed-parameter linear systems and that they be realizable. The restriction to fixed-parameter systems is sometimes not necessary, but when the inputs are stationary random processes, nothing is lost by making this restriction. The restriction of realizability is introduced to guarantee that the results have practical value rather than to simplify the analysis. In fact, the subtleties of the Wiener theory of smoothing filters exist because of this restriction. At times it is assumed that the system is to have been excited over all past time; at other times it is assumed that it has been excited only for a finite time. We shall discuss the smoothing filter under both these conditions.

We have talked only about the smoothing filter, but it is clear that a similar set of considerations applies to any problem in system optimization. Some specification under each of the four conditions mentioned must be made before a problem in the mathematical sense is even defined. Three of these conditions—purpose, nature of the inputs, and freedom of choice allowed—are governed pretty clearly by what the system is to be used for and where, and by theoretical and practical limitations on design techniques. The other condition, criterion of goodness of performance, is also influenced by these things, but it is more intangible. We shall suppose that intuition, backed by experience and the knowledge mentioned, is sufficient to lead to a suitable criterion. If one looks more closely into this matter, he is led to the subject of decision theory,† the

† See, for example, Middleton and Van Meter (I) which contains an extensive bibliography.

basic problem of which is to establish rules for making a decision as to when one thing is better than another. In this chapter we shall use consistently a least-mean-square error criterion, except in Art. 11-8, where we shall use something closely related to it. In Chap. 14, however, we shall again consider what may be regarded as system optimization problems, and various performance criteria will be used.

In the remainder of this chapter we consider a more or less arbitrary collection of problems concerned with optimum linear procedures for smoothing, predicting, and maximizing signal-to-noise ratio. The techniques used are representative of those for a wider class of problems than we actually discuss.

11-2. Smoothing and Predicting of Stationary Inputs Using the Infinite Past (Wiener Theory)

For this problem we assume an input signal $y(t)$ which consists of a wanted signal $s(t)$ added to a noise $n(t)$. Both $s(t)$ and $n(t)$ are taken to be sample functions from real-valued wide-sense stationary random processes with a stationary cross-correlation. We want to find a weighting function $h(t)$ for a fixed-parameter realizable linear filter which acts on the entire past of $y(t)$ in such a way that the output of the filter at time t is a best mean-square approximation to $s(t + \eta)$, $\eta \geq 0$. This is a combined smoothing and prediction problem, and problems of smoothing, and of predicting when there is no noise, are special cases. The solution of this problem is due to Kolmogoroff and to Wiener,[†] and we shall first treat it essentially as Wiener has. It makes no significant difference in the mathematics (although it may in the conceptual foundations) whether $s(t)$ and $n(t)$ are taken to be sample functions of random processes and statistical averages are used (this is the procedure we will follow here) or whether $s(t)$ and $n(t)$ are taken to be individual functions, unknown but with known time correlations, and time averages are used. The reader is referred particularly to a concise development by Levinson[‡] of the theory using the latter assumption.

The input to the filter is

$$y(t) = s(t) + n(t) \tag{11-1}$$

The output for any filter weighting function $h(t)$ is

$$\int_{-\infty}^{\infty} h(t - \tau)y(\tau)\, d\tau = \int_{-\infty}^{\infty} h(\tau)y(t - \tau)\, d\tau \tag{11-2}$$

where $h(t) = 0$, $t < 0$. The average squared error $\mathcal{E}$ is

$$\mathcal{E} = E\left\{ \left[s(t + \eta) - \int_{-\infty}^{\infty} h(\tau)y(t - \tau)\, d\tau \right]^2 \right\} \tag{11-3}$$

[†] Wiener (III).   See p. 59 for a reference to Kolmogoroff.
[‡] Levinson (I).

and it is desired to minimize this if possible by a suitable choice of $h(t)$. There may not actually exist an $h(t)$ which gives a minimum from among the class of all $h(t)$ of realizable filters. This point is discussed later. Since $s(t)$ and $n(t)$ have stationary autocorrelation functions and a stationary cross-correlation function, the mean-squared error expression given by Eq. (11-3) can be expanded to give

$$\mathcal{E} = E\left[s^2(t+\eta)\right] - 2\int_{-\infty}^{\infty} h(\tau)E[s(t+\eta)y(t-\tau)]\,d\tau$$

$$+ \int_{-\infty}^{\infty}\int_{-\infty}^{\infty} h(\tau)h(\mu)E[y(t-\tau)y(t-\mu)]\,d\tau\,d\mu \qquad (11\text{-}4)$$

$$= R_s(0) - 2\int_{-\infty}^{\infty} h(\tau)R_{sy}(\eta+\tau)\,d\tau + \int_{-\infty}^{\infty}\int_{-\infty}^{\infty} h(\tau)h(\mu)R_y(\tau-\mu)\,d\tau\,d\mu$$

We now find a necessary condition which $h(t)$ must satisfy to make $\mathcal{E}$ a minimum. Suppose $g(t)$ is a weighting function for any realizable filter whatsoever. Then $h(t) + \epsilon g(t)$ is the weighting function of a realizable filter, and if $h(t)$ is to provide a minimum mean-squared error, then the expression on the right side of Eq. (11-4) must have at least as great a value when $h(t)$ is replaced by $h(t) + \epsilon g(t)$ as it does with $h(t)$. This must be true for any real ϵ and any $g(t)$ of the class considered. With $h(t)$ replaced by $h(t) + \epsilon g(t)$, the right side of Eq. (11-4) becomes

$$R_s(0) - 2\int_{-\infty}^{\infty} h(\tau)R_{sy}(\eta+\tau)\,d\tau - 2\epsilon\int_{-\infty}^{\infty} g(\tau)R_{sy}(\eta+\tau)\,d\tau$$

$$+ \int_{-\infty}^{\infty}\int_{-\infty}^{\infty} h(\tau)h(\mu)R_y(\tau-\mu)\,d\tau\,d\mu + 2\epsilon\int_{-\infty}^{\infty}\int_{-\infty}^{\infty} h(\tau)g(\mu)R_y(\tau-\mu)$$

$$d\tau\,d\mu + \epsilon^2\int_{-\infty}^{\infty}\int_{-\infty}^{\infty} g(\tau)g(\mu)R_y(\tau-\mu)\,d\tau\,d\mu \qquad (11\text{-}5)$$

For $h(t)$ to give a minimum, this expression minus the right side of Eq. (11-4) must be greater than or equal to zero, i.e.,

$$2\epsilon\left\{\int_{-\infty}^{\infty}\int_{-\infty}^{\infty} h(\tau)g(\mu)R_y(\tau-\mu)\,d\tau\,d\mu - \int_{-\infty}^{\infty} g(\tau)R_{sy}(\eta+\tau)\,d\tau\right\}$$

$$+ \epsilon^2\int_{-\infty}^{\infty}\int_{-\infty}^{\infty} g(\tau)g(\mu)R_y(\mu-\tau)\,d\tau\,d\mu \geq 0 \qquad (11\text{-}6)$$

The last term on the left side of Eq. (11-6) is always nonnegative because $R_y(t)$ is non-negative definite.† If the expression in braces is different from zero, then a suitable choice of ϵ, positive or negative, will make the whole left side of Eq. (11-6) negative. Hence, a necessary condition that the inequality be satisfied is that

$$\int_{-\infty}^{\infty}\int_{-\infty}^{\infty} h(\tau)g(\mu)R_y(\tau-\mu)\,d\tau\,d\mu - \int_{-\infty}^{\infty} g(\tau)R_{sy}(\eta+\tau)\,d\tau = 0 \qquad (11\text{-}7)$$

Equation (11-7) can be written

$$\int_0^{\infty} g(\tau)\left[\int_0^{\infty} h(\mu)R_y(\tau-\mu)\,d\mu - R_{sy}(\eta+\tau)\right]d\tau = 0 \qquad (11\text{-}8)$$

† See Chap. 6, Art. 6-6.

since $g(t)$ and $h(t)$ must vanish for negative values of their arguments. But now Eq. (11-8) can hold for all $g(\tau)$ only if

$$R_{sy}(\tau + \eta) = \int_0^\infty h(\mu) R_y(\tau - \mu)\, d\mu \qquad \tau \geq 0 \qquad (11\text{-}9)$$

This integral equation must be satisfied by $h(t)$ in order that the realizable filter with weighting function $h(t)$ give a minimum mean-squared error prediction for $s(t + \eta)$ amongst the class of all realizable fixed-parameter filters.

We have just shown that for $h(t)$ to satisfy Eq. (11-9) is a necessary condition in order that $h(t)$ provide a minimum; it is also sufficient. For if $h(t)$ satisfies Eq. (11-9), Eq. (11-8) is satisfied for all $g(t)$. Now suppose $f(t)$ is the weighting function of any other realizable filter; we show that $h(t)$ yields a smaller mean-square error than does $f(t)$. Let $g(t) = f(t) - h(t)$. The inequality of Eq. (11-6) is satisfied because Eq. (11-8) is. In particular, Eq. (11-6) is satisfied for this $g(t)$ and for $\epsilon = 1$. But the left-hand side of Eq. (11-6) with $\epsilon = 1$ is simply the difference between the mean-square error using the filter with weighting function $g(t) + h(t) = f(t)$ and that using the filter with weighting function $h(t)$. Hence the error with $h(t)$ is less than or equal to that with any other admissable function.

The problem of finding the optimum smoothing and predicting filter is now very neatly specified by the integral equation (11-9). In Art. 11-4 we shall solve this equation under the restriction that the cross-spectral density $S_{sy}(f)$ be a rational function. But first it seems worthwhile to discuss Eq. (11-9) and its source in some detail, in particular to consider the special cases of pure smoothing and pure prediction and the effect of the realizability requirement.

Infinite-lag Smoothing Filter. Suppose we had not required in the derivation of Eq. (11-9) that the optimum filter be realizable. Then we should have been looking for a weighting function $h(t)$ which would provide a minimum mean-squared error from among the class of all weighting functions, with no requirement that these functions vanish for negative values of their arguments. A little reflection will show that Eq. (11-7) is still a necessary condition for a best $h(t)$ under these circumstances, with $g(t)$ now any weighting function. Hence Eq. (11-8), modified so that the lower limits of both integrals are $-\infty$, is still a necessary condition, and Eq. (11-9) is replaced by the equation

$$R_{sy}(\tau + \eta) = \int_{-\infty}^{\infty} h(\mu) R_y(\tau - \mu)\, d\mu \qquad -\infty < \tau < \infty \qquad (11\text{-}10)$$

which is easily seen to give a sufficient as well as necessary condition that $h(t)$ yield a minimum. Equation (11-10) is readily solved by taking

Fourier transforms. Since the right side is a convolution, one has

$$e^{j\omega\eta}S_{sy}(f) = \int_{-\infty}^{\infty} R_{sy}(\tau + \eta)e^{-j\omega\tau}\,d\tau$$

$$= \int_{-\infty}^{\infty} \int_{-\infty}^{\infty} h(\mu)R_y(\tau - \mu)e^{-j\omega\tau}\,d\mu\,d\tau$$

$$= \int_{-\infty}^{\infty} h(\mu)e^{-j\omega\mu}\,d\mu \int_{-\infty}^{\infty} R_y(\xi)e^{-j\omega\xi}\,d\xi$$

$$= H(j2\pi f)S_y(f)$$

whence
$$H(j2\pi f) = \frac{S_{sy}(f)}{S_y(f)}\,e^{j2\pi f\eta} \tag{11-11}$$

If the signal and noise are uncorrelated and either the signal or the noise has zero mean, Eq. (11-11) becomes

$$H(j2\pi f) = \frac{S_s(f)}{S_s(f) + S_n(f)}\,e^{j2\pi f\eta} \tag{11-12}$$

A filter specified by a system function as given by Eq. (11-12) will not be realizable, but it can be approximated more and more closely by realizable filters as an increasing lag time for the output is allowed.[†] We call the filter given by Eq. (11-11) with $\eta = 0$ the optimum infinite-lag smoothing filter. The solution given by Eq. (11-11) actually has some practical value; it gives approximately the best filter when a long record of input data (signal-plus-noise waveforms) is available before processing (filtering) of the data is begun. In such a situation t enters the problem as a parameter of the recorded data which has no particular connection with real time. In addition, the solution given by Eq. (11-11) is of interest because it has a connection, which will be pointed out later, with the best *realizable* filter for smoothing and predicting. Equation (11-12) has an especially simple interpretation, according to which the best "smoothing" results when the input signal-plus-noise spectrum is most heavily weighted at frequencies for which the ratio of signal spectral density to noise spectral density is largest. The best the filter can do to recover the signal from the noise is to treat favorably those frequency bands which contain largely signal energy and unfavorably those frequency bands which contain largely noise energy.

11-3. Pure Prediction: Nondeterministic Processes[‡]

We shall return in Art. 11-4 to the consideration of smoothing and predicting filters, which must operate on input signals mixed with noise.

---

† See Wallman (I).

‡ The treatment of the pure prediction problem in this article is similar to that in Bode and Shannon (I). It is essentially a heuristic analogue of the rigorous mathematical treatment which deals with linear prediction theory in terms of projections on Hilbert spaces. See, e.g., Doob (II, Chap. XII) and the references there.

In this article, however, we shall consider only the special case in which there is no input noise, so that $y(t) = s(t)$. This is the pure prediction problem. The optimum filter specification which we shall arrive at in this article is contained in the results of Art. 11-4, which are got straightforwardly from Eq. (11-9). Here, however, the pure prediction problem is done in a different way, so as to throw a different light on what is involved in prediction and so as to obtain certain ancillary results.

The only information we have been using about the input signal to the predictor is, first, the fact that the signal is from a wide-sense stationary random process, and second, the autocorrelation function of this random process. The mean-square error depends on the signal input

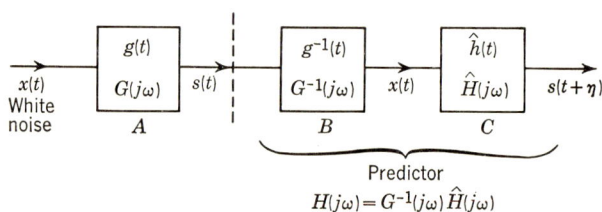

FIG. 11-1. The optimum predicting filter.

only through this autocorrelation function (see Eq. (11-4) with $R_{sy}(t)$ set equal to $R_y(t)$). Hence, if the input-signal random process is replaced by a different random process with the same autocorrelation function, the mean-square error ε for any particular filter is unchanged and the optimum filter is unchanged. This fact provides the basis for the following analysis.

Let us outline briefly the ideas we shall use to find the optimum predicting filter. First, it is imagined that $s(t)$ is replaced at the input to the predicting filter by filtered white noise. That is, a hypothetical filter A is postulated with a system function $G(j\omega)$ chosen so that if A is driven by white noise, its output has exactly the same autocorrelation function (and hence the same spectral density) as has $s(t)$. The output of A is to be the input to the predictor. Then the predictor itself is imagined to be a cascade of two filters, B and C. Filter B is to be the inverse to A and hence is to have system function $G^{-1}(j\omega)$; its purpose is to turn the input signal back into white noise. Filter C is to make the optimum prediction of $s(t + \eta)$ by operating on the white noise; this operation turns out to be quite simple. The system function $H(j\omega)$ of the optimum predictor is then the product of the system functions of the hypothetical filters B and C. See Fig. 11-1.

We shall now treat the problem in more detail. It has to be supposed first that the input signal belongs to that class of random processes which have autocorrelation functions which can be realized with filtered white noise. That is, $R_s(t)$ must be such that for some function $g(t)$, $z(t)$ as defined by

$$z(t) = \int_{-\infty}^{t} g(t - \tau)x(\tau)\,d\tau \tag{11-13}$$

has autocorrelation function $R_z(t) = R_s(t)$, when $x(t)$ is white noise. This means that the system function $G(j2\pi f)$, which is the Fourier transform of $g(t)$, must satisfy†

$$|G(j2\pi f)|^2 = S_s(f) \tag{11-14}$$

This restriction is significant, and we shall discuss it further below. For now, it may simply be remarked that for a large class of wide-sense stationary processes, the spectral density can be factored in terms of the system function of a realizable filter, as in Eq. (11-14). Since for the purposes of the prediction theory there is no point in distinguishing between $s(t)$ and $z(t)$, it will be assumed henceforth that

$$s(t) = \int_{-\infty}^{t} g(t - \tau)x(\tau)\,d\tau \tag{11-15}$$

where $x(t)$ is white noise and the Fourier transform $G(j2\pi f)$ of $g(t)$ satisfies Eq. (11-14).

From Eq. (11-15) we have, for the future value of $s(t)$,

$$s(t + \eta) = \int_{-\infty}^{t+\eta} g(t + \eta - \tau)x(\tau)\,d\tau \tag{11-16}$$

This is the quantity it is desired to predict. Since $s_{t+\eta}$ is a random variable which is composed of a linear superposition of white noise, since at time t part of that white noise is past history, and since the white noise which is yet to occur after time t and before time $t + \eta$ has mean zero and is uncorrelated with what has gone before, it is plausible to suppose that the best linear prediction is got by taking that part of $s_{t+\eta}$ that is composed of white noise which has occurred before time t. Let us now show that this is in fact true. Equation (11-16) can be rewritten

$$s(t + \eta) = \int_{-\infty}^{t} g(t + \eta - \tau)x(\tau)\,d\tau + \int_{t}^{t+\eta} g(t + \eta - \tau)x(\tau)\,d\tau \tag{11-17}$$

$$= \hat{s}(t + \eta) + \int_{t}^{t+\eta} g(t + \eta - \tau)x(\tau)\,d\tau$$

$\hat{s}(t + \eta)$ is exactly that part of $s(t + \eta)$ which is composed of white noise which has occurred before time t; so $\hat{s}(t + \eta)$ is taken to be the prediction

† See Chap. 9, Art. 9-3.

of $s(t + \eta)$, which we want to prove is optimum. The mean-square error of the prediction $\hat{s}(t + \eta)$ is

$$E\{[s_{t+\eta} - \hat{s}_{t+\eta}]^2\} = E\left\{\left[\int_t^{t+\eta} g(t + \eta - \tau)x(\tau) \, d\tau\right]^2\right\} \quad (11\text{-}18)$$

In order to prove that this mean-squared error is a minimum, we shall calculate the mean-squared error using any other linear predicting filter which operates only on the past record of white noise, $x(\tau)$, $\tau \le t$, and show that it cannot be less than that given by Eq. (11-18). Since any linear operation on the past of $s(t)$ is necessarily also a linear operation on $x(\tau)$, $\tau \le t$, it will follow a fortiori that $\hat{s}(t + \eta)$ is as good as any linear prediction which can be made using only the past of $s(t)$.

Let $s'_{t+\eta}$ be another linear prediction, given by

$$s'_{t+\eta} = \int_{-\infty}^{t} f(t - \tau)x(\tau) \, d\tau$$

with some function $f(t)$. Let $h(t) = f(t) - g(t + \eta)$. Then

$$E\{[s_{t+\eta} - s'_{t+\eta}]^2\} = E\left\{\left[\int_{-\infty}^{t+\eta} g(t + \eta - \tau)x(\tau) \, d\tau\right.\right.$$

$$\left.\left. - \int_{-\infty}^{t} g(t + \eta - \tau)x(\tau) \, d\tau - \int_{-\infty}^{t} h(t - \tau)x(\tau) \, d\tau\right]^2\right\}$$

$$= E\left\{\left[\int_{t}^{t+\eta} g(t + \eta - \tau)x(\tau) \, d\tau - \int_{-\infty}^{t} h(t - \tau)x(\tau) \, d\tau\right]^2\right\} \quad (11\text{-}19)$$

When this is expanded and averaged, the cross-product term is zero, for, since $x(\tau)$, $\tau \le t$, is uncorrelated with $x(\tau)$, $\tau > t$, the first integral is uncorrelated with the second. The first integral is just the error for the prediction $\hat{s}_{t+\eta}$, so Eq. (11-19) becomes

$$E\{[s_{t+\eta} - s'_{t+\eta}]^2\} = E\{[s_{t+\eta} - \hat{s}_{t+\eta}]^2\} + E\left\{\left[\int_{-\infty}^{t} h(t - \tau)x(\tau) \, d\tau\right]^2\right\} \quad (11\text{-}20)$$

Since the last term is nonnegative, Eq. (11-20) shows that the prediction error for $s'_{t+\eta}$ is greater than for $\hat{s}_{t+\eta}$. This demonstrates that $\hat{s}_{t+\eta}$ is optimum, as was claimed.

Up to this point, the optimum prediction $\hat{s}(t + \eta)$ has been defined only in terms of the hypothetical white noise $x(t)$ by the equation

$$\hat{s}(t + \eta) = \int_{-\infty}^{t} g(t + \eta - \tau)x(\tau) \, d\tau \quad (11\text{-}21)$$

and not in terms of the input $s(t)$. Let us now obtain an expression for the system function $H(j\omega)$ of the optimum predicting filter, that is, for the filter that will yield $\hat{s}(t + \eta)$ as output when its input is $s(t)$. In Eq. (11-21), if we set

$$\hat{h}(u) = 0 \qquad u < 0$$
$$= g(u + \eta) \qquad u \geq 0 \tag{11-22}$$

then
$$\hat{s}(t + \eta) = \int_{-\infty}^{t} \hat{h}(t - \tau)x(\tau)\, d\tau \tag{11-23}$$

which gives $\hat{s}(t + \eta)$ as filtered white noise. If now $s(t)$ is put into a filter (filter B in Fig. 11-1) which is the inverse to one with impulse response $g(t)$ in order to recover the white noise $x(t)$ at its output, and then $x(t)$ is put into a filter (filter C in Fig. 11-1) with impulse response $\hat{h}(t)$, the over-all action of the two filters is to yield the best prediction $\hat{s}(t + \eta)$ from the input $s(t)$. The predicting filter is therefore obtainable as the cascade of a filter with system function $G^{-1}(j\omega)$ and a filter with system function $\hat{H}(j\omega)$, and the system function of the predicting filter is

$$H(j\omega) = G^{-1}(j\omega)\hat{H}(j\omega) \tag{11-24}$$

$G^{-1}(j\omega)$ is never a proper system function; since $s(t)$ is always smoothed white noise, $g(t)$ is the impulse response of a filter which performs some integration, so the inverse filter must differentiate. This means that $G^{-1}(j\omega)$ does not vanish at infinity. However, the product system function $H(j\omega)$ may or may not vanish at infinity. If it does not, then the best linear prediction cannot be accomplished exactly by a *filter*, that is, by a device characterized by an integral operator as in Eq. (11-23). Even when this happens, the optimum prediction can often be approximated by approximating $H(j\omega)$ to arbitrarily high frequencies. These remarks will be illustrated in the simple example below and by examples in the next article. We could, from Eq. (11-24), write an explicit formula for $H(j\omega)$ in terms of the spectral density of the signal input process; we defer this to the next article, in which the general case of predicting and filtering is considered.

Example 11-3.1. Suppose the signal $s(t)$ is a sample function from a random process with spectral density

$$S_s(f) = \frac{1}{1 + f^2} \tag{11-25}$$

Then if we take for $G(j2\pi f)$

$$G(j2\pi f) = \frac{1}{1 + jf} \tag{11-26}$$

$$|G(j2\pi f)|^2 = \frac{1}{1 + jf}\frac{1}{1 - jf} = S_s(f)$$

so the condition Eq. (11-14) is satisfied. Furthermore

$$g(t) = \frac{1}{2\pi}\int_{-\infty}^{\infty} G(j\omega)e^{j\omega t}\, d\omega$$
$$= 2\pi e^{-2\pi t} \qquad t \geq 0$$
$$= 0 \qquad t < 0 \tag{11-27}$$

so $g(t)$ is the impulse response of a realizable filter, as is required by Eq. (11-15). Now

$$\hat{h}(t) = 0 \qquad t < 0$$
$$= 2\pi e^{-2\pi(t+\eta)} \qquad t \geq 0$$

and

$$\hat{H}(j\omega) = \int_{-\infty}^{\infty} \hat{h}(t)e^{-j\omega t}\,dt$$

$$= 2\pi \int_{0}^{\infty} e^{-2\pi(t+\eta)}e^{-j\omega t}\,dt$$

$$= e^{-2\pi\eta}\frac{1}{1+jf} \qquad (11\text{-}28)$$

Since

$$G^{-1}(j\omega) = 1 + jf$$

we have for the system function of the predicting filter

$$H(j\omega) = (1+jf)\frac{e^{-2\pi\eta}}{1+jf} = e^{-2\pi\eta} \qquad (11\text{-}29)$$

The best predicting filter in this case is simply an attenuator. This example is sufficiently simple that the result given by Eq. (11-29) can be seen intuitively. The spectral density given in Eq. (11-25) can be realized by exciting an R-C filter with time constant $1/2\pi$ by white noise from a zero impedance source, as shown in Fig. 11-2.

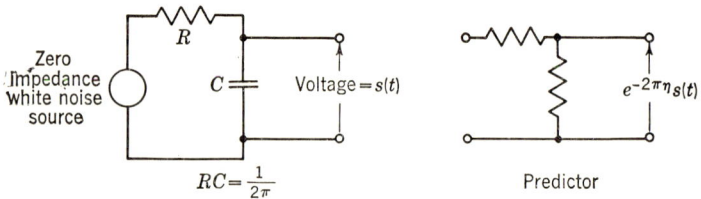

$$RC = \frac{1}{2\pi}$$

Fig. 11-2. Predictor example.

The voltage across the condenser at time t is s_t. The best prediction of this voltage at time $t + \eta$, since $x(\tau)$ has zero mean and is unpredictable for $\tau < t$, is the voltage which would be left after s_t has discharged through the resistance R for η seconds. This is the value which the predicting filter, as specified by Eq. (11-29), will yield.

Integral Equation for Prediction. We have found that $\hat{s}_{t+\eta}$ is the best estimate that can be made for $s_{t+\eta}$ by a linear operation on the past of $s(t)$. If $\hat{s}_{t+\eta}$ can be realized by a linear filter operating on $s(t)$ with weighting function $h(t)$, then it follows, of course, from the minimization done earlier in this section that $h(t)$ must satisfy the integral equation (11-9). It is interesting to observe that the necessity of Eq. (11-9) as a condition on $h(t)$ can be shown directly, however, using $\hat{s}(t + \eta)$. Suppose

$$\hat{s}(t + \eta) = \int_{0}^{\infty} h(\tau)s(t - \tau)\,d\tau \qquad (11\text{-}30)$$

The error is

$$s(t + \eta) - \hat{s}(t + \eta) = s(t + \eta) - \int_{0}^{\infty} h(\tau)s(t - \tau)\,d\tau \qquad (11\text{-}31)$$

but, by Eq. (11-17), the error is a linear function of the white noise $x(\tau)$ only for values of $\tau > t$. Hence the error is uncorrelated with the past of $s(t)$, which is a linear function of $x(\tau)$, $\tau < t$. Thus, using Eq. (11-31) and taking $t = 0$,

$$E\left\{s(-\mu)\left[s(\eta) - \int_0^\infty h(\tau)s(-\tau)\,d\tau\right]\right\} = 0 \qquad \mu > 0$$

Or, expanding, we have the predicting version of Eq. (11-9),

$$R_s(\mu + \eta) = \int_0^\infty h(\tau)R_s(\mu - \tau)\,d\tau \qquad \mu \geq 0 \qquad (11\text{-}32)$$

It may be verified directly that formally the function $h(t)$ defined by

$$h(t) = \int_{-\infty}^\infty e^{i2\pi ft}H(j2\pi f)\,df$$

when $H(j2\pi f)$ is given by Eq. (11-24), satisfies Eq. (11-32).

Gaussian Processes. It has been shown that $\hat{s}_{t+\eta}$ is the best prediction of $s_{t+\eta}$ which can be got by a *linear* superposition of the values of $x(\tau)$, $\tau \leq t$, and hence of the past values of $s(t)$. This of course leaves open the question of whether there are *nonlinear* operations that can be performed on the past values of $s(t)$ which will yield a better mean-square approximation to $\hat{s}_{t+\eta}$. However, if the signal $s(t)$ is a sample function from a gaussian random process, one can show that the best linear mean-square approximation $\hat{s}_{t+\eta}$ is as good as any in a mean-square sense. A justification of this statement is similar to the argument made to show $\hat{s}_{t+\eta}$ is a best linear estimate; the difference lies in the fact that for gaussian random variables, uncorrelatedness implies statistical independence. Suppose $s(t)$ is gaussian, then $x(t)$ is gaussian white noise. Let Y_t be *any* prediction for $s_{t+\eta}$ which depends only on $s(\tau)$, $\tau < t$. Then Y_t is independent of $x(\tau)$, $\tau > t$, and $Z_t = Y_t - s_{t+\eta}$ is also independent of $x(\tau)$, $\tau > t$. The mean-square error of Y_t is

$$\begin{aligned}
E\{[s_{t+\eta} - Y_t]^2\} &= E\{[s_{t+\eta} - \hat{s}_{t+\eta} - Z_t]^2\} \\
&= E\{[s_{t+\eta} - \hat{s}_{t+\eta}]^2\} + E\{[Z_t]^2\} \qquad (11\text{-}33)
\end{aligned}$$

which is greater than or equal to the mean-square error of $\hat{s}_{t+\eta}$.

Deterministic and Nondeterministic Processes. As we mentioned above when we started to consider linear prediction theory using the device of filtered white noise, we were allowing only spectral densities $S_s(f)$ which could be factored, as in Eq. (11-14), into a product of $G(j2\pi f)$ and $G^*(j2\pi f)$, where $G(j2\pi f)$ is the system function of a realizable filter. This restriction is connected with the condition that the random process be nondeterministic. A wide-sense stationary random process is said to be *deterministic* if a future value can be predicted exactly by a linear operation on its past; otherwise, it is *nondeterministic*. An important theorem in prediction theory states that a wide-sense stationary random process is nondeterministic if and only if

$$\int_{-\infty}^{\infty} \left| \frac{\log S(f)}{1+f^2} \right| df \qquad (11\text{-}34)$$

converges, where $S(f)$ is the spectral density of the process.† Now if the process is regarded as generated by white noise passed through a realizable filter,

$$S(f) = |H(j2\pi f)|^2$$

but the condition that a gain function $|H(j2\pi f)|$ be that of a realizable filter is that

$$\int_{-\infty}^{\infty} \left| \frac{\log |H(j2\pi f)|}{1+f^2} \right| df = \frac{1}{2} \int_{-\infty}^{\infty} \left| \frac{\log S(f)}{1+f^2} \right| df \qquad (11\text{-}35)$$

converges. Thus the restriction we imposed was equivalent to restricting the signals to be sample functions from nondeterministic random processes.

Example 11-3.2. Let

$$s(t) = \cos (\omega t + \phi) \qquad \omega = 2\pi f$$

where f and ϕ are independent random variables, where ϕ is uniformly distributed over the interval $0 \leq \phi < 2\pi$, and where f has an even but otherwise arbitrary probability density function $p(f)$. Then $s(t)$ is a stationary random process, every sample function of which is a pure sine wave. It is easily calculated that

$$R_s(\tau) = 2E[\cos \omega\tau]$$

$$= 2 \int_{-\infty}^{\infty} \cos (2\pi f\tau) p(f) \, df$$

But also

$$R_s(\tau) = \int_{-\infty}^{\infty} e^{j2\pi f\tau} S_s(f) \, df = \int_{-\infty}^{\infty} \cos (2\pi f\tau) S_s(f) \, df$$

Hence $S_s(f) = 2p(f)$. We can choose $p(f)$ either to satisfy or not to satisfy the criterion Eq. (11-34), and thus the random process $s(t)$ can be made either nondeterministic or deterministic by suitable choice of $p(f)$. In either case, $s(t + \tau)$ can be predicted perfectly if its past is known, but not by a *linear* operation on its past if it is nondeterministic.

11-4. Solution of the Equation for Predicting and Filtering

We want now to get the solution of the integral equation (11-9) for predicting and filtering, using the infinite past. We shall do this only for the case in which the spectral density $S_y(f)$ of the entire input $y(t)$ is a rational function. Equation (11-9) can be solved under more general conditions, and the method of solution is the same. However, there is

† See Doob (II, p. 584) and Wiener (III, p. 74).

a difficulty in the factorization of the spectral density $S_y(f)$, which is very simple if $S_y(f)$ is rational but calls for more sophisticated function theory if $S_y(f)$ is not rational.† Practically, it is often sufficient to consider only the case of the rational spectral density because of the availability of techniques for approximating arbitrary spectral densities by rational functions.‡

The difficulty in solving Eq. (11-9) arises entirely from the fact that $h(t)$ was required to vanish on half the real line. In fact, we have seen in Art. 11-2 that with this condition waived (infinite-lag filter) the integral equation takes on a form such that it can be solved immediately by taking Fourier transforms. The trick in solving Eq. (11-9) is to factor the spectral density $S_y(f)$ into two parts, one of which is the Fourier transform of a function which vanishes for negative values of its argument, the other of which is the Fourier transform of a function which vanishes for positive values of its argument. We have already used this trick in the preceding article where the solution to the special case for which $R_{sy}(\tau) = R_y(\tau) = R_s(\tau)$ (pure prediction) was obtained in one form in Eq. (11-24). If the system functions $G^{-1}(j\omega)$ and $\hat{H}(j\omega)$ are written in terms of the factors of the spectral density, Eq. (11-24) becomes exactly the solution obtained below for the case $R_{sy}(\tau) = R_y(\tau)$.

First, let us consider the factorization of $S_y(f)$. Since it is assumed that $S_y(f)$ is rational, it may be written

$$S_y(f) = a^2 \frac{(f - w_1) \cdots (f - w_N)}{(f - z_1) \cdots (f - z_M)} \qquad w_n \neq z_m \qquad (11\text{-}36)$$

Since $S_y(f)$ is a spectral density, it has particular properties which imply certain restrictions on the number and location of its poles and zeros:

1. $S_y(f)$ is real for real f. Hence $S_y(f) = S_y^*(f)$, which implies that a^2 is real and all w's and z's with nonzero imaginary parts must occur in conjugate pairs.

2. $S_y(f) \geq 0$. Hence any real root of the numerator must occur with an even multiplicity. (Otherwise the numerator would change sign.)

3. $S_y(f)$ is integrable on the real line. Hence no root of the denominator can be real, and the degree of the numerator must be less than the degree of the denominator, $N < M$.

We can, therefore, split the right side of Eq. (11-36) into two factors, one of which contains all the poles and zeros with positive imaginary parts and the other of which contains all the poles and zeros with negative imaginary parts. Since any real root w_n of the numerator occurs an even number of times, half the factors $(f - w_n)$ may be put with the

† See Wiener (III, Art. 1.7) Levinson (I, Art. 5), or Titchmarsh (II, Art. 11.17).
‡ See, for example, Wiener (III, Art. 1.03).

term containing the poles and zeros with positive imaginary parts and half with the other.   Thus $S_y(f)$ may be written

$$S_y(f) = a\,\frac{(f - w_1)\ \cdots\ (f - w_P)}{(f - z_1)\ \cdots\ (f - z_Q)}\ a\,\frac{(f - w_1^*)\ \cdots\ (f - w_P^*)}{(f - z_1^*)\ \cdots\ (f - z_Q^*)}$$

where $P < Q$, where the z_n, $n = 1, \ldots, Q$, have positive imaginary parts, the w_k, $k = 1, \ldots, P$, have nonnegative imaginary parts, $2P = N$, and $2Q = M$.   Let $G(j\omega)$ be defined by

$$G(j\omega) = G(j2\pi f) = a\,\frac{(f - w_1)\ \cdots\ (f - w_P)}{(f - z_1)\ \cdots\ (f - z_Q)} \qquad (11\text{-}37)$$

Thus $S_y(f) = |G(j2\pi f)|^2$ where $G(j2\pi f) = G(p)$ is a rational function of $p = j2\pi f$ with all its poles and zeros in the left half p plane except possibly for zeros on the imaginary axis.

Define

$$g(t) = \int_{-\infty}^{\infty} e^{j\omega t} G(j2\pi f)\ df$$

and

$$g'(t) = \int_{-\infty}^{\infty} e^{j\omega t} G^*(j2\pi f)\ df$$

Then, since $G(j2\pi f)$ as a function of f has all its poles in the upper half f plane, $g(t) = 0$ for $t < 0$ and $g'(t) = 0$ for $t > 0$.   Since $R_y(t)$ is the inverse Fourier transform of $S_y(f)$, $R_y(t)$ is given by the convolution of $g(t)$ and $g'(t)$,

$$R_y(t) = \int_{-\infty}^{\infty} e^{j\omega t} G(j\omega) G^*(j\omega)\ df$$
$$= \int_{-\infty}^{0} g(t - u) g'(u)\ du \qquad (11\text{-}38)$$

where the upper limit of the integral can be taken to be zero because $g'(t)$ vanishes for $t > 0$.   Intuitively, $G(j\omega)$ is the system function of a realizable linear system which would transform white noise into a random process with autocorrelation function $R_y(t)$.

Let us now introduce a function $A(f)$ defined by

$$S_{sy}(f) = A(f) G^*(j\omega) \qquad (11\text{-}39)$$

[$A(f)$ is $G(j\omega)$ for the case of pure prediction.]   Then, if

$$a(t) = \int_{-\infty}^{\infty} e^{j\omega t} A(f)\ df \qquad (11\text{-}40)$$

we have

$$R_{sy}(\tau) = \int_{-\infty}^{0} a(\tau - u) g'(u)\ du \qquad (11\text{-}41)$$

where the upper limit can again be taken to be zero as in Eq. (11-38).

Substituting $R_y(t)$ as given by Eq. (11-38) and $R_{sy}(t)$ as given by Eq. (11-41) into Eq. (11-9) yields

$$\int_{-\infty}^{0} a(\tau + \eta - u)g'(u)\, du = \int_{0}^{\infty} h(\mu)\, d\mu \int_{-\infty}^{0} g(\tau - \mu - u)g'(u)\, du \qquad \tau > 0$$

or

$$\int_{-\infty}^{0} g'(u)\left\{ a(\tau + \eta - u) - \int_{0}^{\infty} h(\mu)g(\tau - \mu - u)\, d\mu \right\} du = 0 \qquad \tau > 0$$

$$\text{(11-42)}$$

This equation is satisfied if the expression in parentheses vanishes for all $u < 0$ and $\tau > 0$, that is, if the integral equation

$$a(\tau + \eta) = \int_{0}^{\infty} h(\mu)g(\tau - \mu)\, d\mu \qquad \tau > 0 \qquad \text{(11-43)}$$

is satisfied. Equation (11-43) can be solved directly by taking Fourier transforms, whereas the original integral equation (11-9) could not. The reason for the difference is that because $g(t)$ vanishes for $t < 0$, the transform of the right side of Eq. (11-43) factors into a product of transforms, whereas the transform of the right side of Eq. (11-9) will not. We have

$$\int_{0}^{\infty} e^{-j\omega\tau} a(\tau + \eta)\, d\tau = \int_{0}^{\infty} h(\mu)e^{-j\omega\mu}\, d\mu \int_{0}^{\infty} g(\tau - \mu)e^{-j\omega(\tau-\mu)}\, d\tau$$

$$= \int_{0}^{\infty} h(\mu)e^{-j\omega\mu}\, d\mu \int_{-\mu}^{\infty} g(\nu)e^{-j\omega\nu}\, d\nu$$

$$= \int_{0}^{\infty} h(\mu)e^{-j\omega\mu}\, d\mu \int_{0}^{\infty} g(\nu)e^{-j\omega\nu}\, d\nu$$

or $\qquad H(j\omega) = \dfrac{1}{G(j\omega)} \displaystyle\int_{0}^{\infty} e^{-j\omega\tau} a(\tau + \eta)\, d\tau \qquad \text{(11-44)}$

Equation (11-44) gives the solution $H(j\omega)$ for the system function of the optimum filter. Using Eqs. (11-39) and (11-40), we can rewrite Eq. (11-44) in terms of $S_{sy}(f)$ and the factors of $S_y(f)$,

$$H(j\omega) = \frac{1}{G(j\omega)} \int_{0}^{\infty} e^{-j\omega\tau}\, d\tau \int_{-\infty}^{\infty} e^{j\omega'(\tau+\eta)} \frac{S_{sy}(f')}{G^*(j2\pi f')}\, df' \qquad \text{(11-45)}$$

Equation (11-45) is a general formula for the system function of the optimum linear least-mean-square-error predicting and smoothing filter. If $S_y(f)$ and $S_{sy}(f)$ are rational functions, the right side of Eq. (11-45) can always be evaluated straightforwardly, although the work may be tedious. Just as in the case of pure prediction, $H(j\omega)$ may not vanish as $\omega \to \infty$. The remarks made about this possibility in the article on the pure-prediction problem apply here.

The natural special cases of the filtering and prediction problem are predicting when there is no noise present, which we have already discussed, and smoothing without predicting, which we now discuss briefly.

Smoothing. With $\eta = 0$, Eq. (11-45) gives the solution for the system function of the optimum realizable smoothing filter. It is possible to rewrite the right side of Eq. (11-45) so as to simplify somewhat the calculations required in evaluating it in most instances and also so as to point out the relation between it and the solution for the infinite-lag smoothing filter given by Eq. (11-11).

First we point out that if $\phi(f)$ is any rational function (of integrable square) with all its poles in the top half f plane,

$$\int_{-\infty}^{\infty} e^{j2\pi f'\tau}\phi(f')\,df'$$

vanishes for negative values of τ, and hence

$$\int_{0}^{\infty} e^{-j\omega\tau}\,d\tau \int_{-\infty}^{\infty} e^{j\omega'\tau}\phi(f')\,df' = \phi(f) \qquad (11\text{-}46)$$

Also, if $\psi(f)$ is rational with all its poles in the bottom half plane,

$$\int_{0}^{\infty} e^{-j\omega\tau}\,d\tau \int_{-\infty}^{\infty} e^{j\omega'\tau}\psi(f')\,df' = 0 \qquad (11\text{-}47)$$

Thus, given a rational function $\xi(f)$ with no poles on the real axis, if $\xi(f)$ is decomposed into the *sum* of two rational functions $\phi(f) + \psi(f)$, where $\phi(f)$ has all its poles in the top half f plane and $\psi(f)$ in the bottom,

$$\int_{0}^{\infty} e^{-j\omega\tau}\,d\tau \int_{-\infty}^{\infty} e^{j\omega'\tau}\xi(f')\,df' = \phi(f) \qquad (11\text{-}48)$$

This double integral operator, which appears in Eq. (11-45), may be thought of in the present context as a "realizable part of" operator. We introduce the notation for this section, $\phi(f) = [\xi(f)]_+$.

If signal and noise are uncorrelated, $S_{sy}(f) = S_s(f)$. It may be shown rather easily (see Prob. 3) that $S_s(f)/G^*(j2\pi f)$ has no poles on the real f axis when $S_y(f) = S_s(f) + S_n(f)$. Thus, using the special notation just introduced, for the case of smoothing with uncorrelated signal and noise, Eq. (11-45) can be written

$$H(j2\pi f) = \frac{1}{G(j2\pi f)}\left[\frac{S_s(f)}{G^*(j2\pi f)}\right]_+ \qquad (11\text{-}49)$$

To find $H(j2\pi f)$ using Eq. (11-49), a partial-fraction expansion† of $S_s(f)/G^*(j2\pi f)$ is made, but no integrations need be performed.

Equation (11-49) is in a form which permits an intuitive tie-in with the solution for the infinite-lag smoothing filter. From Eq. (11-11) we have, if signal and noise are uncorrelated,

$$H_{-\infty}(j2\pi f) = \frac{S_s(f)}{S_y(f)} = \frac{1}{G(j2\pi f)}\frac{S_s(f)}{G^*(j2\pi f)} \qquad (11\text{-}50)$$

† See Gardner and Barnes (I, p. 159ff.).

where $H_{-\infty}(j2\pi f)$ is the system function of the optimum infinite-lag filter. Equation (11-50) describes a filter which may be thought of as two filters in tandem. Equation (11-49) describes a filter which differs from that of Eq. (11-50) only in that the realizable part of the second filter has been taken.

Example 11-4.1. Suppose that signal and noise are independent and that

$$S_s = \frac{1}{1 + f^2}$$

$$S_n = \frac{Nb^2}{b^2 + f^2} = \frac{N}{1 + (f/b)^2}$$

It is desired to find the system function of the optimum linear smoothing filter. We have

$$S_y(f) = S_s + S_n = \frac{Nb^2(1 + f^2) + b^2 + f^2}{(b + jf)(b - jf)(1 + jf)(1 - jf)}$$

$$= (Nb^2 + 1) \frac{A^2 + f^2}{(b + jf)(b - jf)(1 + jf)(1 - jf)}$$

where $A^2 = [(N + 1)b^2]/(Nb^2 + 1)$.
$S_y(f)$ may be factored into $G(j2\pi f)$, $G^*(j2\pi f)$ where

$$G(j2\pi f) = \frac{b\sqrt{N + 1}}{A} \frac{A + jf}{(b + jf)(1 + jf)}$$

$$G^*(j2\pi f) = \frac{b\sqrt{N + 1}}{A} \frac{A - jf}{(b - jf)(1 - jf)}$$

Then
$$\frac{S_s(f)}{G^*(j2\pi f)} = \frac{A}{b\sqrt{N + 1}} \frac{b - jf}{(1 + jf)(A - jf)}$$

$$= \frac{A}{b\sqrt{N + 1}} \left\{ \frac{b + 1}{A + 1} \frac{1}{1 + jf} + \frac{b - A}{A + 1} \frac{1}{A - jf} \right\}$$

so
$$\left[\frac{S_s(f)}{G^*(j2\pi f)} \right]_+ = \frac{A}{b\sqrt{N + 1}} \frac{b + 1}{A + 1} \frac{1}{1 + jf}$$

and, using Eq. (11-50),

$$H(j2\pi f) = \frac{1}{Nb^2 + 1} \frac{b + 1}{A + 1} \frac{b + jf}{A + jf} \tag{11-51}$$

If we take the limit in Eq. (11-51) as $b \to \infty$, we get the optimum system function for smoothing the signal in white noise of spectral density N. This is

$$H(j2\pi f) = \frac{1}{\sqrt{N + 1} + \sqrt{N}} \frac{1}{\sqrt{N + 1} + j\sqrt{N}f} \tag{11-52}$$

Smoothing and Prediction Error. The mean-square error of any smoothing and predicting filter with impulse response $h(t)$ is given by Eq. (11-4). If it is an optimum filter, then by virtue of Eq. (11-9), the mean-square error is

$$\mathcal{E} = R_s(0) - \int_{-\infty}^{\infty} \int_{-\infty}^{\infty} h(\tau)h(\mu) R_y(\mu - \tau) \, d\mu \, d\tau \tag{11-53}$$

By Parseval's theorem, the double integral can be written in terms of $H(j\omega)$,

$$\int_{-\infty}^{\infty} H(\tau)\, d\tau \int_{-\infty}^{\infty} h(\mu) R_y(\mu - \tau)\, d\mu$$

$$= \int_{-\infty}^{\infty} H(j2\pi f)[H(j2\pi f)S_y(f)]^*\, df \quad (11\text{-}54)$$

$$= \int_{-\infty}^{\infty} |H(j2\pi f)|^2 S_y(f)\, df$$

Then
$$\mathcal{E} = \int_{-\infty}^{\infty} S_s(f)\, df - \int_{-\infty}^{\infty} |H(j2\pi f)|^2 S_y(f)\, df \quad (11\text{-}55)$$

which is probably as convenient a form as any for calculation in the general case. One can substitute for $H(j2\pi f)$ from Eq. (11-46) in Eq. (11-55) and, after some reduction (see Prob. 6), get

$$\mathcal{E} = R_s(0) - \int_0^{\infty} \left| \int_{-\infty}^{\infty} e^{-j\omega'(\tau+\eta)} \frac{S_s(f')}{G^*(j2\pi f')}\, df' \right|^2 d\tau \quad (11\text{-}56)$$

For the case of pure prediction, $S_{sy}(f) = S_y(f) = S_s(f)$ and Eq. (11-56) becomes

$$\mathcal{E} = R_s(0) - \int_0^{\infty} \left| \int_{-\infty}^{\infty} e^{-j\omega'(\tau+\eta)} G(j2\pi f')\, df' \right|^2 d\tau$$

$$= \int_{-\infty}^{\infty} G(j2\pi f) G^*(j2\pi f)\, df - \int_0^{\infty} |g(\tau + \eta)|^2\, d\tau$$

$$= \int_{-\infty}^{\infty} |g(\tau)|^2\, d\tau - \int_0^{\infty} |g(\tau + \eta)|^2\, d\tau$$

where Parseval's theorem has been used. Since $g(t)$ vanishes for $t < 0$, this can be written

$$\mathcal{E} = \int_0^{\infty} |g(\tau)|^2\, d\tau - \int_{\eta}^{\infty} |g(\tau)|^2\, d\tau$$

$$= \int_0^{\eta} |g(\tau)|^2\, d\tau \quad (11\text{-}57)$$

This last result could have been obtained directly from Eq. (11-17).

11-5. Other Filtering Problems Using Least-mean-square Error Criterion

Phillips' Least-mean-square Error Filter. Up to this point, we have asked for the best fixed-parameter linear filter to perform a smoothing or predicting operation. A theoretically less interesting but practically quite useful modification of this approach is to ask for the best filter when the form of the system function of the filter is specified and only a finite number of parameters are left free to be chosen. When the Wiener method, which we have been discussing up to now, is used to find an optimum filter, there may still remain a considerable difficulty, after the system function $H(j\omega)$ is found, in synthesizing a filter whose system function approximates $H(j\omega)$. This is particularly true if servo-mechanisms are involved, as for example in automatic tracking radar systems. On the other hand, if the system function is fixed in form to be that of a practically constructable filter (or servomechanism), then the solution for the best values of the parameters will simply determine the sizes and adjustments of the components of the filter. Of course, by limiting the class of filters to be considered, one in general increases the minimum error which can be achieved. Thus, in principle, quality of performance is sacrificed. Actually, in practical situations the approximations involved in synthesizing a filter to meet the specification of $H(j\omega)$ given by the Wiener theory may degrade the performance so that it is no better than that of a filter designed by the method sketched in this section.

Phillips[†] has described in detail how to determine an optimum smoothing filter when the system function of the smoothing filter is prescribed to be rational and of specified degree in the numerator and denominator. In outline, his method is as follows. The input signal and noise are presumed to be sample functions from stationary random processes. The error, given by

$$s(t) - \int_{-\infty}^{\infty} h(t - \tau)y(\tau) \, d\tau$$

is then also a stationary random process and has a spectral density $S_e(f)$. The error spectral density can be determined straightforwardly from Eq. (11-4) and is (see Prob. 9)

$$S_e(f) = [1 - H(j2\pi f)][1 - H^*(j2\pi f)]S_s(f) + |H(j2\pi f)|^2 S_n(f)$$
$$-[1 - H^*(j2\pi f)]H(j2\pi f)S_{sn}(f) - [1 - H(j2\pi f)]H^*(j2\pi f)S_{ns}(f) \quad (11\text{-}58)$$

The mean-square error is then

$$\mathcal{E} = \int_{-\infty}^{\infty} S_e(f) \, df \quad (11\text{-}59)$$

$H(j2\pi f)$ is taken to be rational with the degrees of the numerator and denominator fixed. Then if $S_s(f)$, $S_n(f)$, $S_{sn}(f)$, $S_{ns}(f)$ are all rational, $S_e(f)$ is rational. The integral of Eq. (11-59) is evaluated by the method of residues with the parameters of $H(j\omega)$ left in literal form. Then these parameters are evaluated so as to minimize $\mathcal{E}$. The reader is referred to the literature[‡] for a thorough discussion of the method and examples.

Extensions and Modifications of the Theory. The theory of best linear predicting and smoothing which has been discussed in the foregoing sections can be extended and modified in various ways. For example, instead of asking for a smoothed or predicted value of $s(t)$, one can ask for a best predicted value of ds/dt or $\int^t s(t) \, dt$ or another linear functional of the signal.[§] The techniques for solving these problems are similar to

---

[†] Phillips (I).
[‡] Phillips (I) and Laning and Battin (I, Art. 5.5).
[§] Wiener (III, Chap. V).

those for the basic smoothing and prediction problem. Smoothing and predicting can be done when there are multiple signals and noises with known autocorrelation and cross-correlations.† The stationarity restriction on the input signal and noise processes may be removed;‡ in this case an integral equation similar to Eq. (11-9) but with the correlations functions of two variables, is obtained. A polynomial with unknown coefficients may be added to the signal.§ The observation interval may be taken to be finite instead of infinite;§ we shall examine this problem in some detail in the next article. For an extensive discussion of the design of linear systems, optimum according to a least-mean-square error criterion, with many examples, the reader is referred to Chaps. 5 through 8 of Laning and Battin (I).

11-6. Smoothing and Predicting with a Finite Observation Time

Let us now consider the optimization problem in which a smoothing and predicting filter is to act on a sample of signal and noise of only finite duration. Here, instead of Eq. (11-2), we have for the output of the filter

$$\hat{s}(t + \eta) = \int_{t-T}^{t} h(t - \tau)y(\tau)\, d\tau = \int_{0}^{T} h(u)y(t - u)\, du \qquad (11\text{-}60)$$

and it is desired to find the $h(t)$ which will minimize

$$E\{[s(t + \eta) - \hat{s}(t + \eta)]^2\}$$

We can now either go through a formal minimization procedure, as at the beginning of Art. 11-2, or take a short cut and argue, as in Eq. (11-24), that in order for the estimate $\hat{s}(t + \eta)$ to be best in a least-mean-square sense, the error must be uncorrelated with $y(\tau)$, $t - T \leq \tau \leq t$; for otherwise a further linear operation on $y(\tau)$ could reduce the error. Thus

$$E\{[\hat{s}(t + \eta) - s(t + \eta)]y(\tau)\} = 0 \qquad t - T \leq \tau \leq t \qquad (11\text{-}61)$$

Substituting from Eq. (11-60), we get

$$\int_{0}^{T} h(u)E[y(t - u)y(\tau)]\, du - E[s(t + \eta)y(\tau)] = 0 \qquad t - T \leq \tau \leq t$$

or
$$\int_{0}^{T} h(u)R_y(v - u)\, du = R_{sy}(v + \eta) \qquad 0 \leq v \leq T \qquad (11\text{-}62)$$

Thus $h(t)$ must satisfy Eq. (11-62) to be the impulse response of the optimum filter, and Eq. (11-62) can readily be shown to be sufficient as well.

---

† Wiener (III, Chap. IV).
‡ Booton (I).
§ Zadeh and Ragazzini (I) and Davis (I).

Equation (11-62) may or may not have a solution. As in the discussion of Eq. (11-9), we shall consider only the case for which the random process $y(t)$ has rational spectral density. Then, if we let $p = j2\pi f$, we can write

$$S_y(f) = \frac{N(p^2)}{D(p^2)} \tag{11-63}$$

where N and D are polynomials of degree n and d, respectively. For convenience, and because later we shall want to consider integral equations like Eq. (11-62) where the right-hand side is not a cross-correlation function, we shall let $R_{sy}(v + \eta)$ be written $z(v)$. Then, adopting this notation and using Eq. (11-63), Eq. (11-62) becomes

$$\int_0^T h(u) \, du \int_{-\infty}^{\infty} e^{p(v-u)} \frac{N(p^2)}{D(p^2)} \, df = z(v) \qquad 0 \le v \le T \tag{11-64}$$

If we operate on both sides of this equation with the differential operator $D(d^2/dv^2)$, a polynomial $D(p^2)$ is built up in the integrand of the left-hand side which cancels the $D(p^2)$ in the denominator and we get

$$\int_0^T h(u) \, du \int_{-\infty}^{\infty} e^{p(v-u)} N(p^2) \, df = D\left(\frac{d^2}{dv^2}\right) z(v) \qquad 0 < v < T \tag{11-65}$$

The polynomial $N(p^2)$ can also be built up by differentiation of the integral, so Eq. (11-65) can be written

$$N\left(\frac{d^2}{dv^2}\right) \left\{ \int_0^T h(u) \, du \int_{-\infty}^{\infty} e^{p(v-u)} \, df \right\} = D\left(\frac{d^2}{dv^2}\right) z(v) \qquad 0 < v < T$$

or
$$N\left(\frac{d^2}{dv^2}\right) h(v) = D\left(\frac{d^2}{dv^2}\right) z(v) \qquad 0 < v < T \tag{11-66}$$

where use has been made of the relation

$$\int_{-\infty}^{\infty} e^{j2\pi f(v-u)} \, df = \delta(v - u)$$

Thus any solution of Eq. (11-64) must satisfy the linear differential equation (11-66). A general solution of Eq. (11-66) has $2n$ undetermined constants. To see if for any choice of these constants the solution of Eq. (11-66) is a solution of Eq. (11-64) and to evaluate the constants, the general solution of Eq. (11-66) may be substituted in Eq. (11-64). It can be shown† that there is some choice of constants for which the solution of Eq. (11-66) for $h(v)$ is a solution of Eq. (11-64) *if and only if the known function $z(v)$ and its derivatives satisfy certain homogeneous*

† See Appendix 2.

boundary conditions at $v = 0$ *and* $v = T$.   By adding to the general solution of Eq. (11-66) the sum

$$\sum_{i=0}^{2d-2n-2} [a_i \delta^{(i)}(v) + b_i \delta^{(i)}(v - T)] \tag{11-67}$$

where the a_i's and b_i's are undetermined and $\delta^{(i)}(x)$ is the ith derivative of the impulse function, a solution (in a purely formal sense) of Eq. (11-64) can always be obtained.†   If derivatives of impulse functions are required in the solution for the weighting function $h(t)$, the significance is, of course, that the optimum filter must perform some differentiation.

Example 11-6.1. Let us consider again Example 11-3, but with the stipulation that the predicting filter is to act on the past of the input only as far back as $t - T$.   It turned out in Example 11-3.1 that although the predicting filter was to be allowed to act on the infinite past, the best filter in fact used none of the past history of the signal at all.   Thus we should get the same optimum filter with the changed condition of this example.   We shall see that this is so.
  We have

$$S(f) = \frac{1}{1 + f^2}$$

$$R_y(t) = R_s(t) = \pi e^{-j2\pi|t|}$$

$$D(p^2) = 1 + f^2 = \frac{1}{4\pi^2}(4\pi^2 - p^2) \qquad p = j2\pi f$$

Then, by Eqs. (11-66) and (11-67),

$$h(v) = \frac{1}{4\pi}\left(4\pi^2 - \frac{d^2}{dt^2}\right) e^{-2\pi(v+\eta)} + a\delta(v) + b\delta(v - T)$$
$$= a\delta(v) + b\delta(v - T)$$

The undetermined coefficients a and b are determined by substituting this expression for $h(v)$ back in the integral equation

$$\int_0^T h(u)e^{-2\pi|v-u|}\, du = e^{-2\pi(v+\eta)} \qquad 0 \le v \le T$$

This gives

$$ae^{-2\pi v} + be^{-2\pi(T-v)} = e^{-2\pi(v+\eta)}$$

or $a = e^{-2\pi\eta}$, $b = 0$.   The weighting function of the optimum predicting filter is then

$$h(u) = \delta(u)e^{-2\pi\eta}$$

which agrees with the result in Example 11-3.1.

Connection with Characteristic-value Problems. The integral equation (11-62) is closely connected with the integral equation

$$\int_0^T \phi(u)R_y(v - u)\, du = \sigma^2\phi(v) \qquad 0 \le v \le T \tag{11-68}$$

† See Appendix 2.

In fact, under certain conditions,† $h(u)$ can be expressed as an infinite series involving the characteristic functions of Eq. (11-68). We now show this formally. Again, let the function on the right-hand side of Eq. (11-62) be denoted $z(v)$. Then Eq. (11-62) is

$$\int_0^T h(u) R_y(v - u) \, du = z(v) \qquad 0 \le v \le T \qquad (11\text{-}69)$$

Let $\{\phi_k(v)\}$ be a set of orthonormal characteristic functions of Eq. (11-68), and let $z(v)$ be expanded in terms of the ϕ_k's,‡

$$z(v) = \sum_{k=0}^{\infty} z_k \phi_k(v) \qquad 0 \le v \le T \qquad (11\text{-}70)$$

where

$$z_k = \int_0^T z(v) \phi_k^*(v) \, dv$$

Then, since by Mercer's theorem

$$R_y(v - u) = \sum_{k=0}^{\infty} \sigma_k^2 \phi_k(v) \phi_k^*(u)$$

the left side of Eq. (11-69) is

$$\sum_{k=0}^{\infty} \lambda_k^2 \phi_k(v) \int_0^T h(u) \phi_k^*(u) du = \sum_{k=0}^{\infty} \sigma_k^2 \phi_k(v) h_k \qquad (11\text{-}71)$$

where h_k is the kth "Fourier coefficient" of $h(u)$ and

$$h(u) = \sum_{k=0}^{\infty} h_k \phi_k(u) \qquad (11\text{-}72)$$

Comparing the series from Eqs. (11-70) and (11-71), we see that the Eq. (11-69) is satisfied if

$$h_k = \frac{z_k}{\sigma_k^2}$$

that is, if

$$h(u) = \sum_{k=0}^{\infty} \frac{z_k}{\sigma_k^2} \phi_k(u) \qquad (11\text{-}73)$$

Applying this result to the filtering equation (11-62), we see that the solution can be written

$$h(u) = \sum_{k=0}^{\infty} \frac{\phi_k(u)}{\sigma_k^2} \int_0^T R_{sy}(v + \eta) \phi_k^*(v) \, dv \qquad (11\text{-}74)$$

† See Appendix 2.
‡ This can always be done for any $z(v)$ of integrable square if $R_y(t)$ is a correlation function of "filtered white noise." See Appendix 2.

Equation (11-74) could have been got by attacking the optimization problem differently. One can expand $y(t)$ in its orthogonal series, Eq. (6-31), in terms of the $\phi_k(t)$ directly, and $h(t)$ in the series given by Eq. (11-72), and calculate the mean-square error in the form of an infinite series and then choose the coefficients h_k so as to minimize the error. When this approach is used, there is no point in requiring stationarity of $y(t)$, for the orthogonal expansion of $y(t)$ is valid even when $y(t)$ is not stationary. Davis† has carried through a solution of a generalized version of the problem by this method. In his treatment he has allowed the signal to include a polynomial of known degree with unknown coefficients.

11-7. Maximizing Signal-to-noise Ratio: The Matched Filter‡

In the preceding sections of this chapter we have been concerned with linear filters intended to recover the form of a signal masked by noise. In some circumstances, for example, in the simple detection of a radar signal, the form of the time function which is the original signal is not important; merely the presence or absence of a signal is. In such a situation, the idea of designing a filter to have maximum output signal-to-noise ratio has considerable appeal, even though the output signal may be a distortion of the input signal. It should be remarked that there is no ambiguity in the notions of output signal and noise from a linear filter. If L denotes the linear operation performed by the filter and the input is

$$y(t) = s(t) + n(t) \tag{11-75}$$

then the output is

$$\begin{aligned} y_o(t) &= L[s + n](t) \\ &= L[s](t) + L[n](t) \end{aligned} \tag{11-76}$$

where it is perfectly reasonable to define $L[s](t)$ to be the output signal $s_o(t)$ and $L[n](t)$ to be the output noise $n_o(t)$.

One case of particular interest, and the one we shall discuss here, is that in which the signal $s(t)$ is a known function of time. It is desired to specify a linear filter to act on the input

$$y(t) = s(t) + n(t)$$

where $n(t)$, the noise, is a sample function from a wide-sense stationary random process, so that the output signal-to-noise ratio

$$\left(\frac{S}{N}\right)_o = \frac{s_o{}^2(t)}{E[n_o{}^2(t)]} \tag{11-77}$$

† Davis (I).
‡ See Zadeh and Ragazzini (II).

is a maximum at some chosen time $t = t_1$. We shall suppose that the filter acts on the input for a finite length of time T. Then

$$s_o(t_1) = \int_0^T h(\tau)s(t_1 - \tau)\, d\tau \qquad (11\text{-}78a)$$

$$n_o(t_1) = \int_0^T h(\tau)n(t_1 - \tau)\, d\tau \qquad (11\text{-}78b)$$

and $$E[n_o{}^2(t_1)] = \int_0^T \int_0^T h(\mu)h(\tau)R_n(\mu - \tau)\, d\mu\, d\tau \qquad (11\text{-}79)$$

Suppose that the maximum output signal-to-noise ratio is $1/\lambda$. Then for any linear filter we have

$$E[n_o{}^2(t_1)] - \lambda s_o{}^2(t_1) \geq 0 \qquad (11\text{-}80)$$

where the equality sign holds only for the output of an optimum filter. Since, if the impulse response $h(t)$ of the filter is multiplied by a constant, the signal-to-noise ratio is not changed, we may suppose the filter gain to be normalized so that $s_o(t_1) = 1$. Let us now derive a condition for the weighting function $h(t)$ of an optimum filter. Let $g(t)$ be any real function with the property that

$$\int_0^T g(\tau)s(t_1 - \tau)\, d\tau = 0 \qquad (11\text{-}81)$$

Then, for any number ϵ,

$$\int_0^T [h(t) + \epsilon g(t)]s(t_1 - \tau)\, d\tau = 1 \qquad (11\text{-}82)$$

For convenience, let us introduce the notation $\sigma^2(h)$ for $E[n_o{}^2(t_1)]$ when the filter has weighting function $h(t)$. Then, from Eq. (11-75) and the remark following, from the normalization

$$\sigma^2(h) - \lambda = 0 \qquad (11\text{-}83)$$

and from Eqs. (11-80) and (11-82),

$$\sigma^2(h + \epsilon g) - \lambda \geq 0 \qquad (11\text{-}84)$$

Subtracting Eq. (11-83) from (11-84) gives

$$\sigma^2(h + \epsilon g) - \sigma^2(h) \geq 0 \qquad (11\text{-}85)$$

for any ϵ and any $g(t)$ satisfying Eq. (11-81). Expanding and cancelling terms gives

$$\epsilon^2 \int_0^T \int_0^T g(\mu)g(\tau)R_n(\tau - \mu)\, d\mu\, d\tau + 2\epsilon \int_0^T \int_0^T h(\mu)g(\tau)R_n(\tau - \mu)\, d\mu\, d\tau \geq 0$$

This inequality is satisfied for all values of ϵ only if the second integral vanishes:

$$\int_0^T g(\tau)\, d\tau \int_0^T h(\mu)R_n(\tau - \mu)\, d\mu = 0 \qquad (11\text{-}86)$$

and Eq. (11-86) is satisfied for all $g(\tau)$ satisfying Eq. (11-81) only if

$$\int_0^T h(\mu) R_n(\tau - \mu)\, d\mu = \alpha s(t_1 - \tau) \qquad 0 \le \tau \le T \qquad (11\text{-}87)$$

where α is any constant.   That this is so can be shown by supposing

$$\int_0^T h(\mu) R_n(\tau - \mu)\, d\mu = a(\tau) \qquad 0 \le \tau \le T$$

where $a(\tau)$ is not a multiple of $s(t_1 - \tau)$, and then letting

$$g(\tau) = a(\tau) - \int_0^T a(u)s(t_1 - u)\, du\, \frac{s(t_1 - \tau)}{\int_0^T s^2(t_1 - u)\, du}$$

It is easily verified that $g(\tau)$ thus defined satisfies Eq. (11-81) and that Eq. (11-86) cannot be satisfied.

Thus Eq. (11-87) is a condition which must be satisfied in order that $h(t)$ be the impulse response of a filter which maximizes the output signal-to-noise ratio.   The value of α does not matter as far as the signal-to-noise ratio is concerned but affects only the normalization.   By substituting $s(t_1 - t)$ from Eq. (11-87) back into Eq. (11-78a), one shows easily that

$$\alpha = \frac{E[n_o^2(t_1)]}{s_o(t_1)} \qquad (11\text{-}88)$$

The reader may verify directly that Eq. (11-87) is a sufficient as well as necessary condition that $h(t)$ be optimum.   The verification is similar to that at the beginning of Art. 11-2.

An interesting limiting case of Eq. (11-87) occurs when the noise is white noise, so that $R_n(t) = N\delta(t)$ where N is the noise power.   Then, taking $\alpha = 1$, Eq. (11-87) becomes

$$N \int_0^T h(\mu)\delta(\tau - \mu)\, d\mu = s(t_1 - \tau) \qquad 0 \le \tau \le T$$

whence

$$h(\tau) = \frac{1}{N} s(t_1 - \tau) \qquad 0 \le \tau \le T$$

Thus, in this special case, the optimum weighting function has the form of the signal run backward starting from the fixed time t_1.   A filter with this characteristic is called a *matched filter*.[†]

Example 11-7.1. Let the signal be a series of rectangular pulses as shown in Fig. 11-3. Let the noise have spectral density

$$S_n(f) = \frac{1}{a^2 + (j2\pi f)^2}$$

Then
$$R_n(t) = \frac{1}{2a} e^{-a|t|} \qquad a > 0$$

[†] Cf. Van Vleck and Middleton (I) and Zadeh and Ragazzini (II).

and the integral equation for the $h(t)$ of the maximum signal-to-noise ratio filter is, for $\alpha = 1$,

$$\int_0^T h(\mu) \frac{e^{-a|\tau-\mu|}}{2a} \, d\mu = s(t_1 - \tau) \qquad 0 \le \tau \le T$$

This equation has the solution (without impulse functions at $\tau = 0$ and $\tau = T$)

$$h(t) = \left(a^2 - \frac{d^2}{dt^2}\right) s(t_1 - t) \tag{11-89}$$

if the conditions

$$
\begin{aligned}
as(t_1) + s'(t_1) &= 0 \\
as(t_1 - T) - s'(t_1 - T) &= 0
\end{aligned} \tag{11-90}
$$

are satisfied. With the time origin chosen as shown in Fig. 11-3 and $t_1 = T$, these

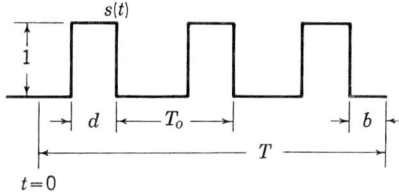

$$t=0$$

FIG. 11-3. Rectangular pulse series.

conditions are satisfied. Then the optimum filter weighting function is

$$
\begin{aligned}
h(t) = {}& a^2 s(T - t) - \delta^{(2)}(t - b) + \delta^{(2)}(t - b - d) \\
& - \cdots + \delta^{(2)}(t - (k - 1)T_0 - b - d)
\end{aligned} \tag{11-91}
$$

Example 11-7.2. Let the noise be the same as in the previous example and the signal be

$$s(t) = \sin^2 \omega_o t$$

Let

$$T = \frac{2\pi n}{\omega_o}$$

Then with $t_1 = T$, the conditions required by Eq. (11-90) are met, and the solution is again given by Eq. (11-89), which becomes in this case

$$
\begin{aligned}
h(t) &= \left(a^2 - \frac{d^2}{dt^2}\right) \sin^2 \omega_o(T - t) \\
&= \frac{a^2}{2} - \left(\frac{a^2}{2} + 2\omega_o^2\right) \cos 2\omega_o(T - t) \qquad 0 \le t \le T
\end{aligned}
$$

11-8. Problems

1. Let $s(t)$ be a signal which is a sample function from a stationary random process with spectral density

$$S_s(f) = \frac{1}{1 + f^4}$$

Show that the least-mean-square error-predicting filter which operates on the infinite past has system function

$$H(j2\pi f) = \exp\left[\frac{-2\pi\eta}{\sqrt{2}}\right] \left\{ \cos \frac{2\pi\eta}{\sqrt{2}} + \sin \frac{2\pi\eta}{\sqrt{2}} + j\sqrt{2} f \sin \frac{2\pi\eta}{\sqrt{2}} \right\}$$

where η is the lead time.

2.† Suppose that a random voltage $e(t)$ is developed across a parallel RLC circuit driven by a white-noise current source $i(t)$, as shown in Fig. 11-4.

    a. Letting $\omega_o{}^2 = 1/LC$ and $Q = R/\omega_o L$, find the spectral density $S_e(f)$ in terms of ω_o, L, and Q, assuming that $S_i(f) = 1$.

    b. Using the normalization $\alpha = j\omega/\omega_o$ and supposing the circuit sharply enough tuned so that $1/Q^2$ is negligible with respect to one, show that approximately

$$G(j\omega) = G(\alpha\omega_o) = \frac{\alpha}{(\alpha + 1/2Q - j)(\alpha + 1/2Q + j)}$$

    c. With the same approximation as in **b**, show that the optimum predicting filter for $e(t)$ (using the infinite past) has system function

$$H(j\omega) = \exp\left(-\frac{\omega_o\eta}{2Q}\right) \sin \omega_o\eta \left[\cot \omega_o\eta - \frac{1}{2Q} - \frac{\omega_o}{j\omega} \right]$$

where η is the prediction lead time.

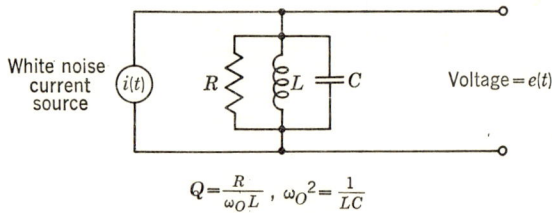

$$Q = \frac{R}{\omega_o L}, \quad \omega_0{}^2 = \frac{1}{LC}$$

FIG. 11-4. Parallel RLC circuit.

3. Show for the case of uncorrelated signal and noise, where $S_y(f) = S_s(f) + S_n(f)$, $S_s(f)$ and $S_n(f)$ any rational spectral densities, that $S_s(f)/G^*(j2\pi f)$ can have no poles on the real f axis.

4. Suppose that signal and noise are independent and that

$$S_s(f) = \frac{1}{1 + (f/a)^2}$$
$$S_n(f) = N$$

(the noise is white noise). Find the optimum linear smoothing filter which operates on the infinite past of the input. Compare the result with the limiting case of Example 11-4.1.

5. Show that for independent signal and noise, the system function of the optimum linear smoothing filter (infinite past) may be written

$$H(j2\pi f) = 1 - \frac{1}{G(j2\pi f)}\left[\frac{S_n(f)}{G^*(j2\pi f)} \right]_+$$

6. Derive Eq. (11-56) of the text from Eq. (11-55).

7. Using Eq. (11-55), calculate the mean-square error of the output of the smoothing filter of Prob. 4 (the filter specified by Eq. (11-52)).

8. Calculate the mean-square error of the output of the predicting filter of Example 11-3.1.

    a. By using Eq. (11-55).

    b. By using Eq. (11-57).

† Lee and Stutt (I).

9. Derive the expression, Eq. (11-58), for the spectral density of the mean-square error of the output of an arbitrary filter with signal-plus-noise input.

10. Let a signal $s(t)$ be from a stationary random process with correlation function $R_s(t) = e^{-|t|} + e^{-2|t|}$. Find the impulse response of the linear predicting filter for this signal, which is optimum among those which operate on the signal only for a time T.

11. Derive Eq. (11-69) of the text directly by expanding $y(t)$ and $h(t)$ in the series

$$\left.\begin{aligned} y(t) &= \sum_{k=0}^{\infty} \sigma_k y_k \phi_k(t) \\ h(t) &= \sum_{k=0}^{\infty} h_k \phi_k(t) \end{aligned}\right\} \quad 0 \le t \le T$$

where the $\phi_k(t)$ are given by Eq. (11-68) and the h_k are unknown coefficients to be determined. Find the mean-square error as an infinite series which involves the h_k, and then determine the h_k so as to minimize it.

12. With signal and noise as given in Prob. 4, find the impulse response function of the optimum linear filter which operates on the input only for a time T. Check the limiting case as $T \to \infty$ with the result of Prob. 4.

13. Find the impulse response of the linear filter which maximizes signal-to-noise ratio for the signal

$$s(t) = \tfrac{1}{4} - \tfrac{1}{6}[e^{2\pi(t-T)} + e^{-2\pi t}] + \tfrac{1}{24}[e^{4\pi(t-T)} + e^{-4\pi t}] \qquad 0 \le t \le T$$

when the spectral density of the noise is

$$S_n(f) = \frac{1}{(1 + f^2)(4 + f^2)}$$

CHAPTER 12

NONLINEAR DEVICES: THE DIRECT METHOD

The problem which will concern us in this chapter and the next is that of determining the statistical properties of the output of a nonlinear device,† e.g., a detector or a limiting amplifier. In these chapters, we shall restrict our discussion to those nonlinear devices for which the output $y(t)$ at a given instant of time can be expressed as a function of the input $x(t)$ at the same instant of time. That is, we shall assume that we can write

$$y(t) = g[x(t)] \qquad (12\text{-}1)$$

where $g(x)$ is a single-valued function of x. We thus rule out of consideration those nonlinear devices which contain energy-storage elements, since such elements generally require that the present value of the output be a function not only of the present value of the input but also of the past history of the input.

12-1. General Remarks

The problem at hand can be stated as follows: knowing the (single-valued) *transfer characteristic* $y = g(x)$ of a nonlinear device and the statistical properties of its input, what are the statistical properties of its output? Basically, this is simply a problem of a transformation of variables, the elements of which were discussed in earlier chapters. For example, it follows from the results of Art. 3-6 that if $A(Y)$ is the set of points in the sample space of the input random variable x_t which corresponds to the set of points $(-\infty < y_t \le Y)$ in the sample space of the output random variable y, then

$$P(y_t \le Y) = P[x_t \,\varepsilon\, A(Y)] \qquad (12\text{-}2)$$

and the probability density function of the output random variable is given by

$$p(Y) = \frac{dP[x_t \,\varepsilon\, A(Y)]}{dY} \qquad (12\text{-}3)$$

wherever the derivative exists. It also follows from the results of Art. 3-6

† Cf. Burgess (I), and Rice (I, Arts. 4.1, 4.2, 4.5, and 4.7).

that if the probability density function $p_1(x_t)$ of the input random variable exists and is continuous almost everywhere, and if the transfer characteristic provides a one-to-one mapping from x to y, then the probability density function $p_2(y)$ of the output random variable is given by

$$p_2(y_t) = p_1(x_t) \left| \frac{dx}{dy} \right| \tag{12-4}$$

Although this equation may not always apply (e.g., when the transfer characteristic is constant over an interval in x as for a half-wave detector), Eq. (12-3) always does.

Further, it follows from Art. 4-1 that averages with respect to the output can always be obtained by averaging with respect to the input. Thus

$$E[f(y_t)] = E\{f[g(x_t)]\} = \int_{-\infty}^{+\infty} f[g(x_t)]p(x_t)\, dx_t \tag{12-5}$$

The nth moment of the output is therefore given by

$$E(y_t{}^n) = \int_{-\infty}^{+\infty} g^n(x_t)p(x_t)\, dx_t \tag{12-6}$$

Similarly, the autocorrelation function of the output is

$$R_y(t_1,t_2) = \int_{-\infty}^{+\infty} \int_{-\infty}^{+\infty} g(x_1)g(x_2)p(x_1,x_2)\, dx_1\, dx_2 \tag{12-7}$$

where $x_1 = x_{t_1}$ and $x_2 = x_{t_2}$.

The problem of determining the statistical properties of the output of a nonlinear device can hence be solved in principle by applying the various results which we obtained earlier on the transformation of variables. We shall apply those results in this chapter to a study of two nonlinear devices of some practical importance:† the full-wave square-law detector and the half-wave linear detector. However, when we try to use the method adopted in this chapter to solve more complicated problems, considerable analytical difficulties often arise. In these cases, a less direct method involving the use of the Fourier transform of the transfer characteristic may be used. The transform method of analysis will be discussed in Chap. 13.

12-2. The Square-law Detector

By *full-wave square-law detector* we mean a full-wave square-law device with the transfer characteristic

$$y = ax^2 \tag{12-8}$$

---

† It should be noted that we have already studied the envelope detector, in essence, in Arts. 8-5, 8-6, and 9-5.

where a is a scaling constant, followed by a low-pass or averaging filter. Such a detector is shown schematically in Fig. 12-1; the full-wave square-law transfer characteristic is shown in Fig. 12-2. The square-law detector is analytically the simplest of the various nonlinear devices we shall

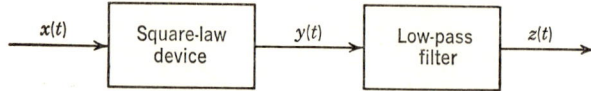

$$x(t) \longrightarrow \boxed{\begin{array}{c}\text{Square-law}\\\text{device}\end{array}} \xrightarrow{y(t)} \boxed{\begin{array}{c}\text{Low-pass}\\\text{filter}\end{array}} \xrightarrow{z(t)}$$

FIG. 12-1. The square-law detector.

study. It is of interest to us, however, not only because of the simplicity of its analysis but also because of its practical importance. The determination of the statistical properties of the detector output can be made most easily by determining the statistical properties first of the output of the square-law device and then of the low-pass-filter output. In this article, we shall consider first a general input and then an input with gaussian statistics; the case of a sine wave plus narrow-band gaussian noise input will be studied in the next article.

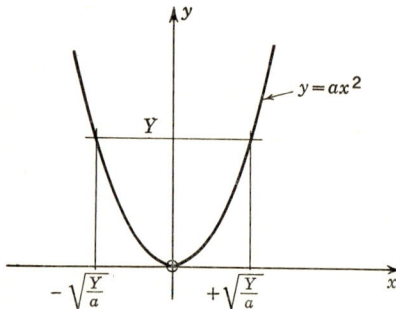

FIG. 12-2. The full-wave square-law transfer characteristic.

The probability density function of the output of a full-wave square-law *device* was derived in Art. 3-6 and is, from Eqs. (3-45) and (3-46),

$$p_2(y_t) = \begin{cases} \dfrac{p_1(x_t = +\sqrt{y_t/a}) + p_1(x_t = -\sqrt{y_t/a})}{2\sqrt{ay_t}} & \text{for } y_t \geq 0 \\ 0 & \text{otherwise} \end{cases} \quad (12\text{-}9)$$

When the input probability density function is an even function, this becomes

$$p_2(y_t) = \begin{cases} \dfrac{p_1(x_t = +\sqrt{y_t/a})}{\sqrt{ay_t}} & \text{for } y_t \geq 0 \\ 0 & \text{otherwise} \end{cases} \quad (12\text{-}10)$$

The determination of the probability density function of the output of the low-pass filter was discussed in Art. 9-5. However, for an arbitrary filter characteristic, the density function obtained in general form there is so unwieldy as to make calculation of averages difficult. The calculation of the autocorrelation function, which requires second-order distributions, is even more difficult. It is therefore useful to specialize to the case of a narrow-band gaussian noise input and an idealized type of

low-pass filter which simplifies the calculations enough to permit more extensive results.   This special case will be considered in the next section.
    The nth moment of the output of the square-law device is

$$E(y_t{}^n) = a^n \int_{-\infty}^{+\infty} x_t{}^{2n} p_1(x_t) \, dx_t = a^n E(x_t{}^{2n}) \qquad (12\text{-}11)$$

The autocorrelation function of the output of the square-law device is

$$R_y(t_1,t_2) = a^2 E(x_1{}^2 x_2{}^2) \qquad (12\text{-}12)$$

which becomes a function of $t_1 - t_2$ when the input is stationary.   It is not feasible to go much beyond these simple results without specifying the statistical properties of the detector input.
    Gaussian Input. Suppose now that the detector input $x(t)$ is a sample function of a real gaussian random process with zero mean.   In this case,

$$p_1(x_t) = \frac{1}{\sqrt{2\pi}\,\sigma_{x_t}} \exp\left(-\frac{x_t{}^2}{2\sigma_{x_t}{}^2}\right) \qquad (12\text{-}13)$$

It then follows from Eq. (12-10) that the probability density function of the output of the square-law *device* is

$$p_2(y_t) = \begin{cases} \dfrac{1}{\sqrt{2\pi a y_t}\,\sigma_{x_t}} \exp\left(-\dfrac{y_t}{2a\sigma_{x_t}{}^2}\right) & \text{for } y_t \geq 0 \\ 0 & \text{otherwise} \end{cases} \qquad (12\text{-}14)$$

which is a chi-squared density function.
    If the detector input is a narrow-band gaussian random process, we can write, from Eq. (8-73),

$$x(t) = V(t) \cos\left[\omega_c t + \phi(t)\right] \qquad (12\text{-}15)$$

where $f_c = \omega_c/2\pi$ is the center frequency of the input spectral density, $V(t) \geq 0$ is the envelope of the input, and $0 \leq \phi(t) \leq 2\pi$ is the input phase.   The output of the square-law device then is

$$y(t) = \frac{aV^2(t)}{2} + \frac{aV^2(t)}{2} \cos\left[2\omega_c t + 2\phi(t)\right] \qquad (12\text{-}16)$$

The first term of this expression has a spectral density centered on zero frequency, whereas the second term has a spectral density centered on $2f_c$.   If the bandwidth of the input is narrow compared to its center frequency, these two spectral densities will not overlap.†   On passing $y(t)$ through a *low-pass zonal filter* (i.e., a nonrealizable filter which will pass without distortion the low-frequency part of its input and filter out completely the high-frequency part), we obtain for the filter output

$$z(t) = \frac{aV^2(t)}{2} \qquad (12\text{-}17)$$

† Cf. Fig. 12-4.

Since $V(t)$ is the envelope of a narrow-band gaussian random process, it has a Rayleigh probability density function:

$$p(V_t) = \begin{cases} \dfrac{V_t}{\sigma_{x_t}{}^2} \exp\left(- \dfrac{V_t{}^2}{2\sigma_{x_t}{}^2} \right) & \text{for } V_t \geq 0 \\ 0 & \text{otherwise} \end{cases} \qquad (12\text{-}18)$$

from Eq. (8-85). The probability density function of the filter output is therefore

$$p_3(z_t) = \begin{cases} \dfrac{1}{a\sigma_{x_t}{}^2} \exp\left(- \dfrac{z_t}{a\sigma_{x_t}{}^2} \right) & \text{for } z_t \geq 0 \\ 0 & \text{otherwise} \end{cases} \qquad (12\text{-}19)$$

which is an exponential density function. The probability density function of the normalized input $\xi = x/\sigma_x$, the normalized output of the square-law device $\eta = y/a\sigma_x{}^2$, and the normalized filter output $\zeta = z/a\sigma_x{}^2$ are shown in Fig. 12-3.

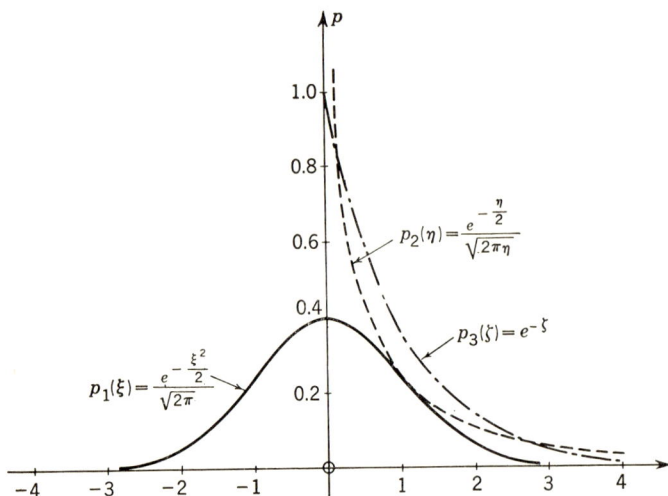

Fig. 12-3. Full-wave square-law detector probability density functions.

The nth moments of the output of the square-law device can be obtained by substituting into Eq. (12-11) the previously derived values (Eq. (8-11)) for the even moments of a gaussian random variable. Thus we obtain

$$E(y_t{}^n) = a^n \cdot 1 \cdot 3 \cdot 5 \cdots (2n - 1)\sigma_{x_t}{}^{2n} \qquad (12\text{-}20)$$

In particular,

$$\begin{aligned} E(y_t) &= a\sigma_{x_t}{}^2 \\ E(y_t{}^2) &= 3a^2\sigma_{x_t}{}^4 = 3E^2(y_t) \\ \sigma^2(y_t) &= 2a^2\sigma_{x_t}{}^4 = 2E^2(y_t) \end{aligned} \qquad (12\text{-}21)$$

and hence

When the detector input is a narrow-band gaussian random process and the output filter is a low-pass zonal filter, the nth moment of the filter output is,[†] using Eq. (12-19),

$$E(z_t{}^n) = \int_0^\infty \frac{z_t{}^n}{a\sigma_{x_t}{}^2} \exp\left(-\frac{z_t}{a\sigma_x{}^2}\right) dz_t \qquad (12\text{-}22)$$

$$= n!\,a^n\sigma_{x_t}{}^{2n}$$

Hence

$$E(z_t) = a\sigma_{x_t}{}^2 = E(y_t)$$

$$E(z_t{}^2) = 2a^2\sigma_{x_t}{}^4 \qquad (12\text{-}23)$$

and

$$\sigma^2(z_t) = a^2\sigma_{x_t}{}^4 = E^2(z_t)$$

Thus we see from Eqs. (12-21) and (12-23) that the means of the outputs of the square-law device and the low-pass filter are both equal to a times the variance of the input and that the variance of the output of the low-pass zonal filter is one-half the variance of the output of the square-law device.

The autocorrelation function of the output of the full-wave square-law device in response to a gaussian input is, from Eqs. (12-12) and (8-121),

$$R_y(t_1,t_2) = a^2 E^2(x_t{}^2) + 2a^2 E^2(x_1 x_2) \qquad (12\text{-}24a)$$

which becomes, on setting $\tau = t_1 - t_2$ and $\sigma_x = \sigma_{x_t}$, all t,

$$R_y(\tau) = a^2\sigma_x{}^4 + 2a^2 R_x{}^2(\tau) \qquad (12\text{-}24b)$$

when the input random process is stationary. The spectral density of the output of the square-law device, given by the Fourier transform of $R_y(\tau)$, is

$$S_y(f) = a^2\sigma_x{}^4\delta(f) + 2a^2 \int_{-\infty}^{+\infty} R_x{}^2(\tau)e^{-j2\pi f\tau} d\tau$$

since, from Eq. (A1-16), the Fourier transform of a constant is an impulse function. Now

$$\int_{-\infty}^{+\infty} R_x{}^2(\tau)e^{-j2\pi f\tau}\, d\tau = \int_{-\infty}^{+\infty} S_x(f')\, df' \int_{-\infty}^{+\infty} R_x(\tau)\exp[-j2\pi(f-f')\tau]\, d\tau$$

$$= \int_{-\infty}^{+\infty} S_x(f')S_x(f-f')\, df'$$

Hence

$$S_y(f) = a^2\sigma_x{}^4\delta(f) + 2a^2 \int_{-\infty}^{+\infty} S_x(f')S_x(f-f')\, df' \qquad (12\text{-}25)$$

Thus the output spectral density is composed of two parts: an impulse part

$$S_{\bar{y}}(f) = a^2\sigma_x{}^4\delta(f) \qquad (12\text{-}26a)$$

which corresponds to the output mean value, and a part

$$S_y(f)_r = 2a^2 \int_{-\infty}^{+\infty} S_x(f')S_x(f-f')\, df' \qquad (12\text{-}26b)$$

which corresponds to the random variations of the output.

† Cf. Dwight (I, Eq. 861.2).

A feeling for the above results can perhaps best be obtained by assuming some simple form for the input spectral density. Accordingly, let the input spectral density have a constant value A over a narrow band of width B centered on some frequency f_c, where $0 < B < f_c$:

$$S_x(f) = \begin{cases} A & \text{for } f_c - \dfrac{B}{2} < |f| < f_c + \dfrac{B}{2} \\ 0 & \text{otherwise} \end{cases} \qquad (12\text{-}27)$$

This spectral density is shown in Fig. 12-4a. In this case,

$$\sigma_x{}^2 = \int_{-\infty}^{+\infty} S_x(f)\, df = 2AB \qquad (12\text{-}28)$$

The impulsive part of the spectral density of the output of the square-law device is therefore

$$S_{\bar{y}}(f) = 4a^2 A^2 B^2 \delta(f) \qquad (12\text{-}29)$$

and is shown in Fig. 12-4b. The mean and the variance of the output of the square-law device are, from Eqs. (12-21),

$$E(y_t) = 2aAB \quad \text{and} \quad \sigma^2(y_t) = 8a^2 A^2 B^2 \qquad (12\text{-}30)$$

and the mean and variance of the output of the low-pass zonal filter are, from Eqs. (12-23),

$$E(z_t) = 2aAB \quad \text{and} \quad \sigma^2(z_t) = 4a^2 A^2 B^2 \qquad (12\text{-}31)$$

The spectral density of the fluctuating part of the output of the square-law device is formed, as shown by Eq. (12-26b), by convolving the input spectral density $S_x(f)$ with itself. Here we have

$$S_y(f)_r = \begin{cases} 4a^2 A^2(B - |f|) & \text{for } 0 < |f| \leq B \\ 2a^2 A^2(B - |\,|f| - 2f_c|) & \text{for } 2f_c - B < |f| < 2f_c + B \\ 0 & \text{otherwise} \end{cases} \qquad (12\text{-}32)$$

which is also plotted in Fig. 12-4b. The spectral density of the output of the square-law device is thus nonzero only in the narrow bands centered about zero frequency and twice the center frequency of the input spectral density. The low-pass zonal filter passes that part about zero frequency and stops the part about $\pm 2f_c$. Hence

$$S_z(f) = S_{\bar{z}}(f) + S_z(f)_r \qquad (12\text{-}33)$$

where

$$S_{\bar{z}}(f) = 4a^2 A^2 B^2 \delta(f) \qquad (12\text{-}34a)$$

and

$$S_z(f)_r = \begin{cases} 4a^2 A^2(B - |f|) & \text{for } 0 < |f| < B \\ 0 & \text{otherwise} \end{cases} \qquad (12\text{-}34b)$$

These spectral densities are shown in Fig. 12-4c. From the above results, we see that the bandwidths of the outputs of the square-law device and

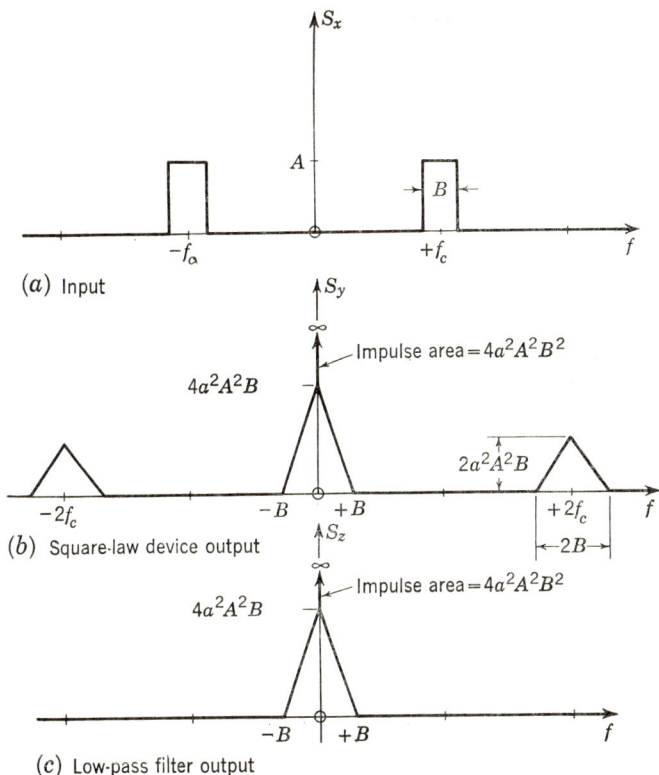

FIG. 12-4. Spectral densities for a full-wave square-law detector in response to a narrow-band gaussian input.

the low-pass filter are twice the bandwidth of the detector input. A moment's reflection should convince one that these results are consistent with those obtained in the elementary case of sine-wave inputs.

12-3. The Square-law Detector: Signal-plus-Noise Input

Let us now consider the case in which the input to the full-wave square-law detector is the sum of a signal $s(t)$ and a noise $n(t)$:

$$x(t) = s(t) + n(t) \tag{12-35}$$

where $s(t)$ and $n(t)$ are sample functions from statistically independent random processes with zero means.

Since the output of the square-law device is

$$y(t) = a[s(t) + n(t)]^2 = a[s^2(t) + 2s(t)n(t) + n^2(t)] \tag{12-36}$$

and the input processes are independent, the mean of the output is

$$E(y_t) = a[E(s_t{}^2) + E(n_t{}^2)] \tag{12-37a}$$

which, when the input processes are stationary, becomes

$$E(y) = a(\sigma_s{}^2 + \sigma_n{}^2) \tag{12-37b}$$

where $\sigma_s = \sigma(s_t)$ and $\sigma_n = \sigma(n_t)$ for all t. The mean-square value of the square-law device output is, in general,

$$E(y_t{}^2) = a^2[E(s_t{}^4) + 6E(s_t{}^2)E(n_t{}^2) + E(n_t{}^4)] \tag{12-38a}$$

which becomes

$$E(y^2) = a^2[E(s^4) + 6\sigma_s{}^2\sigma_n{}^2 + E(n^4)] \tag{12-38b}$$

when the inputs are stationary.

The autocorrelation function of the output of the square-law device is

$$R_y(t_1,t_2) = E(y_1 y_2) = a^2 E[(s_1 + n_1)^2 (s_2 + n_2)^2]$$

Hence

$$R_y(t_1,t_2) = a^2 \left[\begin{array}{l} E(s_1{}^2 s_2{}^2) + 4E(s_1 s_2)E(n_1 n_2) \\ + E(s_1{}^2)E(n_2{}^2) + E(s_2{}^2)E(n_1{}^2) + E(n_1{}^2 n_2{}^2) \end{array} \right] \tag{12-39a}$$

When the input random processes are stationary, we get, on setting $\tau = t_1 - t_2$,

$$R_y(\tau) = a^2[R_{s^2}(\tau) + 4R_s(\tau)R_n(\tau) + 2\sigma_s{}^2\sigma_n{}^2 + R_{n^2}(\tau)] \tag{12-39b}$$

where $R_s(\tau)$ and $R_n(\tau)$ are the signal and noise autocorrelation functions, respectively, and where

$$R_{s^2}(\tau) = E(s_1{}^2 s_2{}^2) \qquad \text{and} \qquad R_{n^2}(\tau) = E(n_1{}^2 n_2{}^2) \tag{12-40}$$

The autocorrelation function of the output of the square-law device therefore contains three types of terms:

$$R_y(\tau) = R_{sxs}(\tau) + R_{sxn}(\tau) + R_{nxn}(\tau) \tag{12-41}$$

in which the term

$$R_{sxs}(\tau) = a^2 R_{s^2}(\tau) \tag{12-42a}$$

is due to the interaction of the signal with itself, the term

$$R_{nxn}(\tau) = a^2 R_{n^2}(\tau) \tag{12-42b}$$

is due to the interaction of the noise with itself, and the term

$$R_{sxn}(\tau) = 4a^2 R_s(\tau)R_n(\tau) + 2a^2\sigma_s{}^2\sigma_n{}^2 \tag{12-42c}$$

is due to the interaction of the signal with the noise. Of these terms we may say that only the sxs term (which would be present if there were no noise) relates to the desired signal output; the sxn and nxn terms relate to the output noise.

The spectral density of the output of the square-law device may be obtained by taking the Fourier transform of $R_y(\tau)$. Thus we get

$$S_y(f) = S_{sxs}(f) + S_{sxn}(f) + S_{nxn}(f) \qquad (12\text{-}43)$$

where

$$S_{sxs}(f) = a^2 \int_{-\infty}^{+\infty} R_{s^2}(\tau) \exp(-j2\pi f\tau)\, d\tau \qquad (12\text{-}44a)$$

$$S_{nxn}(f) = a^2 \int_{-\infty}^{+\infty} R_{n^2}(\tau) \exp(-j2\pi f\tau)\, d\tau \qquad (12\text{-}44b)$$

and

$$S_{sxn}(f) = 4a^2 \int_{-\infty}^{+\infty} R_s(\tau) R_n(\tau) \exp(-j2\pi f\tau)\, d\tau + 2a^2 \sigma_s{}^2 \sigma_n{}^2 \delta(f) \qquad (12\text{-}44c)$$

$$= 4a^2 \int_{-\infty}^{+\infty} S_n(f') S_s(f - f')\, df' + 2a^2 \sigma_s{}^2 \sigma_n{}^2 \delta(f)$$

in which $S_s(f)$ and $S_n(f)$ are the spectral densities of the input signal and noise, respectively.   The occurrence of the *sxn* term shows that the output noise increases in the presence of an input signal.   Although this result has been obtained here for a full-wave square-law detector, we shall see later that a similar result holds for any nonlinear device.

Sine Wave Plus Gaussian Noise Input. In the preceding section we obtained various statistical properties of the output of a full-wave square-law detector in response to a signal-plus-noise input of a general nature. Suppose now that the input noise $n(t)$ is a sample function of a stationary real gaussian random process with zero mean and that the input signal is a sine wave of the form

$$s(t) = P \cos(\omega_c t + \theta) \qquad (12\text{-}4$$

where P is a constant and θ is a random variable uniformly distributed over the interval $0 \leq \theta \leq 2\pi$ and independent of the input noise process.

It follows from Eq. (12-24b), since the input noise is gaussian, that the *nxn* part of the autocorrelation function of the output of the full-wave square-law device is

$$R_{nxn}(\tau) = 2a^2 R_n{}^2(\tau) + a^2 \sigma_n{}^4 \qquad (12\text{-}46)$$

The corresponding spectral density then is, from Eq. (12-25),

$$S_{nxn}(f) = 2a^2 \int_{-\infty}^{+\infty} S_n(f') S_n(f - f')\, df' + a^2 \sigma_n{}^4 \delta(f) \qquad (12\text{-}47)$$

We now require certain properties of the input signal in order to determine the remaining parts of the autocorrelation function of the output of the square-law device.   First, the autocorrelation function of the input signal is

$$R_s(t_1,t_2) = P^2 E[\cos(\omega_c t_1 + \theta) \cos(\omega_c t_2 + \theta)]$$

$$= \frac{P^2}{2} \cos[\omega_c(t_1 - t_2)] + \frac{P^2}{2}\frac{1}{2\pi} \int_0^{2\pi} \cos[\omega_c(t_1 + t_2) + 2\theta]\, d\theta$$

$$= \frac{P^2}{2} \cos[\omega_c(t_1 - t_2)]$$

We can therefore write

$$R_s(\tau) = \frac{P^2}{2} \cos \omega_c \tau \tag{12-48}$$

where $\tau = t_1 - t_2$. The spectral density of the input signal then is, using Eq. (A1-18),

$$S_s(f) = \frac{P^2}{4}[\delta(f - f_c) + \delta(f + f_c)] \tag{12-49}$$

where $f_c = \omega_c/2\pi$. From Eqs. (12-42c) and (12-48), we have

$$R_{sxn}(\tau) = 2a^2P^2R_n(\tau) \cos \omega_c \tau + a^2P^2\sigma_n^2 \tag{12-50}$$

The corresponding spectral density is

$$S_{sxn}(f) = a^2P^2[S_n(f - f_c) + S_n(f + f_c)] + a^2P^2\sigma_n^2\delta(f) \tag{12-51}$$

which follows from Eqs. (12-44c) and (12-49).

Next, the autocorrelation function of the square of the input signal is,

$$E(s_1{}^2s_2{}^2) = P^4E[\cos^2(\omega_c t_1 + \theta) \cos^2(\omega_c t_2 + \theta)]$$
$$= \frac{P^4}{4} + \frac{P^4}{8} \cos 2\omega_c(t_1 - t_2)$$

The sxs part of the autocorrelation function of the output of the square-law device is, therefore,

$$R_{sxs}(\tau) = \frac{a^2P^4}{4} + \frac{a^2P^4}{8} \cos 2\omega_c \tau \tag{12-52}$$

where $\tau = t_1 - t_2$. The corresponding spectral density is

$$S_{sxs}(f) = \frac{a^2P^4}{4} \delta(f) + \frac{a^2P^4}{16}[\delta(f - 2f_c) + \delta(f + 2f_c)] \tag{12-53}$$

To summarize, when the input to a full-wave square-law device consists of a sine wave plus a gaussian noise with zero mean, the autocorrelation function of the device output is, from Eqs. (12-46), (12-50), and (12-52),

$$R_y(\tau) = a^2\left(\frac{P^2}{2} + \sigma_n^2\right)^2 + 2a^2R_n{}^2(\tau) + 2a^2P^2R_n(\tau) \cos \omega_c \tau + \frac{a^2P^4}{8} \cos 2\omega_c \tau \tag{12-54}$$

and the output spectral density is

$$S_y(f) = a^2\left(\frac{P^2}{2} + \sigma_n^2\right)^2 \delta(f) + 2a^2 \int_{-\infty}^{+\infty} S_n(f')S_n(f - f') \, df'$$
$$+ a^2P^2[S_n(f - f_c) + S_n(f + f_c)] + \frac{a^2P^4}{16}[\delta(f - 2f_c) + \delta(f + 2f_c)] \tag{12-55}$$

which follows from Eqs. (12-47), (12-51), and (12-53).

The first term in Eq. (12-54) is simply the mean of the output of the square-law device,

$$E(y) = a\left(\frac{P^2}{2} + \sigma_n{}^2\right) \tag{12-56}$$

The mean square of the output, obtained by evaluating the autocorrelation function at $\tau = 0$, is

$$E(y^2) = 3a^2\left(\frac{P^4}{8} + P^2\sigma_n{}^2 + \sigma_n{}^4\right) \tag{12-57}$$

The variance of the output of the square-law device therefore is

$$\sigma_y{}^2 = 2a^2\left(\frac{P^4}{16} + P^2\sigma_n{}^2 + \sigma_n{}^4\right) \tag{12-58}$$

Again we shall try to obtain a feeling for the analytic results by assuming a simple form for the input noise spectral density. As in Art. 12-2, let the input noise spectral density have a constant value A over a narrow band of width B centered on a frequency f_c, where $0 < B < f_c$. The total input spectral density then is

$$S_x(f) = \frac{P^2}{4}\left[\delta(f - f_c) + \delta(f + f_c)\right]$$

$$+ \begin{cases} A & \text{for } f_c - \dfrac{B}{2} < |f| < f_c + \dfrac{B}{2} \\ 0 & \text{otherwise} \end{cases} \tag{12-59}$$

and is shown in Fig. 12-5a.

Next, let us consider the various terms which make up the output spectral density. We observed above that the noise terms at the square-law-device output resulting from the interaction of the input noise with itself (the $n×n$ terms) are the same as the total-output-noise terms arising in the case of noise alone as an input. Thus, from Eqs. (12-29) and (12-32),

$$S_{nxn}(f) = 4a^2 A^2 B^2 \delta(f)$$

$$+ \begin{cases} 4a^2 A^2(B - |f|) & \text{for } 0 < |f| < B \\ 2a^2 A^2(B - ||f| - 2f_c|) & \text{for } 2f_c - B < |f| < 2f_c + B \\ 0 & \text{otherwise} \end{cases} \tag{12-60}$$

This spectral density is shown in Fig. 12-4b.

Equation (12-53) shows that the output-spectral-density terms resulting from the interaction of the input signal with itself (the $s×s$ terms) consist of three impulses, one located at zero frequency and a pair located at $\pm 2f_c$. These terms are shown in Fig. 12-5b.

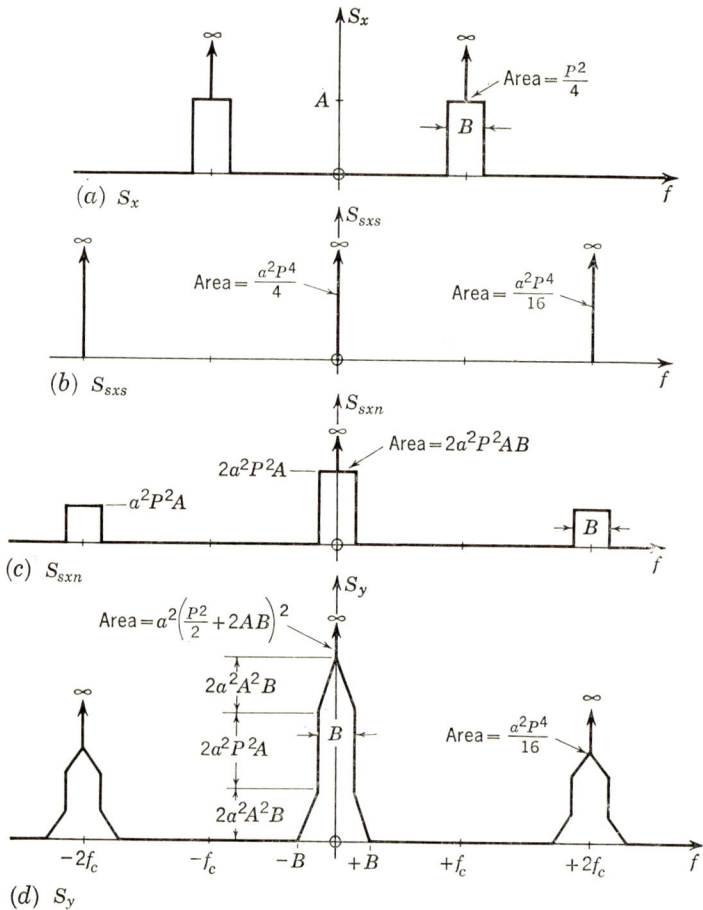

FIG. 12-5. Spectral densities for a full-wave square-law device in response to a sine wave plus gaussian input.

The output-spectral-density terms resulting from the interaction of the input signal with the input noise (the *sxn* terms) are shown by Eq. (12-51) to consist of an impulse at zero frequency plus a pair of terms resulting from the shifting of the input-noise spectral density by $\pm f_c$. Hence

$$S_{sxn}(f) = 2a^2P^2AB\delta(f)$$

$$+ \begin{cases} 2a^2P^2A & \text{for } 0 < |f| \le \dfrac{B}{2} \\[2mm] a^2P^2A & \text{for } 2f_c - \dfrac{B}{2} < |f| < 2f_c + \dfrac{B}{2} \\[2mm] 0 & \text{otherwise} \end{cases} \quad (12\text{-}61)$$

This spectral density is shown in Fig. 12-5c. The total-output spectral density is then given by the sum of the sxs, sxn, and nxn terms and is shown in Fig. 12-5d.

As in Art. 12-2, we can follow a full-wave square-law *device* by a low-pass zonal filter to obtain a full-wave square-law *detector*. The output spectral density of the square-law detector is given by those terms in Eqs. (12-53), (12-60), and (12-61) which are in the vicinity of zero frequency. Thus

$$S_z(f) = a^2 \left(\frac{P^2}{2} + 2AB \right)^2 \delta(f)$$

$$+ \begin{cases} 2a^2P^2A & \text{for } 0 < |f| \leq \dfrac{B}{2} \\ 0 & \text{otherwise} \end{cases} \qquad (12\text{-}62)$$

$$+ \begin{cases} 4a^2A^2(B - |f|) & \text{for } 0 < |f| \leq B \\ 0 & \text{otherwise} \end{cases}$$

The second term in this equation is the result of the interaction between the input signal and noise; the third is the result of the interaction of the input noise with itself. The relative importance of these two terms can be determined by comparing their total areas, $\sigma^2(z_{sxn})$ and $\sigma^2(z_{nxn})$, respectively. From Eq. (12-62),

$$\frac{\sigma^2(z_{sxn})}{\sigma^2(z_{nxn})} = \frac{2a^2P^2AB}{4a^2A^2B^2} = \frac{P^2}{2AB} \qquad (12\text{-}63)$$

Now the input signal-to-noise power ratio (i.e., the ratio of the input signal and noise variances) is

$$\left(\frac{S}{N} \right)_i = \frac{\sigma_s^2}{\sigma_n^2} = \frac{P^2/2}{2AB} \qquad (12\text{-}64)$$

Hence
$$\frac{\sigma^2(z_{sxn})}{\sigma^2(z_{nxn})} = 2 \left(\frac{S}{N} \right)_i \qquad (12\text{-}65)$$

Thus, as the input signal-to-noise power ratio increases, the output noise becomes more and more due to the interaction of the input signal and noise and less and less to the interaction of the input noise with itself.

Modulated Sine Wave Plus Gaussian Noise Input. In the preceding section, we assumed that the detector input signal was a pure sine wave. Let us now consider the input signal to be a randomly amplitude-modulated sine wave:

$$s(t) = P(t) \cos (\omega_c t + \theta) \qquad (12\text{-}66)$$

where θ is uniformly distributed over the interval $0 \leq \theta \leq 2\pi$ and where $P(t)$ is a sample function of a stationary real random process which is statistically independent of θ and of the input noise to the detector.

(The analysis which follows is also valid when $P(t)$ is periodic but contains no frequencies commensurable with $f_c = \omega_c/2\pi$).

We shall assume, as before, that the input noise is a sample function of a stationary real gaussian process with zero mean. The nxn terms of the square-law device output autocorrelation function and spectral density are therefore given again by Eqs. (12-46) and (12-47), respectively.

The autocorrelation function of the input signal is

$$R_s(\tau) = E(P_t P_{t+\tau}) E\{\cos(\omega_c t + \theta)\cos[\omega_c(t+\tau)+\theta]\} \quad (12\text{-}67)$$
$$= \tfrac{1}{2} R_P(\tau)\cos\omega_c\tau$$

where $R_P(\tau)$ is the autocorrelation function of the input signal-modulating process. The spectral density of the input signal then is

$$S_s(f) = \tfrac{1}{4}\int_{-\infty}^{+\infty} S_P(f')[\delta(f - f_c - f') + \delta(f + f_c - f')]\,df' \quad (12\text{-}68)$$
$$= \tfrac{1}{4}[S_P(f - f_c) + S_P(f + f_c)]$$

where $S_P(f)$ is the spectral density of the input signal-modulating process. It therefore follows, from Eqs. (12-42c) and (12-67), that the sxn portion of the autocorrelation function of the square-law device output is

$$R_{sxn}(\tau) = 2a^2 R_P(\tau) R_n(\tau)\cos\omega_c\tau + a^2 R_P(0)\sigma_n{}^2 \quad (12\text{-}69)$$

The corresponding spectral density is

$$S_{sxn}(f) = 2a^2\int_{-\infty}^{+\infty} R_P(\tau) R_n(\tau)\cos 2\pi f_c\tau \exp(-j2\pi f_c\tau)\,d\tau + a^2 R_P(0)\sigma_n{}^2\delta(f)$$

The integral in this equation can be evaluated as one-half the sum of the Fourier transform of the product $R_P(\tau) R_n(\tau)$ evaluated at $f - f_c$ and at $f + f_c$. Since the Fourier transform of $R_P(\tau) R_n(\tau)$ is the convolution of the corresponding spectral densities, we get

$$S_{sxn}(f) = a^2\int_{-\infty}^{+\infty} S_n(f')[S_P(f - f_c - f') + S_P(f + f_c - f')]\,df'$$
$$+ a^2 R_P(0)\sigma_n{}^2\delta(f) \quad (12\text{-}70)$$

The sxs part of the autocorrelation function of the output of the square-law device is

$$R_{sxs}(\tau) = a^2 E(P_t{}^2 P_{t+\tau}{}^2) E\{\cos^2(\omega_c t + \theta)\cos^2[\omega_c(t+\tau)+\theta]\}$$
$$= \frac{a^2}{4} R_{P^2}(\tau) + \frac{a^2}{8} R_{P^2}(\tau)\cos 2\omega_c\tau \quad (12\text{-}71)$$

where $R_{P^2}(\tau)$ is the autocorrelation function of the square of the input signal-modulation process. The sxs part of the spectral density of the output of the square-law device is therefore

$$S_{sxs}(f) = \frac{a^2}{4} S_{P^2}(f) + \frac{a^2}{16}[S_{P^2}(f - 2f_c) + S_{P^2}(f + 2f_c)] \quad (12\text{-}72)$$

where $S_{P^2}(f)$ is the spectral density of the square of the input signal-modulation process.

Let us now compare these results with those obtained when the input sine wave is unmodulated. First of all, the nxn portions of the output spectral densities are the same in both cases. Next, a comparison of Eqs. (12-51) and (12-70) shows that the nonimpulsive part of the sxn portion of the unmodulated output spectral density is convolved with the spectral density of $P(t)$ in the case of modulation. Finally, a comparison of Eqs. (12-53) and (12-72) shows that the impulses in the case of no modulation are replaced by terms containing the spectral density of $P^2(t)$ when there is modulation. The over-all effect of the modulation is therefore to cause a spreading in frequency of the sxn and sxs parts of the output spectral density.

Signal-to-noise Ratios. As the final topic in our study of the full-wave square-law detector, let us consider the relation between the output and input signal-to-noise power ratios.

From our expression for the sxs part of the autocorrelation function of the output of the square-law *device*, Eq. (12-71), we see that the signal power at the *detector* output is

$$S_o = \frac{a^2}{4} R_{P^2}(0) = \frac{a^2}{4} E(P^4) \tag{12-73}$$

We can express S_o in terms of the input signal power

$$S_i = R_s(0) = \tfrac{1}{2} R_P(0) = \tfrac{1}{2} E(P^2) \tag{12-74}$$

as

$$S_o = a^2 k_P S_i^2 \tag{12-75}$$

where

$$k_P = \frac{E(P^4)}{E^2(P^2)} \tag{12-76}$$

is a function only of the form of the probability distribution of the modulating signal. Since k_P remains constant when S_i is varied, e.g., when $P(t)$ changes to $\alpha P(t)$, Eq. (12-75) shows that the output signal power varies as the square of the input signal power—hardly an unexpected result.

The noise power at the detector output is, from Eqs. (12-46) and (12-69),

$$\begin{aligned} N_o &= \tfrac{1}{2} 2a^2 R_n{}^2(0) + \tfrac{1}{2} 2a^2 R_P(0) R_n(0) \\ &= a^2[\sigma_n{}^4 + \sigma_n{}^2 E(P^2)] \end{aligned} \tag{12-77}$$

where the factors of $\tfrac{1}{2}$ arise because half the noise power at the output of the square-law device is centered about zero frequency and half is centered about twice the carrier frequency.† This result can be expressed in terms of the input signal power S_i and the input noise power $N_i = \sigma_n{}^2$ as

$$N_o = a^2 N_i{}^2 \left(1 + 2 \frac{S_i}{N_i} \right) \tag{12-78}$$

† Cf. Figs. 12-4 and 12-5.

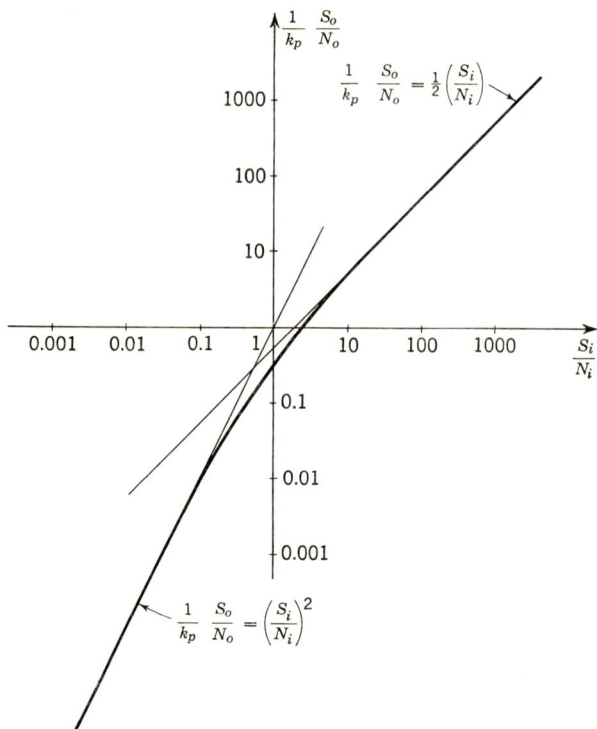

Fig. 12-6. Output signal-to-noise power ratio versus input signal-to-noise power ratio for a full-wave square-law detector.

This result corresponds to that obtained previously in the case of an unmodulated sine wave, Eq. (12-65).

The ratio of the output signal and noise powers is therefore

$$\frac{S_o}{N_o} = k_P \frac{(S_i/N_i)^2}{1 + 2S_i/N_i} \tag{12-79}$$

which is plotted in Fig. 12-6. When the input signal-to-noise power ratio is very large we get, approximately,

$$\frac{S_o}{N_o} = \frac{k_P}{2} \frac{S_i}{N_i} \tag{12-80}$$

When the input signal-to-noise power ratio is very small, we get, approximately,

$$\frac{S_o}{N_o} = k_P \left(\frac{S_i}{N_i}\right)^2 \tag{12-81}$$

Thus the output signal-to-noise power ratio varies directly as the input signal-to-noise power ratio for large values of the latter and as the

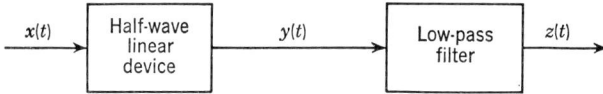

FIG. 12-7. The half-wave linear detector.

square of the input signal-to-noise power ratio for small values of the latter. This result shows the *small-signal suppression effect* of a detector. Although we obtained it here for the full-wave square-law detector, we shall show in Chap. 13 that it is a property of detectors in general.

12-4. The Half-wave Linear Detector

For our second application of the direct method of analysis of non-linear devices, let us study the half-wave linear detector.† This detector consists of a half-wave linear device with the transfer characteristic

$$y = \begin{cases} bx & \text{when } x \geq 0 \\ 0 & \text{when } x < 0 \end{cases} \quad (12\text{-}82)$$

where b is a scaling constant, followed by a low-pass, or averaging, filter. Such a detector is shown schematically in Fig. 12-7; the half-wave linear transfer characteristic is shown in Fig. 12-8.

The probability distribution function of the output of the half-wave linear device is

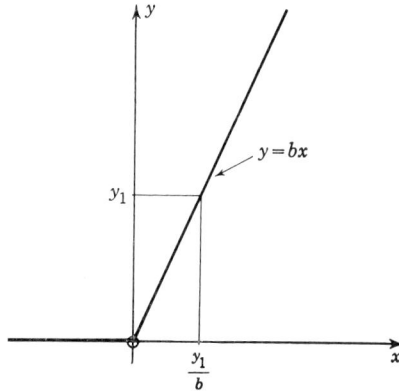

FIG. 12-8. The half-wave linear transfer characteristic.

$$P(y_t \leq y_1) = \begin{cases} 0 & \text{for } y_1 < 0 \\ P\left(x_t \leq \dfrac{y_1}{b}\right) & \text{for } y_1 \geq 0 \end{cases}$$

as may be seen from Fig. 12-8. This equation can be expressed in terms of the input probability density function as

$$P(y_t \leq y_1) = \begin{cases} 0 & \text{for } y_1 < 0 \\ P(x_t < 0) + \displaystyle\int_0^{y_1/b} p_1(x_t)\, dx_t & \text{for } y_1 \geq 0 \end{cases} \quad (12\text{-}83)$$

By differentiating Eq. (12-83) with respect to y_1, we get

$$p_2(y_t) = P(x_t < 0)\delta(y_t) + \frac{1}{b} p_1\left(x_t = \frac{y_t}{b}\right) U(y_t) \quad (12\text{-}84)$$

where $U(y_t)$ is the unit step function defined by Eq. (A1-6).

† Cf. Rice (I, Arts. 4.2 and 4.7) and Burgess (I).

The various moments of the output can be obtained by substituting Eq. (12-82) into Eq. (12-6), giving

$$E(y_t{}^n) = b^n \int_0^\infty x_t{}^n p_1(x_t)\, dx_t \tag{12-85}$$

In order to proceed further, we generally need to specify the particular form of the input probability density function. However, when the input probability density function is an even function, we can express the even-order moments of the device output directly in terms of the even-order moments of the detector input. For, in this case,

$$E(y_t{}^{2n}) = b^{2n} \int_0^\infty x_t{}^{2n} p_1(x_t)\, dx_t = \frac{b^{2n}}{2} \int_{-\infty}^{+\infty} x_t{}^{2n} p_1(x_t)\, dx_t$$

Hence
$$E(y_t{}^{2n}) = \frac{b^{2n}}{2} E(x_t{}^{2n}) \tag{12-86}$$

where $p_1(x_t) = p_1(-x_t)$.

The autocorrelation function of the output of the half-wave linear device is, from Eqs. (12-7) and (12-82),

$$R_y(t_1, t_2) = b^2 \int_0^{+\infty} \int_0^{+\infty} x_1 x_2 p(x_1, x_2)\, dx_1\, dx_2 \tag{12-87}$$

where $x_1 = x_{t_1}$ and $x_2 = x_{t_2}$.

Gaussian Input. Suppose now that the detector input $x(t)$ is a sample function of a real gaussian random process with zero mean. In this case, the probability density function of the detector input is given by Eq. (12-13) and the probability density function of the output of the half-wave linear device is, from Eq. (12-84),

$$p_2(y_t) = \frac{1}{2} \delta(y_t) + \frac{U(y_t)}{\sqrt{2\pi}\, b\sigma(x_t)} \exp\left[-\frac{y_t{}^2}{2b^2\sigma^2(x_t)} \right] \tag{12-88}$$

The probability density function of $\eta = y_t / b\sigma(x_t)$, along with that of $\xi = x_t / \sigma(x_t)$ is plotted in Fig. 12-9.

Since $p_1(x_t)$ is an even function here, the even-order moments of the output can be obtained most directly from Eq. (12-86). Thus, on using Eq. (8-11b),

$$E(y_t{}^{2n}) = \tfrac{1}{2} b^{2n} \sigma^{2n}(x_t)[1 \cdot 3 \cdot 5 \cdots (2n-1)] \tag{12-89}$$

where $n = 1, 2, \ldots$. The odd-order moments of the device output can be determined by substituting Eq. (12-13) into Eq. (12-85), giving

$$E(y_t{}^{2m+1}) = \frac{b^{2m+1}}{\sqrt{2\pi}\, \sigma(x_t)} \int_0^\infty x_t{}^{2m+1} \exp\left[-\frac{x_t{}^2}{2\sigma^2(x_t)} \right] dx_t$$

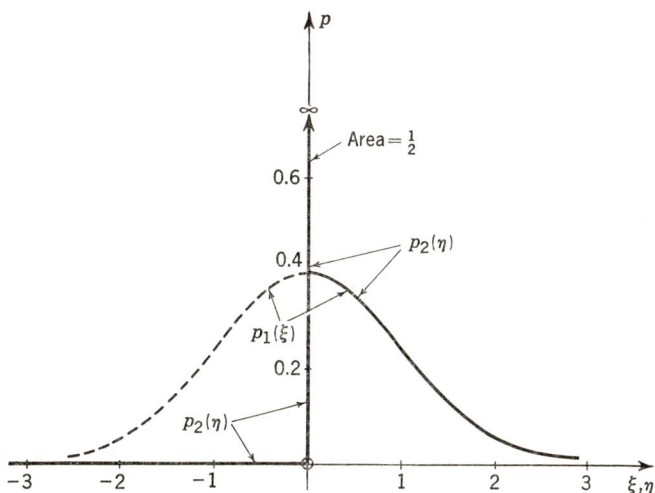

FIG. 12-9. Half-wave linear-device probability density functions.

where $m = 0, 1, 2, \ldots$ On defining $z = x^2$, this becomes

$$E(y_t^{2m+1}) = \frac{b^{2m+1}}{2\sqrt{2\pi}\,\sigma(x_t)} \int_0^\infty z^m \exp\left[-\frac{z}{2\sigma^2(x_t)}\right] dz$$

Hence†

$$E(y_t^{2m+1}) = \frac{2^m m!}{\sqrt{2\pi}} b^{2m+1}\sigma^{2m+1}(x_t) \qquad (12\text{-}90)$$

In particular, the mean value of the output of the half-wave linear device is

$$E(y_t) = \frac{b\sigma(x_t)}{\sqrt{2\pi}} \qquad (12\text{-}91)$$

and the variance is

$$\sigma^2(y_t) = E(y_t^2) - E^2(y_t)$$
$$= \frac{1}{2}\left(1 - \frac{1}{\pi}\right) b^2\sigma^2(x_t) = 0.3408 b^2\sigma^2(x_t) \qquad (12\text{-}92)$$

The autocorrelation function of the output of the half-wave linear device is found by substituting for $p(x_1,x_2)$ in Eq. (12-87) the joint probability density function of two gaussian random variables with zero means.   When the input process is stationary, we get, using Eq. (8-22),

$$R_y(\tau) = \frac{b^2}{2\pi\sigma_x^2[1 - \rho_x^2(\tau)]^{1/2}} \int_0^\infty \int_0^\infty x_1 x_2$$
$$\exp\left\{-\frac{x_1^2 + x_2^2 - 2\rho_x(\tau)x_1 x_2}{2\sigma_x^2[1 - \rho_x^2(\tau)]}\right\} dx_1\, dx_2$$

† See Dwight (I, Eq. 861.2).

where $\sigma_x = \sigma(x_t)$ for all t and $\rho_x(\tau) = E(x_1 x_2)/\sigma_x^2$. In order to facilitate evaluation of the double integral in this equation, we define

$$z = \frac{x}{\{2\sigma_x^2[1 - \rho_x^2(\tau)]\}^{1/2}}$$

We may then write

$$R_y(\tau) = \frac{2}{\pi} b^2 \sigma_x^2 [1 - \rho_x^2(\tau)]^{3/2} \int_0^\infty \int_0^\infty z_1 z_2$$

$$\exp[-z_1^2 - z_2^2 + 2\rho_x(\tau) z_1 z_2] \, dz_1 \, dz_2$$

Now it has been shown that†

$$\int_0^\infty dx \int_0^\infty dy \; xy \exp(-x^2 - y^2 - 2xy \cos \phi) = \tfrac{1}{4} \csc^2 \phi (1 - \phi \cot \phi)$$

$$(12\text{-}93)$$

when $0 \le \phi \le 2\pi$. Therefore, on defining $-\rho_x(\tau) = \cos \phi$ (since $|\rho_x(\tau)| \le 1$), we obtain

$$R_y(\tau) = \frac{b^2 \sigma_x^2}{2\pi} \{[1 - \rho_x^2(\tau)]^{1/2} + \rho_x(\tau) \cos^{-1}[-\rho_x(\tau)]\} \qquad (12\text{-}94)$$

as the autocorrelation function of the output of a half-wave linear device in response to a stationary real gaussian random process input.

In view of the practical difficulties involved in taking the Fourier transform of Eq. (12-94) to get the spectral density of the output of the half-wave linear device, it is convenient to express the output autocorrelation function in series form before taking its transform. The power-series expansion of the arc cosine is‡

$$\cos^{-1}(-\rho) = \frac{\pi}{2} + \rho + \frac{\rho^3}{2 \cdot 3} + \frac{3\rho^5}{2 \cdot 4 \cdot 5} + \frac{3 \cdot 5\rho^7}{2 \cdot 4 \cdot 6 \cdot 7}$$

$$+ \cdots, \qquad -1 \le \rho \le 1 \quad (12\text{-}95)$$

The power-series expansion of $[1 - \rho^2]^{1/2}$ is§

$$[1 - \rho^2]^{1/2} = 1 - \frac{\rho^2}{2} - \frac{\rho^4}{2 \cdot 4} - \frac{3\rho^6}{2 \cdot 4 \cdot 6} - \cdots, \qquad -1 \le \rho \le 1 \quad (12\text{-}96)$$

On substituting these expansions in Eq. (12-94) and collecting terms, we find that the first few terms of a series expansion of $R_y(\tau)$ are

$$R_y(\tau) = \frac{b^2 \sigma_x^2}{2\pi} \left[1 + \frac{\pi}{2} \rho_x(\tau) + \frac{1}{2} \rho_x^2(\tau) + \frac{1}{24} \rho_x^4(\tau) + \frac{1}{80} \rho_x^6(\tau) + \cdots \right]$$

$$(12\text{-}97)$$

† Rice (I, Eq. 3.5-4).
‡ Dwight (I, Eq. 502).
§ Dwight (I, Eq. 5.3).

Since $|\rho_x(\tau)| \leq 1$, the terms in $\rho_x(\tau)$ above the second power are relatively unimportant. To a reasonable approximation, then,

$$R_y(\tau) = \frac{b^2\sigma_x{}^2}{2\pi} + \frac{b^2}{4} R_x(\tau) + \frac{b^2}{4\pi\sigma_x{}^2} R_x{}^2(\tau) \qquad (12\text{-}98)$$

We know from Eq. (12-91) that the first term in the series expansion of $R_y(\tau)$ is the square of the mean of y_t. Therefore the rest of the series expansion evaluated at $\tau = 0$ must give the variance. Using Eq. (12-98) we then have, approximately, for the variance of y_t

$$\sigma_y{}^2 = \frac{b^2\sigma_x{}^2}{4}\left(1 + \frac{1}{4\pi}\right) = 0.3295b^2\sigma_x{}^2 \qquad (12\text{-}99)$$

A comparison of this result with the exact value given in Eq. (12-92) shows that the approximate form of the output autocorrelation function, Eq. (12-98), gives a value for the output variance which is in error by only about 3 per cent. Since the approximation afforded by Eq. (12-98) becomes better as $R_x(\tau)$ decreases, $[R_y(\tau) - \bar{y}_r{}^2]$ as calculated from Eq. (12-98) is never in error by more than about 3 per cent.

An approximate expression for the spectral density of the output of the half-wave linear device can now be obtained by taking the Fourier transform of Eq. (12-98). Thus

$$S_y(f) = \frac{b^2\sigma_x{}^2}{2\pi} \delta(f) + \frac{b^2}{4} S_x(f) + \frac{b^2}{4\pi\sigma_x{}^2} \int_{-\infty}^{+\infty} S_x(f')S_x(f - f')\, df' \quad (12\text{-}100)$$

where $S_x(f)$ is the spectral density of the input process. As in the study of the square-law detector, let us assume that the input process has the narrow-band rectangular spectrum

$$S_x(f) = \begin{cases} A & \text{for } f_c - \dfrac{B}{2} < |f| < f_c + \dfrac{B}{2} \\ 0 & \text{otherwise} \end{cases} \qquad (12\text{-}101)$$

In this case, $\sigma_x{}^2 = 2AB$ and, from Eq. (12-100),

$$S_y(f) = \frac{b^2AB}{\pi} \delta(f) + \begin{cases} \dfrac{b^2A}{4} & \text{for } f_c - \dfrac{B}{2} < |f| < f_c + \dfrac{B}{2} \\ 0 & \text{otherwise} \end{cases}$$

$$+ \begin{cases} \dfrac{b^2A}{4\pi}\left(1 - \dfrac{|f|}{B}\right) & \text{for } 0 < |f| < B \\ \dfrac{b^2A}{8\pi}\left(1 - \dfrac{1}{B}| |f| - 2f_c|\right) & \text{for } 2f_c - B < |f| < 2f_c + B \\ 0 & \text{otherwise} \end{cases} \qquad (12\text{-}102)$$

Suppose now that the half-wave linear device is followed by a low-pass zonal filter. The spectral density of the filter output will be nonzero only in the region about zero frequency. Hence

$$S_z(f) = \frac{b^2AB}{\pi}\,\delta(f) + \begin{cases} \dfrac{b^2A}{4\pi}\left(1 - \dfrac{|f|}{B}\right) & \text{for } 0 < |f| < B \\ 0 & \text{otherwise} \end{cases} \qquad (12\text{-}103)$$

The various spectra for this case are shown in Fig. 12-10.

It is interesting to compare the above results for the half-wave linear detector with those obtained previously for the full-wave square-law

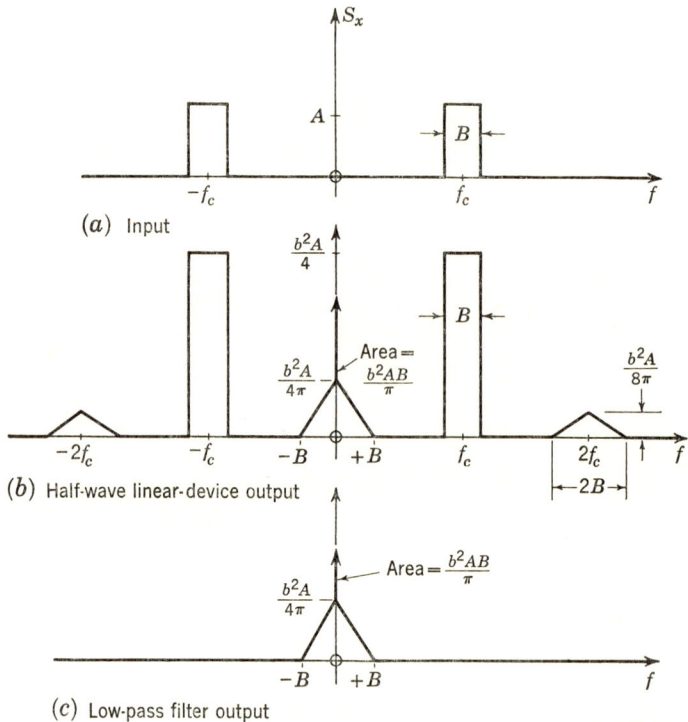

(a) Input

(b) Half-wave linear-device output

(c) Low-pass filter output

FIG. 12-10. Spectral densities for a half-wave linear detector in response to a narrow-band gaussian input.

detector. From a comparison of Figs. 12-4 and 12-10, it may be seen that the spectra for both detectors have the same shape in the band centered on zero frequency and on twice the center frequency of the input spectrum. The area of each band, however, is proportional to the square of the input variance in the case of the square-law detector and to the variance itself for the linear detector. A further difference is that the output of the half-wave linear *device* has an additional noise band coincident in frequency with the input band.   These results are based, of course, on the validity of the approximate relation, Eq. (12-98), for the autocorrelation function of the output of the half-wave linear device.

We have gone here about as far with the analysis of the half-wave linear detector as it is convenient to go with the direct method of analysis. A study of the behavior of this detector in response to an input consisting of a signal plus a noise will be deferred to the next chapter.

12-5. Problems

1. Let the input to a full-wave square-law device whose transfer characteristic is $y = ax^2$ be a sample function $x(t)$ of a stationary real gaussian random process with mean m and variance σ^2.

a. Determine the probability density function of y.
b. Determine the mean and variance of y.
c. Plot the result of **a** for the case in which $m = 5\sigma$.

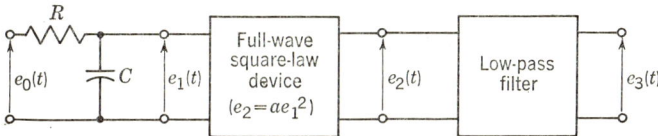

FIG. 12-11. A square-law detector.

2. Consider the system shown schematically in Fig. 12-11.   Let the input $e_0(t)$ be a sample function of a stationary real gaussian random process with zero mean and spectral density

$$S_0(f) = A >> 2kTR$$

a. Determine the autocorrelation function of e_2.
b. Determine the spectral density of e_2.
c. Sketch the autocorrelation functions and spectral densities of e_0, e_1, and e_2.

3.† Consider the system shown schematically in Fig. 12-11 and let the input be as in Prob. 2.   Let the unit impulse response of the low-pass filter be

$$h(t,T) = \begin{cases} \alpha e^{-\alpha t} & \text{for } 0 \leq t \leq T \\ 0 & \text{otherwise} \end{cases} \tag{12-104}$$

where α is a constant.

a. Determine the mean m_3 of e_3.
b. Determine the variance $\sigma_3{}^2$ of e_3.
c. Determine the ratio (σ_3/M_3) and its limiting value as $T \to \infty$.

4. Consider the system shown schematically in Fig. 12-11 and let the input be as in Prob. 2.   Let the low-pass filter be an integrating filter with the unit impulse response

$$h(t,T) = \begin{cases} 1 & \text{for } 0 \leq t \leq T \\ 0 & \text{otherwise} \end{cases} \tag{12-105}$$

a. Determine m_3.
b. Determine σ_3.
c. Determine (σ_3/m_3) and its limiting value as $T \to \infty$.

† Cf. Probs. 9 and 10 of Chap. 9.

5. Let the input to a full-wave square-law device be of the form

$$x(t) = P(1 + m \cos \omega_m t) \cos \omega_c t + n(t) \qquad (12\text{-}106)$$

where P and m are constants, where $0 < m < 1$ and $\omega_m < < \omega_c$, and where $n(t)$ is a sample function of a stationary real guassian random process with zero mean. The spectral density of the noise has a constant value N_0 for $\omega_c - 2\omega_m \leq |\omega| \leq \omega_c + 2\omega_m$ and is zero elsewhere.

    a. Determine the output spectral density and sketch.

The desired output signal is the sinusoidal wave of angular frequency ω_m.

    b. Determine the power ratio of the output low-frequency distortion to the desired output.

    c. Determine the power ratio of the desired output signal to the entire low-frequency noise.

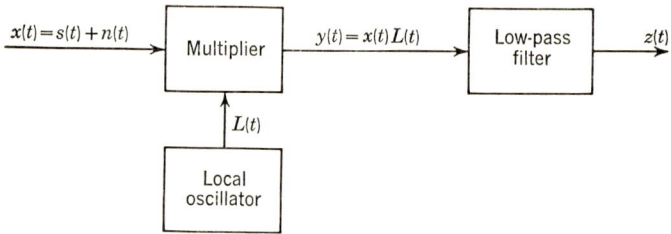

FIG. 12-12. A synchronous detector.

6. Consider the synchronous detector shown schematically in Fig. 12-12. Let the input signal be a randomly modulated sine wave of the form

$$s(t) = P(t) \cos \omega_c t \qquad (12\text{-}107)$$

where $P(t)$ is a sample function of a stationary real random process with the spectral density, where $f_0 < < f_c$,

$$S_s(f) = \begin{cases} S_0 & \text{for } 0 \leq |f| \leq \dfrac{f_0}{2} \\ 0 & \text{otherwise} \end{cases} \qquad (12\text{-}108)$$

and which is statistically independent of the input noise process. Let the local-oscillator output be

$$L(t) = A \cos \omega_c t \qquad (12\text{-}109)$$

where A is a constant. Further, let the output filter be a low-pass zonal filter which passes all frequencies $0 \leq f \leq f_0$ unaltered and which attenuates completely all other frequencies.

    a. Suppose that the input noise $n(t)$ is a sample function of a stationary real random process with zero mean and spectral density

$$S_n(f) = \begin{cases} N_0 & \text{for } f_c - \dfrac{f_0}{2} \leq |f| \leq f_c + \dfrac{f_0}{2} \\ 0 & \text{otherwise} \end{cases} \qquad (12\text{-}110)$$

      Determine the signal-to-noise power ratio at the output of the low-pass filter in terms of the input signal-to-noise power ratio.

b. Let the input noise be of the form

$$n(t) = \nu(t) \cos \omega_c t \tag{12-111}$$

where $\nu(t)$ is a sample function of a stationary real random process with zero mean and spectral density:

$$S_n(f) = \begin{cases} N_0 & \text{for } 0 \le |f| \le \dfrac{f_0}{2} \\ 0 & \text{otherwise} \end{cases} \tag{12-112}$$

Determine the signal-to-noise power ratio at the output of the low-pass filter in terms of the input signal-to-noise power ratio.

7. Consider the synchronous detector shown schematically in Fig. 12-12. Let the input noise $n(t)$ be a sample function of a stationary real gaussian random process with zero mean and spectral density:

$$S_n(f) = \frac{\sigma_n{}^2}{2(2\pi\sigma^2)^{1/2}} \left\{ \exp\left[-\frac{(f - f_c)^2}{2\sigma^2} \right] + \exp\left[-\frac{(f + f_c)^2}{2\sigma^2} \right] \right\} \tag{12-113}$$

where $\sigma \ll f_c$. Further, let the input signal be of the form

$$s(t) = P \cos(\omega_c t + \theta) \tag{12-114}$$

where P is a constant and θ is a uniformly distributed (over $0 \le \theta < 2\pi$) random variable which is independent of the input noise process, and let the local oscillator output be of the form

$$L(t) = A \cos \omega_c t$$

where A is a constant.

a. Determine the spectral density of the multiplier output.
b. Sketch the spectral densities of the input signal plus noise, and the multiplier output.

8. Let the low-pass output filter in Prob. 7 be a simple R-C network with a half-power bandwidth $\Delta f \ll \sigma$. Derive an expression for the signal-to-noise power ratio at the filter output in terms of the filter half-power bandwidth and for the input signal-to-noise power ratio.

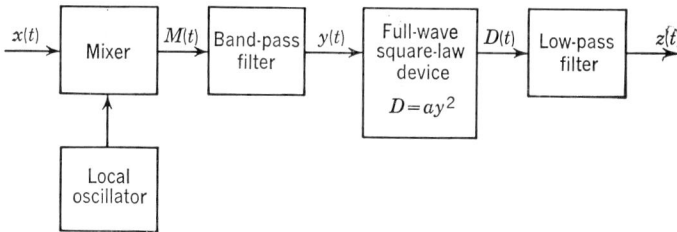

FIG. 12-13. A receiver.

9. Consider the receiver shown schematically in Fig. 12-13. Let the input be of the form

$$x(t) = P(t) \cos \omega_c t \tag{12-115}$$

where $P(t)$ is a sample function of a stationary real gaussian random process with zero mean and spectral density

$$S_p(f) = \begin{cases} 1 & \text{for } 0 \le |f| \le 0.05f_c \\ 0 & \text{otherwise} \end{cases} \tag{12-116}$$

The output of the mixer is

$$M(t) = [1 + x(t)]g(t) \tag{12-117}$$

where $g(t)$ is the periodic function of time shown in Fig. 12-14 and where ω_0 in that figure is $\frac{5}{4}$ of ω_c.   The band-pass filter has the system function

$$H_{BP}(f) = \begin{cases} 1 & \text{for } 0.1f_c \le |f| \le 0.4f_c \\ 0 & \text{otherwise} \end{cases} \tag{12-118}$$

and the low-pass filter has the system function

$$H_{LP}(f) = \begin{cases} 1 & \text{for } 0 \le |f| \le 0.15f_c \\ 0 & \text{otherwise} \end{cases} \tag{12-119}$$

a. Determine the spectral density of the mixer output in terms of the input spectral density and sketch.
b. Determine the spectral density of the low-pass filter output and sketch.
c. Determine the mean and variance of the low-pass filter output.

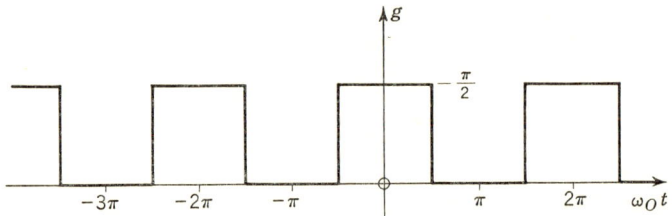

FIG. 12-14. $g(t)$

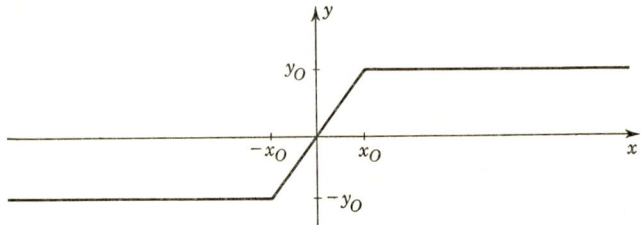

FIG. 12-15. The limiter transfer characteristic.

10. The transfer characteristic of a given symmetrical limiter is shown in Fig. 12-15.

a. Derive an expression for the probability density function of the limiter output in terms of the probability density function of the limiter input.
b. Evaluate **a** when the input random process is a stationary real gaussian random process with a zero mean.
c. Determine the mean and variance of the limiter output for part **b**.
d. If the input probability density function is even about a mean of zero but otherwise arbitrary, determine the probability density function of the limiter output for the case of the ideal limiter (i.e., $y_0 > 0$ but $x_0 = 0$).
e. Evaluate the mean and variance for the limiter output in part **c**.

CHAPTER 13

NONLINEAR DEVICES: THE TRANSFORM METHOD

When the direct method of analysis of nonlinear devices is applied to problems other than those discussed in the preceding chapter, analytic difficulties often arise in the evaluation of the various integrals involved. In many interesting cases, such difficulties can be avoided through the use of the transform method of analysis, which we shall discuss in this chapter.† We shall first introduce the Fourier (or Laplace) transform of the transfer characteristic of a nonlinear device; then we shall apply it to the determination of the autocorrelation function of the device output and, finally, study the specific case of a νth-law detector. As in Chap. 12, we shall restrict our attention to those nonlinear devices for which the device output at time t can be expressed as some single-valued function of the device input at time t.

13-1. The Transfer Function

Let $y = g(x)$ be the transfer characteristic of the nonlinear device in question. If $g(x)$ and its derivative are sectionally continuous and if $g(x)$ is absolutely integrable, i.e., if

$$\int_{-\infty}^{+\infty} |g(x)| \, dx < +\infty \tag{13-1}$$

then the Fourier transform $f(jv)$ of the transfer characteristic exists, where

$$f(jv) = \int_{-\infty}^{+\infty} g(x)e^{-jvx} \, dx \tag{13-2}$$

and the output of the nonlinear device may be expressed in terms of its input by the inverse Fourier transformation‡

$$y = g(x) = \frac{1}{2\pi} \int_{-\infty}^{+\infty} f(jv)e^{jxv} \, dv \tag{13-3}$$

We shall call $f(jv)$ the *transfer function* of the nonlinear device. The representation of a nonlinear device by its transfer function was intro-

† Cf. Middleton (II) and Rice (I, Part 4).
‡ See, for example, Churchill (I, p. 89ff.).

duced by Bennett and Rice[†] and was apparently first applied to the study of noise in nonlinear devices, more or less simultaneously, by Bennett,[‡] Rice,[§] and Middleton.[¶]

In many interesting cases (e.g., the half-wave linear device), the transfer characteristic is not absolutely integrable, its Fourier transform does not exist, and Eq. (13-3) cannot be used. However, the definition of the transfer function may often be extended to cover such cases, and results similar to Eq. (13-3) can be obtained. For example,[††] suppose that $g(x)$ is zero for $x < 0$, that it and its derivative are sectionally continuous, and that $g(x)$ is of exponential order as $x \to \infty$; i.e.,

$$|g(x)| \leq M_1 e^{u_1 x} \qquad \text{for } x \geq 0 \tag{13-4}$$

where M_1 and u_1 are constants. The function

$$\gamma(x) = g(x)e^{-u'x}$$

where $u' > u_1$, then, is absolutely integrable, since

$$|\gamma(x)| \leq M_1 e^{-(u'-u_1)x}$$

and its Fourier transform does exist. We can therefore write, from Eq. (13-3),

$$\gamma(x) = \frac{1}{2\pi} \int_{-\infty}^{+\infty} e^{jxv} \, dv \left[\int_0^\infty g(\xi)e^{-u\xi}e^{-jv\xi} \, d\xi \right]$$

and hence

$$g(x) = \frac{1}{2\pi} \int_{-\infty}^{+\infty} e^{(u'+jv)x} \, dv \left[\int_0^\infty g(\xi)e^{-(u'+jv)\xi} \, d\xi \right]$$

If now we introduce the complex variable $w = u + jv$, the integral with respect to the real variable v may be replaced by a contour integral along the line $w = u' + jv$ in the w plane, and we can write

$$g(x) = \frac{1}{2\pi j} \int_{u'-j\infty}^{u'+j\infty} f(w)e^{xw} \, dw \tag{13-5}$$

where $u' > u$ and

$$f(w) = \int_0^\infty g(x)e^{-wx} \, dx \tag{13-6}$$

The transfer function of the nonlinear device in this case may therefore

---

† Bennett and Rice (I).
‡ Bennett (I).
§ Rice (I, Art. 4.8).
¶ Middleton (II) and (III).
†† Cf. Section 57 of Churchill (II).

be defined as the *unilateral Laplace transform*,† Eq. (13-6), of the transfer characteristic, and the device output can be obtained from the inverse Laplace transformation, Eq. (13-5).

In many problems, the transfer characteristic does not vanish over a semi-infinite interval, and the above transform relations cannot be used. It often happens, however, that the transfer characteristic does satisfy the usual continuity conditions and is of exponential order for both positive and negative values of x; i.e.,

$$|g(x)| \leq M_2 e^{u_2 x} \quad \text{for } x > 0$$

and
$$|g(x)| \leq M_3 e^{-u_3 x} \quad \text{for } x < 0$$

where M_2, u_2, M_3, and u_3 are constants. In such a case, let us define the half-wave transfer characteristics‡

$$g_+(x) = \begin{cases} g(x) & \text{when } x > 0 \\ 0 & \text{when } x \leq 0 \end{cases} \tag{13-7a}$$

and
$$g_-(x) = \begin{cases} 0 & \text{when } x \geq 0 \\ g(x) & \text{when } x < 0 \end{cases} \tag{13-7b}$$

Then
$$g(x) = g_+(x) + g_-(x) \tag{13-7c}$$

Now $g_+(x)$ and $g_-(x)$ do satisfy the conditions of the preceding paragraph and hence do have unilateral Laplace transforms, say $f_+(w)$ and $f_-(w)$, respectively, where

$$f_+(w) = \int_0^\infty g_+(x) e^{-wx} \, dx \tag{13-8a}$$

which converges for $u > u_2$, and

$$f_-(w) = \int_{-\infty}^0 g_-(x) e^{-wx} \, dx \tag{13-8b}$$

which converges for $u < u_3$. The transfer function of the given device may then be thought of as the pair $f_+(w)$ and $f_-(w)$. Since $g_+(x)$ can be obtained from $f_+(w)$ by a contour integral of the type in Eq. (13-5) and $g_-(x)$ can similarly be obtained from $f_-(w)$, it follows from Eq. (13-7c) that the complete transfer characteristic is given by the equation

$$g(x) = \frac{1}{2\pi j} \int_{\mathbf{C}_+} f_+(w) e^{xw} \, dw + \frac{1}{2\pi j} \int_{\mathbf{C}_-} f_-(w) e^{xw} \, dw \tag{13-9}$$

where the contour $\mathbf{C}_+$ may be taken to be the line $w = u' + jv$, $u' > u_2$, and $\mathbf{C}_-$ to be the line $w = u'' + jv$, $u'' < u_3$.

If in the above $u_2 < u_3$, $f_+(w)$ and $f_-(w)$ have overlapping regions of

† Churchill (II, Chap. VI).
‡ Cf. Titchmarsh (II, Art. 1.3).

convergence in the w plane and we can accordingly define the transfer function of the device to be the *bilateral Laplace transform*,[†]

$$f(w) = f_+(w) + f_-(w) = \int_{-\infty}^{+\infty} g(x)e^{-wx}\, dx \qquad (13\text{-}10)$$

which converges for $u_2 < u < u_3$. In this case, the inversion contours may both be taken to lie in the overlap region, and hence can be the same.

13-2. νth-Law Devices

An important class of nonlinear devices is that based on the half-wave νth-law transfer characteristic

$$\gamma(x) = \begin{cases} ax^\nu & \text{when } x > 0 \\ 0 & \text{when } x \le 0 \end{cases} \qquad (13\text{-}11)$$

where a is a scaling constant and ν is some nonnegative real number. Particular examples are the half-wave νth-law device with the transfer characteristic $g(x) = \gamma(x)$, the full-wave (even) νth-law device with the transfer characteristic

$$\begin{aligned} g_e(x) &= \gamma(x) + \gamma(-x) \\ &= \begin{cases} ax^\nu & \text{when } x > 0 \\ 0 & \text{when } x = 0 \\ a(-x)^\nu & \text{when } x < 0 \end{cases} \end{aligned} \qquad (13\text{-}12)$$

and the full-wave (odd) νth-law device with the transfer characteristic

$$\begin{aligned} g_o(x) &= \gamma(x) - \gamma(-x) \\ &= \begin{cases} ax^\nu & \text{when } x > 0 \\ 0 & \text{when } x = 0 \\ -a(-x)^\nu & \text{when } x < 0 \end{cases} \end{aligned} \qquad (13\text{-}13)$$

Plots of several such transfer characteristics are shown in Fig. 13-1. The half-wave and full-wave (even) νth-law devices are often used in detectors, and the full-wave (odd) νth-law device occurs in some nonlinear amplifiers.

The Laplace transform of the half-wave νth-law transfer characteristic is

$$\phi(w) = a \int_0^\infty x^\nu e^{-wx}\, dx$$

The integral on the right converges only when $\Re(w) > 0$. The corre-

[†] Cf. van der Pol and Bremmer (I, Art. II.5).

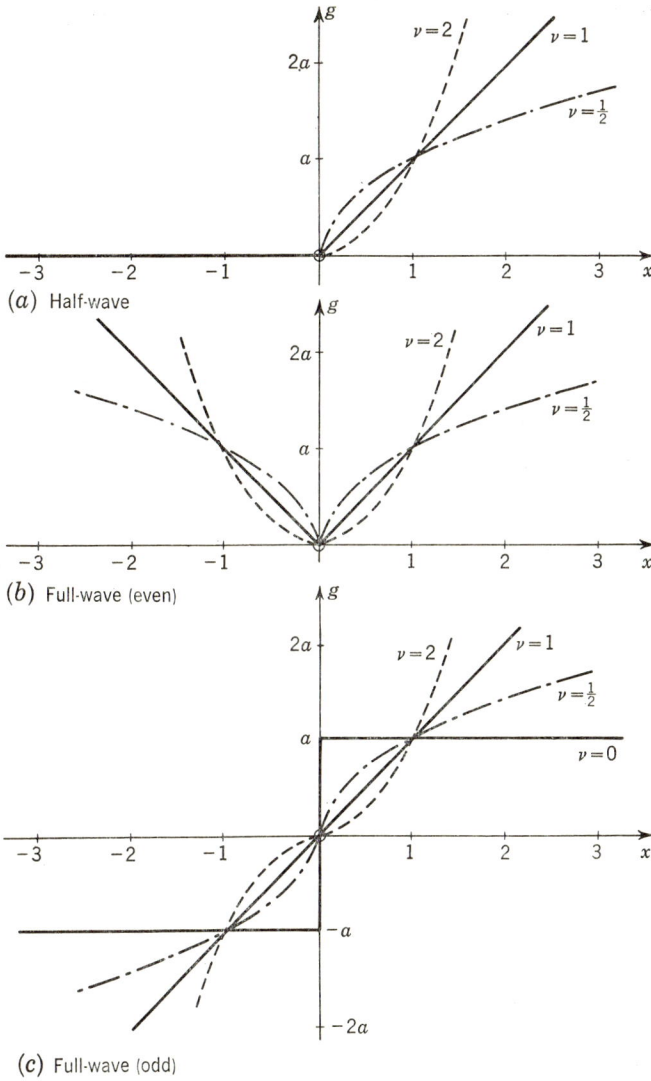

(c) Full-wave (odd)

FIG. 13-1. νth-Law transfer characteristics.

sponding inverse transformation contour C_+ must therefore lie to the right of the $w = jv$ axis.   Letting $t = wx$,

$$\phi(w) = \frac{a}{w^{\nu+1}} \int_0^\infty t^\nu e^{-t} \, dt$$

Hence
$$\phi(w) = \frac{a\Gamma(\nu + 1)}{w^{\nu+1}} \qquad (13\text{-}14)$$

where $\Gamma(z)$ is the gamma function[†] defined by the integral

$$\Gamma(z) = \int_0^\infty e^{-t}t^{z-1}\,dt \tag{13-15}$$

where $\Re(z) > 0$. It then follows that the transfer function of the half-wave νth-law device is

$$f(w) = \phi(w) \tag{13-16}$$

Since

$$a \int_{-\infty}^0 (-x)^\nu e^{-wx}\,dx = a \int_0^\infty t^\nu e^{-(-w)t}\,dt = \phi(-w)$$

the transfer function pair of the full-wave (even) νth-law device is

$$f_{e_+}(w) = \phi(w) \qquad \text{and} \qquad f_{e_-}(w) = \phi(-w) \tag{13-17}$$

and that of the full-wave (odd) νth-law device is

$$f_{o_+}(w) = \phi(w) \qquad \text{and} \qquad f_{o_-}(w) = -\phi(-w) \tag{13-18}$$

The integral for $\phi(-w)$ converges only when $\Re(w) < 0$; the corresponding inverse transformation contour $\mathbf{C}_-$ must therefore lie to the left of the $w = jv$ axis. We shall subsequently choose $\mathbf{C}_+$ to be the line $w = \epsilon + jv$ and $\mathbf{C}_-$ to be the line $w = -\epsilon + jv$, where $\epsilon > 0$ and $-\infty < v < +\infty$, as shown in Fig. 13-2.

Narrow-band Inputs. Suppose that the input to a half-wave νth-law device is a narrow-band wave of the form

$$x(t) = V(t)\cos\theta(t)$$
$$= V(t)\cos[\omega_c t + \phi(t)] \tag{13-19}$$

where $V(t) \geq 0$. The output of that device can then be obtained by substituting in Eq. (13-5) for $x(t)$ from Eq. (13-19) and for $f(w)$ from Eq. (13-16). Thus

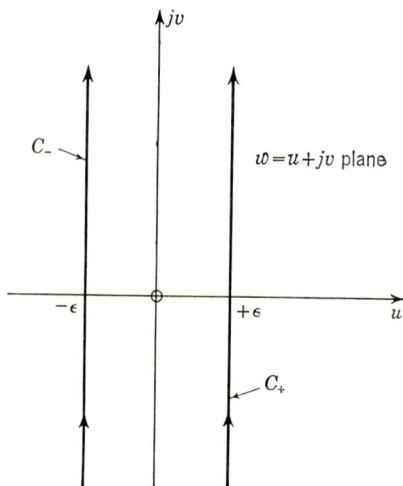

FIG. 13-2. νth-Law inversion contours.

$$y = \frac{1}{2\pi j}\int_{\mathbf{C}_+} f(w)e^{wV\cos\theta}\,dw = \frac{a\Gamma(\nu+1)}{2\pi j}\int_{\epsilon-j\infty}^{\epsilon+j\infty}\frac{e^{wV\cos\theta}}{w^{\nu+1}}\,dw$$

The exponential can be expanded using the Jacobi-Anger formula[‡]

$$\exp(z\cos\theta) = \sum_{m=0}^\infty \varepsilon_m I_m(z)\cos m\theta \tag{13-20}$$

[†] Whittaker and Watson (I, Art. 12.2) or Magnus and Oberhettinger (I, Chap. I).
[‡] Magnus and Oberhettinger (I, Chap. 3, Art. 1).

where ε_m is the Neumann factor $\varepsilon_0 = 1$, $\varepsilon_m = 2$ $(m = 1,2, \ldots)$ and $I_m(z)$ is a modified Bessel function of the first kind.   Hence

$$y = \sum_{m=0}^{\infty} a\Gamma(\nu + 1) \cos m\theta \, \frac{\varepsilon_m}{2\pi j} \int_{\epsilon-j\infty}^{\epsilon+j\infty} \frac{I_m(wV)}{w^{\nu+1}} \, dw$$

$$= \sum_{m=0}^{\infty} a\Gamma(\nu + 1) V^\nu \cos m\theta \, \frac{\varepsilon_m}{2\pi j} \int_{\delta-j\infty}^{\delta+j\infty} \frac{I_m(\zeta)}{\zeta^{\nu+1}} \, d\zeta$$

on letting $\zeta = wV$ and $\delta = \epsilon V$.   Therefore, on defining the coefficients

$$C(\nu,m) = \frac{\varepsilon_m a\Gamma(\nu + 1)}{2\pi j} \int_{\delta-j\infty}^{\delta+j\infty} \frac{I_m(\zeta)}{\zeta^{\nu+1}} \, d\zeta \qquad (13\text{-}21)$$

it follows that the output of a half-wave νth-law device in response to a narrow-band input may be expressed as

$$y(t) = \sum_{m=0}^{\infty} C(\nu,m) V^\nu(t) \cos [m\omega_c t + m\phi(t)] \qquad (13\text{-}22)$$

We have thus decomposed the device output into a sum of harmonic terms.   Each harmonic is envelope modulated by the νth power of the input envelope, and the mth harmonic is phase-modulated by m times the input phase modulation.

Suppose, next, that a narrow-band wave is applied to the input of a full-wave (even) νth-law device.   The output can be obtained by substituting in Eq. (13-9) for $x(t)$ from Eq. (13-19) and for $f_{e_+}(w)$ and $f_{e_-}(w)$ from Eq. (13-17).   In this way, we find that

$$y_e = \frac{a\Gamma(\nu + 1)}{2\pi j} \int_{\epsilon-j\infty}^{\epsilon+j\infty} \frac{e^{wV\cos\theta}}{w^{\nu+1}} \, dw + \frac{a\Gamma(\nu + 1)}{2\pi j} \int_{-\epsilon-j\infty}^{-\epsilon+j\infty} \frac{e^{wV\cos\theta}}{(-w)^{\nu+1}} \, dw$$

$$= \frac{a\Gamma(\nu + 1)}{2\pi j} \int_{\epsilon-j\infty}^{\epsilon+j\infty} \frac{e^{wV\cos\theta} + e^{-wV\cos\theta}}{w^{\nu+1}} \, dw$$

Hence, on expanding the exponentials with Eq. (13-20) and using

$$I_m(-z) = (-1)^m I_m(z) \qquad (13\text{-}23)$$

we get

$$y_e = \sum_{m=0}^{\infty} [1 + (-1)^m] a\Gamma(\nu + 1) V^\nu \cos m\theta \, \frac{\varepsilon_m}{2\pi j} \int_{\delta-j\infty}^{\delta+j\infty} \frac{I_m(\zeta)}{\zeta^{\nu+1}} \, d\zeta$$

Thus we can express the output of a full-wave (even) νth-law device in response to a narrow-band input as

$$y_e(t) = \sum_{\substack{m=0 \\ (m\ even)}}^{\infty} 2C(\nu,m)V^\nu(t)\cos[m\omega_c t + m\phi(t)] \tag{13-24}$$

where $C(\nu,m)$ is given by Eq. (13-21). In a similar manner it may also be shown that the output of a full-wave (odd) νth-law device in response to a narrow-band input can be expressed as

$$y_o(t) = \sum_{\substack{m=1 \\ (m\ odd)}}^{\infty} 2C(\nu,m)V^\nu(t)\cos[m\omega_c t + m\phi(t)] \tag{13-25}$$

Hence, whereas the output of a half-wave νth-law device generally contains all harmonics of its input, the output of a full-wave (even) νth-law device contains only even harmonics (including zero frequency) of the input, and the output of a full-wave (odd) νth-law device contains only odd harmonics of the input.

Evaluation of the Coefficients $C(\nu,m)$.† In order to evaluate the coefficients $C(\nu,m)$, let us first consider the integral of $I_m(\zeta)/\zeta^{\nu+1}$ around the contour shown in Fig. 13-3. Where $\zeta = \xi + j\eta$, let

$$I_1 = \int_{\delta-j\beta}^{\delta+j\beta} \frac{I_m(\zeta)\,d\zeta}{\zeta^{\nu+1}} \qquad \zeta = \delta + j\eta$$

$$I_2 = \int_{\delta+j\beta}^{0+j\beta} \frac{I_m(\zeta)\,d\zeta}{\zeta^{\nu+1}} \qquad \zeta = \xi + j\beta$$

$$I_3 = \int_{-j\beta}^{+j\beta} \frac{I_m(\zeta)\,d\zeta}{\zeta^{\nu+1}} \qquad \zeta = j\eta$$

and $$I_4 = \int_{0-j\beta}^{\delta-j\beta} \frac{I_m(\zeta)\,d\zeta}{\zeta^{\nu+1}} \qquad \zeta = \xi - j\beta \tag{13-26}$$

Then, as $\beta \to \infty$, I_1 tends to the integral in Eq. (13-21).
Since‡

$$I_m(z) = \left(\frac{z}{2}\right)^m \sum_{n=0}^{\infty} \frac{(z/2)^{2n}}{n!\Gamma(m+n+1)} \tag{13-27}$$

† Cf. Watson (I, Art. 13.24).
‡ Watson (I, Art. 3.7) or Magnus and Oberhettinger (I, Chap. 3, Art. 1).

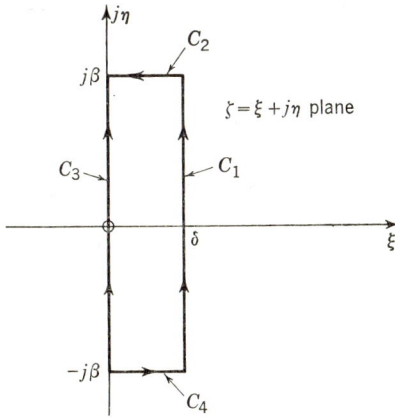

FIG. 13-3. Contour of integration.

the Bessel function $I_m(z)$ varies as z^m for small values of z. If, then, we assume for the moment that $m > \nu + 1$, the singularity of $I_m(\zeta)/\zeta^{\nu+1}$ at the origin vanishes, and that function becomes analytic inside of, and on, the contour of Fig. 13-3. It therefore follows from the Cauchy theorem[†] that

$$I_1 + I_2 - I_3 + I_4 = 0 \qquad (13\text{-}28)$$

Consider now the integrals I_2 and I_4. An asymptotic expansion of $I_m(z)$ for large z is[‡]

$$I_m(z) = \frac{e^z}{(2\pi z)^{\frac{1}{2}}} \sum_{n=0}^{\infty} \frac{(-1)^n \Gamma(m + n + \frac{1}{2})}{n!\,\Gamma(m - n + \frac{1}{2})(2z)^n} \qquad (13\text{-}29)$$

For large values of $|z|$, then,

$$I_m(z) = \frac{e^z}{(2\pi z)^{\frac{1}{2}}} \qquad (13\text{-}30)$$

approximately. Hence, for large values of β,

$$I_2 = \frac{e^{j\beta}}{\sqrt{2\pi}} \int_\delta^0 \frac{e^\xi \, d\xi}{(\xi + j\beta)^{\nu+\frac{3}{2}}}$$

and, when $\nu + \frac{3}{2} > 0$,

$$|I_2| \le \frac{\delta e^\delta}{\sqrt{2\pi}\,\beta^{\nu+\frac{3}{2}}} \to 0 \qquad \text{as } \beta \to \infty$$

† Churchill (II, Sec. 48) or Titchmarsh (I, Art. 2.33).
‡ Watson (I, Sec. 7.23).

It therefore follows that $I_2 \to 0$, and similarly that $I_4 \to 0$, as $\beta \to \infty$, and hence that

$$I = \int_{\delta-j\infty}^{\delta+j\infty} \frac{I_m(\zeta)\ d\zeta}{\zeta^{\nu+1}} = \int_{-j\infty}^{+j\infty} \frac{I_m(\zeta)\ d\zeta}{\zeta^{\nu+1}} \tag{13-31}$$

On the imaginary axis, $\zeta = j\eta$ and $d\zeta = j\ d\eta$.   Hence

$$I = j^{(m-\nu)} \int_{-\infty}^{+\infty} \frac{J_m(\eta)\ d\eta}{\eta^{\nu+1}}$$

since

$$I_m(j\eta) = j^m J_m(\eta) \tag{13-32}$$

Now, setting $\eta = -t$,

$$\int_{-\infty}^{0} \frac{J_m(\eta)\ d\eta}{\eta^{\nu+1}} = -(-1)^{(m-\nu)} \int_{0}^{\infty} \frac{J_m(t)\ dt}{t^{\nu+1}}$$

Therefore

$$I = j^{(m-\nu)}[1 - (-1)^{(m-\nu)}] \int_{0}^{\infty} \frac{J_m(t)\ dt}{t^{\nu+1}}$$

$$= 2j \sin\ (m - \nu)\ \frac{\pi}{2} \int_{0}^{\infty} \frac{J_m(t)\ dt}{t^{\nu+1}} \tag{13-33}$$

The integral in this equation is Weber's infinite integral† and is

$$\int_{0}^{\infty} \frac{J_m(t)\ dt}{t^{\nu+1}} = \frac{\Gamma\left(\dfrac{m-\nu}{2}\right)}{2^{\nu+1}\Gamma\left(1 - \dfrac{m+\nu}{2}\right)} \tag{13-34}$$

under the conditions that $\nu + \frac{1}{2} > 0$ and $m > \nu$ (which are met by our assumptions so far).   Since‡

$$\Gamma(z)\Gamma(1 - z) = \frac{\pi}{\sin\ \pi z} \tag{13-35}$$

it follows from Eqs. (13-22), (13-33), (13-34), and (13-35) that for m even,

$$C(\nu,m) = \frac{\varepsilon_m a\Gamma(\nu + 1)}{2^{\nu+1}\Gamma\left(1 - \dfrac{m-\nu}{2}\right)\Gamma\left(1 + \dfrac{m+\nu}{2}\right)} \tag{13-36}$$

when $m > \nu + 1$ and $\nu + \frac{1}{2} > 0$.   Since, for fixed ν, $C(\nu,m)/\varepsilon_m$ is a single-valued analytic function of m for all m, the theory of analytic continuation§ permits us to remove the restriction $m > \nu + 1$.   The

† Watson (I, Art. 13.24).
‡ Magnus and Oberhettinger (I, Chap. 1, Art. 1).
§ Churchill (II, Sec. 50) or Titchmarsh (I, Arts. 4.1 through 4.4).

remaining condition $\nu + \frac{1}{2} > 0$ is automatically satisfied by our original assumption that ν is nonnegative.

Since $|\Gamma(-n)| = \infty$, $n = 0, 1, 2 \ldots$, it follows from Eq. (13-36) that the coefficients $C(\nu,m)$ are zero whenever $(m - \nu)/2$ is a positive integer. This occurs, for example, when m is even and greater than ν if ν is an even integer, and when m is odd and greater than ν if ν is an odd integer. It therefore follows from Eq. (13-24) that if ν is an even integer, the output harmonics of a full-wave (even) νth-law device vanish whenever $m > \nu$; similarly it follows from Eq. (13-25) that if ν is an odd integer, the output harmonics of a full-wave (odd) νth-law device vanish whenever $m > \nu$.

νth-Law Detectors and Nonlinear Amplifiers. If a half-wave or full-wave (even) νth-law device is followed by a low-pass zonal filter, a νth-law detector is formed. The output of the half-wave νth-law detector is, from Eq. (13-22),

$$z(t) = C(\nu,0)V^\nu(t) \tag{13-37a}$$

and the output of a full-wave νth-law detector is, from Eq. (13-24),

$$z_e(t) = 2C(\nu,0)V^\nu(t) \tag{13-37b}$$

where, from Eq. (13-36),

$$C(\nu,0) = \frac{a\Gamma(\nu + 1)}{2^{\nu+1}\Gamma^2\left(1 + \dfrac{\nu}{2}\right)} \tag{13-38}$$

In particular, the output of a half-wave linear detector is

$$z_1(t) = C(1,0)V(t) = \frac{a}{\pi}V(t)$$

and the output of a full-wave square-law detector is†

$$z_2(t) = 2C(2,0)V^2(t) = \frac{a}{2}V^2(t)$$

The half-wave linear detector is thus an envelope detector.

If a full-wave (odd) νth-law device is followed by a band-pass zonal filter centered on $f_c = \omega_c/2\pi$ (i.e., a filter whose system function is unity over the frequency interval about f_c and zero elsewhere) a νth-law nonlinear amplifier is formed. The output of such an amplifier is, from Eq. (13-25),

$$z_o(t) = 2C(\nu,1)V^\nu(t)\cos[\omega_c t + \phi(t)] \tag{13-39a}$$

where, from Eq. (13-36),

$$C(\nu,1) = \frac{a\Gamma(\nu + 1)}{2^\nu\Gamma[(\nu + 1)/2]\Gamma[(\nu + 3)/2]} \tag{13-39b}$$

† Cf. Art. 12-2, Eq. (12-17).

When $\nu = 0$, the output of the full-wave (odd) νth-law device can assume only the values $\pm a$, as shown by Fig. 13-1c. The device is then called an *ideal limiter*, and the cascade of such a device with a band-pass zonal filter, centered on the input carrier frequency, is called an *ideal band-pass limiter*. When the input to an ideal band-pass limiter is a narrow-band wave, as in Eq. (13-19), the output is, from Eq. (13-39),

$$z_o(t) = 2C(0,1) \cos [\omega_c t + \phi(t)]$$
$$= \frac{4a}{\pi} \cos [\omega_c t + \phi(t)]$$

Hence, when the input to an ideal band-pass limiter is a narrow-band wave, the output is a purely phase-modulated wave; the phase modulation of the output is identical to that of the input.

13-3. The Output Autocorrelation Function and Spectral Density

The autocorrelation function of the output of a nonlinear device can be stated as follows in terms of the transfer function of that device, from Eq. (13-5),†

$$R_y(t_1,t_2) = \int_{-\infty}^{+\infty} \int_{-\infty}^{+\infty} g(x_1)g(x_2)p(x_1,x_2)\, dx_1\, dx_2$$
$$= \frac{1}{(2\pi j)^2} \int_C f(w_1)\, dw_1 \int_C f(w_2)\, dw_2 \int_{-\infty}^{+\infty} \int_{-\infty}^{+\infty} p(x_1,x_2)$$
$$\exp(w_1 x_1 + w_2 x_2)\, dx_1\, dx_2$$

The double integral with respect to x_1 and x_2 is, from Eq. (4-25), the joint characteristic function of x_1 and x_2 expressed as a function of the complex variables w_1 and w_2. Hence

$$R_y(t_1,t_2) = \frac{1}{(2\pi j)^2} \int_C f(w_1)\, dw_1 \int_C f(w_2)\, dw_2\, M_x(w_1,w_2) \quad (13\text{-}40)$$

Equation (13-40) is the fundamental equation of the transform method of analysis of nonlinear devices in response to random inputs. The remainder of this chapter is concerned with the evaluation of this expression for various specific devices and inputs.

In many problems, the device input is the sum of a signal and a noise:

$$x(t) = s(t) + n(t) \quad (13\text{-}41)$$

† We assume here for compactness that the transfer function may in fact be expressed either as a unilateral Laplace transform of the transfer characteristic, as in Eq. (13-6), or as a bilateral Laplace transform as in Eq. (13-7). For those cases in which the transfer function must be expressed as a transform pair, as in Eq. (13-8), each inversion contour in this equation must be replaced by a pair of contours, as in Eq. (13-9).

where $s(t)$ and $n(t)$ are sample functions of statistically independent random processes. In these cases the input characteristic function factors, and Eq. (13-40) becomes

$$R_y(t_1,t_2) = \frac{1}{(2\pi j)^2} \int_C f(w_1) \, dw_1 \int_C f(w_2) \, dw_2 \, M_s(w_1,w_2) M_n(w_1,w_2) \quad (13\text{-}42)$$

where $M_s(w_1,w_2)$ is the joint characteristic function of s_1 and s_2, and $M_n(w_1,w_2)$ is the joint characteristic function of n_1 and n_2.

Gaussian Input Noise. When the input noise is a sample function of a real gaussian random process with zero mean, we have, from Eq. (8-23),

$$M_n(w_1,w_2) = \exp\left\{\frac{1}{2}\left[\sigma_1{}^2 w_1{}^2 + 2R_n(t_1,t_2)w_1 w_2 + \sigma_2{}^2 w_2{}^2\right]\right\} \quad (13\text{-}43)$$

where $\sigma_1 = \sigma(n_1)$, $\sigma_2 = \sigma(n_2)$, and $R_n(t_1,t_2) = E(n_1 n_2)$. The output autocorrelation function then becomes

$$R_y(t_1,t_2) = \frac{1}{(2\pi j)^2} \int_C f(w_1) \exp\left(\frac{\sigma_1{}^2 w_1{}^2}{2}\right) dw_1 \int_C f(w_2) \exp\left(\frac{\sigma_2{}^2 w_2{}^2}{2}\right) dw_2$$
$$\exp[R_n(t_1,t_2)w_1 w_2] M_s(w_1,w_2)$$

If now $\exp[R_n(t_1,t_2)w_1 w_2]$ and $M_s(w_1,w_2)$ could each be factored as a product of a function of w_1 times a function of w_2, or as a sum of such products, then the double integral in the above equation could be evaluated as a product of integrals. That the exponential term can be so factored may be seen by expanding it in a power series:

$$\exp[R_n(t_1,t_2)w_1 w_2] = \sum_{k=0}^{\infty} \frac{R_n{}^k(t_1,t_2)w_1{}^k w_2{}^k}{k!} \quad (13\text{-}44)$$

The autocorrelation function of the output of the nonlinear device can therefore be written as

$$R_y(t_1,t_2) = \sum_{k=0}^{\infty} \frac{R_n{}^k(t_1,t_2)}{k!(2\pi j)^2} \int_C f(w_1)w_1{}^k \exp\left(\frac{\sigma_1{}^2 w_1{}^2}{2}\right) dw_1$$
$$\int_C f(w_2)w_2{}^k \exp\left(\frac{\sigma_1{}^2 w_1{}^2}{2}\right) dw_2 \, M_s(w_1,w_2) \quad (13\text{-}45)$$

when the input noise is gaussian. In order to proceed further, the characteristic function of the input signal must be specified.

Sine-wave Signals. Suppose next that the input signal is an amplitude-modulated sine wave, i.e., that

$$s(t) = P(t) \cos \theta(t) = P(t) \cos [\omega_c t + \phi] \quad (13\text{-}46)$$

where $P(t)$ is a sample function of a low-frequency random process (i.e., one whose spectral density is nonzero only in a region about zero frequency which is narrow compared to f_c) and where the random variable ϕ is uniformly distributed over the interval $0 \leq \phi < 2\pi$ and independent of both the input modulation and noise. The characteristic function of such a signal is

$$M_s(w_1,w_2) = E[\exp(w_1P_1 \cos \theta_1 + w_2P_2 \cos \theta_2)]$$

The exponential can be expanded using the Jacobi–Anger formula, Eq. (13-20), to give

$$M_s(w_1,w_2) = \sum_{m=0}^{\infty} \sum_{n=0}^{\infty} \varepsilon_m\varepsilon_n E[I_m(w_1P_1)I_n(w_2P_2)]E(\cos m\theta_1 \cos n\theta_2)$$

Since

$$E(\cos m\theta_1 \cos n\theta_2) = E[\cos m(\omega_c t_1 + \phi) \cos n(\omega_c t_2 + \phi)]$$
$$= \begin{cases} 0 & \text{when } n \neq m \\ \dfrac{1}{\varepsilon_m} \cos m\omega_c\tau & \text{when } n = m \end{cases}$$

where $\tau = t_1 - t_2$, it follows that

$$M_s(w_1,w_2) = \sum_{m=0}^{\infty} \varepsilon_m E[I_m(w_1P_1)I_m(w_2P_2)] \cos m\omega_c\tau \qquad (13\text{-}47)$$

when the input signal is an amplitude-modulated sine wave.

The autocorrelation function of the output of a nonlinear device in response to a sine-wave signal and a gaussian noise may now be obtained by substituting Eq. (13-47) into Eq. (13-45). Thus, if we define the function

$$h_{mk}(t_i) = \frac{1}{2\pi j} \int_C f(w)w^k I_m[wP_i] \exp\left[\frac{\sigma_i^2 w^2}{2}\right] dw \qquad (13\text{-}48)$$

where $P_i = P(t_i)$, $\sigma_i = \sigma[n(t_i)]$, and the autocorrelation function

$$R_{mk}(t_1,t_2) = E[h_{mk}(t_1)h_{mk}(t_2)] \qquad (13\text{-}49)$$

where the averaging is with respect to the input signal modulation, it follows that the output autocorrelation function can be expressed as

$$R_y(t_1,t_2) = \sum_{m=0}^{\infty} \sum_{k=0}^{\infty} \frac{\varepsilon_m}{k!} R_{mk}(t_1,t_2)R_n{}^k(t_1,t_2) \cos m\omega_c\tau \qquad (13\text{-}50)$$

When both the input signal modulation and noise are stationary, Eq. (13-50) becomes

$$R_y(\tau) = \sum_{m=0}^{\infty} \sum_{k=0}^{\infty} \frac{\varepsilon_m}{k!} R_{mk}(\tau)R_n{}^k(\tau) \cos m\omega_c\tau \qquad (13\text{-}51)$$

If, in addition, the input signal is an unmodulated sine wave

$$P \cos (\omega_c t + \phi)$$

we get

$$R_y(\tau) = \sum_{m=0}^{\infty} \sum_{k=0}^{\infty} \frac{\varepsilon_m h_{mk}{}^2}{k!} R_n{}^k(\tau) \cos m\omega_c\tau \qquad (13\text{-}52)$$

since in this case the coefficients $h_{mk}(t_1)$ and $h_{mk}(t_2)$ are constant and equal.

Output Signal and Noise Terms. For the moment, consider the case in which the input noise is an unmodulated sine wave. The output auto-correlation function is then given by Eq. (13-52). Let us now expand that result and examine the various terms:

$$R_y(\tau) = h_{00}{}^2 + 2 \sum_{m=1}^{\infty} h_{m0}{}^2 \cos m\omega_c\tau + \sum_{k=1}^{\infty} \frac{h_{0k}{}^2}{k!} R_n{}^k(\tau)$$

$$+ 2 \sum_{m=1}^{\infty} \sum_{k=1}^{\infty} \frac{h_{mk}{}^2}{k!} R_n{}^k(\tau) \cos m\omega_c\tau \qquad (13\text{-}53)$$

The first term in this equation corresponds to the constant part of the device output. The set of terms $(m \geq 1, k = 0)$ corresponds to the periodic part of the output and is primarily due to the interaction of the input signal with itself. The remaining terms correspond to the random variations of the output, i.e., the output noise. Of these remaining terms, those in the set $(m = 0, k \geq 1)$ are due mainly to the interaction of the input noise with itself, and those in the set $(m \geq 1, k \geq 1)$ to the inter-action of the input signal with the input noise.

If we express the output of the nonlinear device in terms of its mean, its periodic components, and its random part as

$$y(t) = m_y + \sum_{m=1}^{\infty} A_m \cos (m\omega_c t + \phi_m) + \eta(t) \qquad (13\text{-}54)$$

the autocorrelation function of the output may be written as

$$R_y(\tau) = m_y{}^2 + \tfrac{1}{2} \sum_{m=1}^{\infty} A_m{}^2 \cos m\omega_c\tau + R_\eta(\tau) \qquad (13\text{-}55)$$

where $R_\eta(\tau) = E(\eta_1\eta_2)$. A comparison of Eqs. (13-53) and (13-55) shows that the output mean value and the amplitudes of the output periodic components can be expressed directly in terms of the coefficients h_{mk}. Thus

$$m_y = h_{00} \qquad (13\text{-}56)$$

and

$$A_m = 2h_{m0} \qquad m \geq 1 \qquad (13\text{-}57)$$

In addition, the autocorrelation function of the random part of the output may be expressed as

$$R_\eta(\tau) = R_{(NXN)}(\tau) + R_{(SXN)}(\tau) \qquad (13\text{-}58)$$

where

$$R_{(NXN)}(\tau) = \sum_{k=1}^{\infty} \frac{h_{0k}^2}{k!} R_n^k(\tau) \qquad (13\text{-}59)$$

represents that part of the output noise due mainly to the interaction of the input noise with itself, and where

$$R_{(SXN)}(\tau) = 2 \sum_{m=1}^{\infty} \sum_{k=1}^{\infty} \frac{h_{mk}^2}{k!} R_n^k(\tau) \cos m\omega_c\tau \qquad (13\text{-}60)$$

represents that part of the output noise that is due to the interaction of the input signal with the input noise.† The expansion of the output autocorrelation function leading to Eq. (13-52) has thus enabled us to isolate the output mean, the output periodic components, and the (NXN) and (SXN) parts of the output noise.

These results were obtained under the assumption of a stationary input noise and an unmodulated sine wave as an input signal. However, a similar splitting up of the output autocorrelation function is possible in the general case, and we can write

$$R_y(t_1,t_2) = R_{(SXS)}(t_1,t_2) + R_{(NXN)}(t_1,t_2) + R_{(SXN)}(t_1,t_2) \qquad (13\text{-}61)$$

where we have defined, from Eq. (13-50),

$$R_{(SXS)}(t_1,t_2) = \sum_{m=0}^{\infty} \varepsilon_m R_{m0}(t_1,t_2) \cos m\omega_c\tau \qquad (13\text{-}62)$$

$$R_{(NXN)}(t_1,t_2) = \sum_{k=1}^{\infty} \frac{1}{k!} R_{0k}(t_1,t_2) R_n^k(t_1,t_2) \qquad (13\text{-}63)$$

and

$$R_{(SXN)}(t_1,t_2) = 2 \sum_{m=1}^{\infty} \sum_{k=1}^{\infty} R_{mk}(t_1,t_2) R_n^k(t_1,t_2) \cos m\omega_c\tau \qquad (13\text{-}64)$$

It should be noted that, strictly speaking, all the above terms are functions of the input signal-modulation process.

Just which of the various (SXS) terms in Eq. (13-62) represents, in fact, the output signal depends entirely upon the use of the nonlinear device. For example, if the device is used as a detector, the desired out-

---

† Note that $R_{(SXN)}(\tau) = 0$ when $P(t) = 0$, since $I_m(0) = 0$, $m \geq 1$, in Eq. (13-48).

put signal is centered on zero frequency. In this case, the signal part of the output autocorrelation function would be

$$R_{S_o}(t_1,t_2) = R_{00}(t_1,t_2) \qquad (13\text{-}65)$$

On the other hand, when the device is a nonlinear amplifier,

$$R_{S_o}(t_1,t_2) = 2R_{10}(t_1,t_2)\cos\omega_c\tau \qquad (13\text{-}66)$$

since then the desired output signal is centered on the input carrier frequency, i.e., on f_c.

13-4. The Output Spectral Density

The spectral density of the output of a nonlinear device may be obtained, as usual, by taking the Fourier transform of the output autocorrelation function. Consider first the case in which the input to the nonlinear device is an unmodulated sine wave plus a stationary gaussian noise. In this case, the output autocorrelation function is given by Eq. (13-53). Hence

$$S_y(f) = h_{00}^2\,\delta(f) + \sum_{m=1}^{\infty} h_{m0}^2[\delta(f + mf_c) + \delta(f - mf_c)] + \sum_{k=1}^{\infty} \frac{h_{0k}^2}{k!}\,{}_kS_n(f)$$

$$+ \sum_{m=1}^{\infty}\sum_{k=1}^{\infty} \frac{h_{mk}^2}{k!}\,[{}_kS_n(f + mf_c) + {}_kS_n(f - mf_c)] \qquad (13\text{-}67)$$

where we have defined ${}_kS_n(f)$ to be the Fourier transform of $R_n{}^k(\tau)$:

$$_kS_n(f) = \int_{-\infty}^{+\infty} R_n{}^k(\tau)\exp(-j2\pi f\tau)\,d\tau \qquad (13\text{-}68)$$

The first term in Eq. (13-67) is an impulse located at zero frequency corresponding to the mean value of the output. The set of impulses located at $\pm mf_c$ correspond to the periodic components of the output, and the remaining terms are due to the output noise. As before, the output noise terms may be separated into two parts: the part representing the interaction of the input noise with itself,

$$S_{(NXN)}(f) = \sum_{k=1}^{\infty} \frac{h_{0k}^2}{k!}\,{}_kS_n(f) \qquad (13\text{-}69)$$

and the part representing the interaction of the input signal and noise,

$$S_{(SXN)}(f) = \sum_{m=1}^{\infty}\sum_{k=1}^{\infty} \frac{h_{mk}^2}{k!}\,[{}_kS_n(f + mf_c) + {}_kS_n(f - mf_c)] \qquad (13\text{-}70)$$

These two equations are, of course, simply the Fourier transforms of the corresponding autocorrelation functions as given by Eqs. (13-59) and (13-60), respectively.

Let us now determine the relation between $_kS_n(f)$ and $S_n(f)$. When $k = 1$, it follows from Eq. (13-68) that

$$_1S_n(f) = S_n(f) \tag{13-71}$$

Further, when $k \geq 2$, $R_n{}^k(\tau)$ can be factored to give

$$_kS_n(f) = \int_{-\infty}^{+\infty} R_n{}^{k-1}(\tau) R_n(\tau) \exp(-j2\pi f\tau) \, d\tau$$

Hence $_kS_n(f)$ may be expressed as a convolution of $_{k-1}S_n(f)$ with $S_n(f)$:

$$_kS_n(f) = \int_{-\infty}^{+\infty} {}_{k-1}S_n(f')S_n(f - f') \, df' \tag{13-72}$$

By repeated application of this recursion formula, we get

$$_kS_n(f) = \int_{-\infty}^{+\infty} \cdots \int_{-\infty}^{+\infty} S_n(f_{k-1})S_n(f_{k-2} - f_{k-1}) \cdots$$
$$S_n(f - f_1) \, df_{k-1} \cdots df_1 \tag{13-73}$$

Thus the spectral density $_kS_n(f)$ may be expressed as the $(k - 1)$ fold convolution of the input noise spectral density with itself.

Consider next the case in which the input signal is an amplitude-modulated sine wave and in which the input signal modulation and input noise are stationary. The output autocorrelation function in this case, obtained by setting $\tau = t_1 - t_2$ in Eq. (13-61), is

$$R_y(\tau) = R_{(SXS)}(\tau) + R_{(NXN)}(\tau) + R_{(SXN)}(\tau) \tag{13-74}$$

where, from Eqs. (13-62), (13-63), and (13-64),

$$R_{(SXS)}(\tau) = \sum_{m=0}^{\infty} \varepsilon_m R_{m0}(\tau) \cos m\omega_c\tau \tag{13-75}$$

$$R_{(NXN)}(\tau) = \sum_{k=1}^{\infty} \frac{1}{k!} R_{0k}(\tau) R_n{}^k(\tau) \tag{13-76}$$

and $$R_{(SXN)}(\tau) = 2 \sum_{m=1}^{\infty} \sum_{k=1}^{\infty} R_{mk}(\tau) R_n{}^k(\tau) \cos m\omega_c\tau \tag{13-77}$$

The output spectral density then is

$$S_y(f) = S_{(SXS)}(f) + S_{(NXN)}(f) + S_{(SXN)}(f) \tag{13-78}$$

where the various spectral densities in this equation are the Fourier transforms of the corresponding autocorrelation functions in Eq. (13-74).

If now we define $S_{mk}(f)$ to be the Fourier transform of the autocorrelation function of the coefficient $h_{mk}(t)$:

$$S_{mk}(f) = \int_{-\infty}^{+\infty} R_{mk}(\tau) \exp(-j2\pi f\tau) \, d\tau \qquad (13\text{-}79)$$

the various components of the output spectral density may be expressed as

$$S_{(SXS)}(f) = \sum_{m=0}^{\infty} \frac{\varepsilon_m}{2} [S_{m0}(f + mf_c) + S_{m0}(f - mf_c)] \qquad (13\text{-}80)$$

$$S_{(NXN)}(f) = \sum_{k=1}^{\infty} \frac{1}{k!} \int_{-\infty}^{+\infty} S_{0k}(f')_k S_n(f - f') \, df' \qquad (13\text{-}81)$$

and

$$S_{(SXN)}(f) = \sum_{m=1}^{\infty} \sum_{k=1}^{\infty} \frac{1}{k!} \int_{-\infty}^{+\infty} S_{mk}(f')[_k S_n(f + mf_c - f')$$
$$+ {}_k S_n(f - mf_c - f')] \, df' \qquad (13\text{-}82)$$

These spectral densities are, in essence, the convolution with $S_{mk}(f)$ of those obtained previously when the input signal was assumed to be an unmodulated sine wave.

13-5. Narrow-band Inputs

Suppose that the input to a nonlinear device is the sum of an unmodulated sine wave and a stationary gaussian noise with a narrow-band spectral density which is concentrated about the carrier frequency of the sine wave f_c as shown in Fig. 13-4. The spectral density of the device output is then given by Eq. (13-67).

Let us now examine the noise part of the output spectral density. In particular, consider the spectral density $_k S_n(f)$. The first term $_1 S_n(f)$ is the same as the noise part

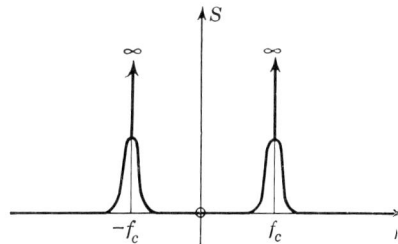

FIG. 13-4. A narrow-band input spectral density.

of the input spectral density, as shown by Eq. (13-71). The terms $k \geq 2$ are given by successive applications of the convolution recursion formula, Eq. (13-72). Thus $_2 S_n(f)$ is the convolution of $_1 S_n(f)$ with $S_n(f)$, i.e., the convolution of $S_n(f)$ with itself. The spectral density $_2 S_n(f)$ therefore has its maximum value at zero frequency and has smaller peaks centered on twice the carrier frequency, as shown in Fig. 13-5b.

It follows from the convolution integral that each of the peaks in $_2S_n(f)$ has the same shape, although the shape of these peaks may well differ from that of the peaks in $S_n(f)$. This process of convolution with $S_n(f)$ may be continued to build up $_kS_n(f)$ for any $k \geq 2$, as shown by Eq. (13-73). Typical plots for $k = 1, 2, 3, 4$ are shown in Fig. 13-5. A moment's reflection on the convolution process shows that when k is odd, the peaks of $_kS_n(f)$ are located on odd multiples of the carrier frequency and extend out to $\pm k f_c$. Similarly, when k is even, the peaks of $_kS_n(f)$ are located on even multiples of f_c, including zero frequency, and again extend out to $\pm k f_c$.

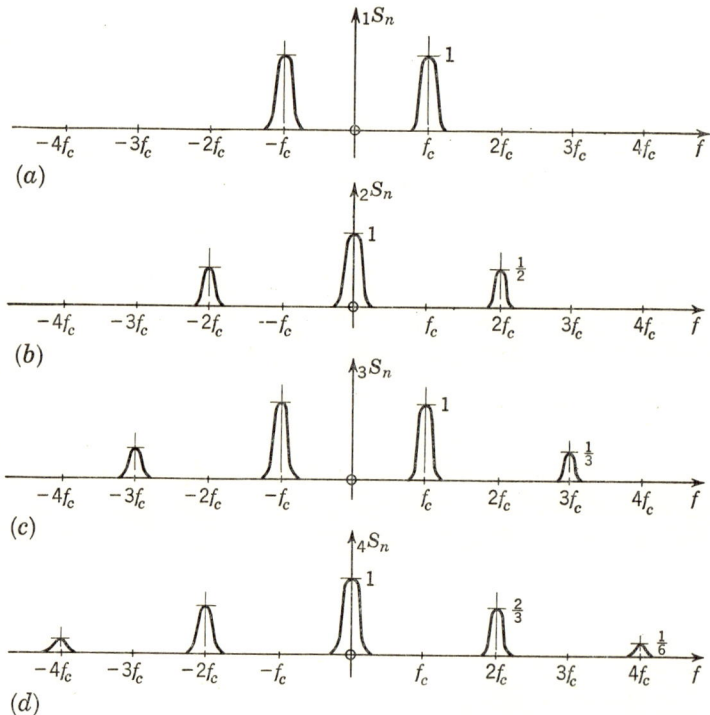

Fig. 13-5. Plots of the spectral density $_kS_n(f)$ for several values of k showing the relative values of the various spectral bands. (a) $_1S_n$. (b) $_2S_n$. (c) $_3S_n$. (d) $_4S_n$.

As shown by Eq. (13-69), the kth (NXN) term of the output spectral density is $_kS_n(f)$ times the factor $h_{0k}{}^2/k!$. Figure 13-5 therefore pictures, except for those factors, the first four (NXN) terms of the output spectral density. Each (SXN) term of the output spectral density has the form $[_kS_n(f + mf_c) + _kS_n(f - mf_c)]$ multiplied by $h_{mk}{}^2/k!$. The plots of the various (SXN) terms may therefore be constructed (except for the scaling constants) from the plots of Fig. 13-5 simply by translating each plot in that figure by $\pm mf_c$. The resultant plots for the case $(m = 2, k = 2)$ are shown in Fig. 13-6.

The above results apply to the case in which the input sine wave is unmodulated. When the input signal is an amplitude-modulated sine wave, it follows, from Eqs. (13-80) through (13-82), that plots of the various terms of the resultant output spectral density are substantially the same as those shown in Figs. 13-5 and 13-6, except for a spreading of each spectral band due to the modulation process.

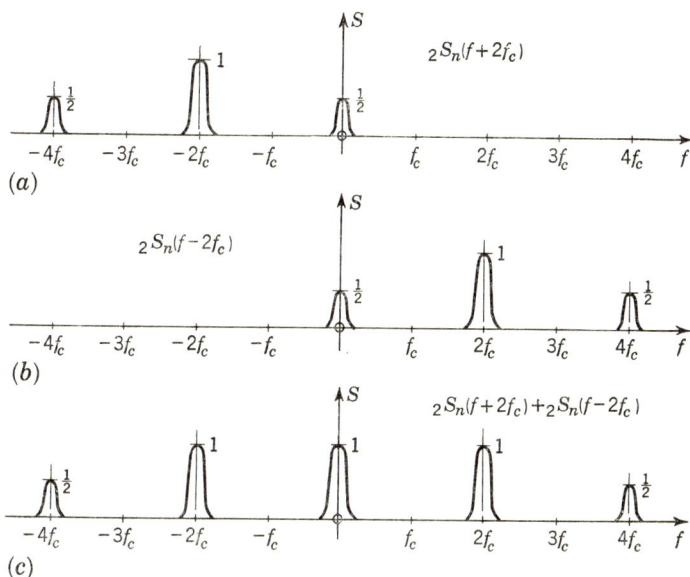

FIG. 13-6. Plots showing the construction of $_2S_n(f + 2f_c) + _2S_n(f - 2f_c)$.

It follows from the formulas for the various (NXN) and (SXN) terms of the output spectral density and from Figs. 13-5 and 13-6 that the complete output noise spectral density consists of a narrow band of noise centered on each multiple of the input carrier frequency. The relative magnitudes of the various bands depend upon the nature of the coefficients h_{mk}. Just which of the various spectral regions is of importance depends entirely on the use of the nonlinear device in question, e.g., whether the device is used as a detector or as a nonlinear amplifier.

An Expansion of $R_n{}^k(\tau)$. As we saw in Chap. 8, the autocorrelation function of a narrow-band stationary real gaussian random process may be expressed as†

$$R_n(\tau) = R_\nu(\tau) \cos [\omega_c \tau + \theta(\tau)] \tag{13-83}$$

† Cf. Art. 8-5 and Prob. 10 of Chap. 8.

where
$$R_\nu(\tau) = [R_c{}^2(\tau) + R_{cs}{}^2(\tau)]^{\frac{1}{2}} \qquad (13\text{-}84a)$$

$$\theta(\tau) = \tan^{-1}\left[\frac{R_{cs}(\tau)}{R_c(\tau)}\right] \qquad (13\text{-}84b)$$

$$R_c(\tau) = 2\int_0^\infty S_n(f)\cos 2\pi(f - f_c)\tau\, df \qquad (13\text{-}85a)$$
and
$$R_{cs}(\tau) = 2\int_0^\infty S_n(f)\sin 2\pi(f - f_c)\tau\, df \qquad (13\text{-}85b)$$

The spectral density $S_\nu(f)$, i.e., the Fourier transform of $R_\nu(\tau)$, is concentrated in a small region about zero frequency, and the spectral density corresponding to $R(\tau) = \cos[\omega_c\tau + \theta(\tau)]$ is concentrated in a narrow band about f_c. Equation (13-83) may therefore be used to expand $R_n{}^k(\tau)$ so as to place in evidence the various spectral-band components.

Since†

$$\cos^k \theta = \sum_{r=0}^{\frac{k-2}{2}} \frac{k!}{(k-r)!r!2^{k-1}} \cos(k-2r)\theta + \frac{k!}{\left(\dfrac{k}{2}!\right)^2 2^k} \qquad (13\text{-}86a)$$

when k is even, and

$$\cos^k \theta = \sum_{r=0}^{\frac{k-1}{2}} \frac{k!}{(k-r)!r!2^{k-1}} \cos(k-2r)\theta \qquad (13\text{-}86b)$$

when k is odd, it follows from Eq. (13-83) that

$$R_n{}^k(\tau) = \begin{cases} \dfrac{R_\nu{}^k(\tau)}{2^{k-1}}\left[\displaystyle\sum_{r=0}^{\frac{k-2}{2}} \frac{k!}{(k-r)!r!} \cos(k-2r)[\omega_c\tau + \theta(\tau)] + \frac{k!}{2\left(\dfrac{k}{2}!\right)^2}\right] \\[4pt] \qquad\qquad\qquad\qquad\qquad\qquad\qquad\text{for } k \text{ even} \quad (13\text{-}87) \\[8pt] \dfrac{R_\nu{}^k(\tau)}{2^{k-1}}\displaystyle\sum_{r=0}^{\frac{k-1}{2}} \frac{k!}{(k-r)!r!} \cos(k-2r)[\omega_c\tau + \theta(\tau)] \\[4pt] \qquad\qquad\qquad\qquad\qquad\qquad\qquad\text{for } k \text{ odd} \end{cases}$$

The autocorrelation function of the output of a nonlinear device in response to an amplitude-modulated sine wave plus a narrow-band gaussian noise can now be obtained by substituting Eq. (13-87) into either Eq. (13-51) or (13-50), depending on whether the signal modulation is stationary or not.

† Dwight (I, Eq. 404).

Detectors. Suppose that the nonlinear device in question is followed by a low-pass zonal filter to form a detector. Because of this filter, only those terms in the expansion of the spectral density of the device output which are concentrated about zero frequency can contribute to the detector output. Let us determine the noise terms in the detector output.

The development leading to the plots in Fig. 13-5 shows that the only $_kS_n(f)$ having spectral bands about zero frequency are those for which k is even; the only (NXN) terms of the device output which contribute to the detector output are therefore those for which k is even. Hence, from Eq. (13-63),

$$R'_{(NXN)}(t_1,t_2) = \sum_{\substack{k=2 \\ (k \text{ even})}}^{\infty} \frac{1}{k!} R_{0k}(t_1,t_2) R_n{}^k(\tau)$$

which becomes, on expanding $R_n{}^k(\tau)$ from Eq. (13-87),

$$R'_{(NXN)}(t_1,t_2) = \sum_{\substack{k=2 \\ (k \text{ even})}}^{\infty} \frac{1}{\left(\frac{k}{2}!\right)^2 2^k} R_{0k}(t_1,t_2) R_\nu{}^k(\tau)$$

$$+ \sum_{\substack{k=2 \\ (k \text{ even})}}^{\infty} \sum_{r=0}^{\frac{k-2}{2}} \frac{R_{0k}(t_1,t_2) R_\nu{}^k(\tau)}{(k-r)!r!2^{k-1}} \cos(k-2r)[\omega_c\tau + \theta(\tau)]$$

Since both $R_{0k}(\tau)$ and $R_\nu{}^k(\tau)$ correspond to low-frequency fluctuations, the terms in the first series correspond to spectral terms concentrated about zero frequency, whereas the terms in the double series correspond to spectral terms centered on $2f_c \le (k-2r)f_c \le kf_c$, $k \ge 2$. Only the first set contributes to the detector output $z(t)$; hence

$$R_{z(NXN)}(t_1,t_2) = \sum_{\substack{k=2 \\ (k \text{ even})}}^{\infty} \frac{1}{\left(\frac{k}{2}!\right)^2 2^k} R_{0k}(t_1,t_2) R_\nu{}^k(\tau) \qquad (13\text{-}88)$$

is the (NXN) portion of the detector output autocorrelation function.

The (SXN) portion of the detector output autocorrelation function may be obtained similarly. Examination of Figs. 13-5 and 13-6 shows that the noise bands in $_kS_n(f)$ are spaced by twice the carrier frequency and that these bands extend only over the frequency interval from $-kf_c$ to $+kf_c$. It follows that the noise bands in $_kS_n(f+mf_c)$ and $_kS_n(f-mf_c)$

are also spaced by twice f_c and that these spectral densities have bands about zero frequency only when $(m + k)$ is even and $m \leq k$. The only (SXN) terms of the autocorrelation function of the device output which can contribute to the detector output are therefore given by (from Eq. (13-64)):

$$R'_{(SXN)}(t_1,t_2) = 2 \sum_{\substack{k=1 \\ (m+k \text{ even})}}^{\infty} \sum_{m=1}^{k} \frac{1}{k!} R_{mk}(t_1,t_2) R_n{}^k(\tau) \cos m\omega_c\tau$$

The (SXN) terms of the autocorrelation function of the detector output may now be determined by substituting in this equation for $R_n{}^k(\tau)$ from Eq. (13-87) and picking out the low-frequency terms. In this way we get

$$R_{z(SXN)}(t_1,t_2) = \sum_{\substack{k=1 \\ (m+k \text{ even})}}^{\infty} \sum_{m=1}^{k} \frac{R_{mk}(t_1,t_2) R_v{}^k(\tau)}{\left(\dfrac{k+m}{2}\right)! \left(\dfrac{k-m}{2}\right)! 2^{k-1}} \qquad (13\text{-}89)$$

The sum of Eqs. (13-88) and (13-89) gives the autocorrelation function of the detector output noise. If the detector low-pass output filter is not a zonal filter, the methods of Chap. 9 may be used, in conjunction with Eqs. (13-88) and (13-89), to find the actual autocorrelation function of the detector output noise.

Nonlinear Amplifiers. As a second example, let us consider a nonlinear band-pass amplifier. The amplifier nonlinearity may be undesired and due to practical inability to obtain a sufficiently large dynamic range, or it may actually be desired as in a "logarithmic" radar receiver or an FM limiter. In either case, it is important to determine the effect of the nonlinearity. As in the preceding sections of this article, we shall assume that the system input is the sum of an amplitude-modulated sine wave and a stationary gaussian noise whose spectral density is concentrated in the vicinity of the carrier frequency of the input signal.

Since the amplifier is nonlinear, it may generate signal and noise components in the vicinity of zero frequency, the input carrier frequency f_c, and all harmonics of f_c. Since the system is supposed to be an "amplifier," however, we are interested only in those components in the vicinity of f_c itself. For convenience, let us represent the nonlinear amplifier by a nonlinear device in cascade with a band-pass zonal filter; the system function of the filter is assumed to be unity over the frequency interval centered about f_c and zero elsewhere. Other filter characteristics can be accounted for by applying the methods of Chap. 9.

The techniques of the preceding several sections may now be used to sort out those terms in the autocorrelation function of the output of the nonlinear device which contribute to the amplifier output. In this way it can be shown that

$$R_{z(NXN)}(t_1,t_2) = \sum_{\substack{k=1 \\ (k\ \text{odd})}}^{\infty} \frac{R_{0k}(t_1,t_2)R_\nu^k(\tau)}{\left(\dfrac{k+1}{2}\right)!\left(\dfrac{k-1}{2}\right)!} \frac{\cos\,[\omega_c\tau + \theta(\tau)]}{2^{k-1}} \qquad (13\text{-}90)$$

is the (NXN) part of the autocorrelation function of the amplifier output and that

$$R_{z(SXN)}(t_1,t_2) = \sum_{\substack{k=1 \\ (m+k\ \text{odd})}}^{\infty} \sum_{m=1}^{k+1} \frac{(k+1)R_{mk}(t_1,t_2)R_\nu^k(\tau)}{\left(\dfrac{k+m+1}{2}\right)!\left(\dfrac{k-m+1}{2}\right)!}$$

$$\frac{\cos\,[\omega_c\tau + \theta(\tau)]}{2^{k-1}} \qquad (13\text{-}91)$$

is the (SXN) portion of that autocorrelation function.

13-6. νth-Law Detectors†

The results of the several preceding articles are general in that the transfer characteristic of the nonlinear device in question has been left unspecified. Let us now consider a specific example, the νth-law detector. We shall assume that the detector consists either of a half-wave νth-law nonlinear device or of a full-wave (even) νth-law device, followed by a low-pass zonal filter. The transfer characteristics and transfer functions of the νth-law devices were discussed in Art. 13-2.

Let the input to a νth-law detector consist of the sum of an amplitude-modulated sine wave and a gaussian noise. If the nonlinear device is a half-wave νth-law device, the autocorrelation function of its output is given by Eq. (13-50) where the coefficient $h_{mk}(t_i)$ in $R_{mk}(t_1,t_2)$ is

$$h_{mk}(t_i) = \frac{a\Gamma(\nu + 1)}{2\pi j} \int_{C_+} w^{k-\nu-1} I_m(wP_i) \exp\left(\frac{\sigma_i^2 w^2}{2}\right) dw \qquad (13\text{-}92)$$

which follows from Eq. (13-48) on substituting for $f(w)$ from Eq. (13-16). If, on the other hand, the nonlinear device is a full-wave (even) νth-law device, it follows from Eqs. (13-17) and (13-18) that

$$h_{mk}(t_i)_e = \begin{cases} 0 & \text{for } m+k \text{ odd} \\ 2h_{mk}(t_i) & \text{for } m+k \text{ even} \end{cases} \qquad (13\text{-}93)$$

where $h_{mk}(t_i)$ is given by Eq. (13-92). When the input noise is narrow-band, it follows from the preceding article that the only coefficients $h_{mk}(t_i)$ which contribute to the detector output are those for which $m+k$ is even.

† Cf. Rice (I, Sec. 4.10) and Middleton (II, Sec. 4).

The only difference between the autocorrelation function of the output of a half-wave νth-law detector and that of a full-wave νth-law detector is therefore a factor of $2^2 = 4$.†

Evaluation of the $h_{mk}(t_i)$.‡ Let us now evaluate the coefficients $h_{mk}(t_i)$ as given by Eq. (13-92) for the νth-law devices. The method to be used is essentially the same as that used in Art. 13-2 to evaluate the coefficients $C(\nu,m)$.

Consider the integral of

$$w^{k-\nu-1}I_m(wP_i)\,\exp(\sigma_i{}^2w^2/2)$$

around the rectangular contour shown in Fig. 13-7. Let I_1 be the integral along the line $w = \epsilon + jv$ from $v = -\beta$ to $v = +\beta$; let I_2 be the integral along the line

$$w = u + j\beta$$

from $u = \epsilon$ to $u = 0$; let I_3 be the integral along the line $w = jv$ from $v = -\beta$ to $v = +\beta$; and let I_4 be the integral along $w = u - j\beta$ from $u = 0$ to $u = \epsilon$. Since, from Eq. (13-27), the Bessel function $I_m(z)$ varies as z^m for small z, and since the exponential varies as z^0 for small z, there is no singularity of the integrand at the origin if we assume for the moment that $m + k - \nu - 1 > 0$. The integrand then becomes analytic inside, and on, the contour of Fig. 13-7, and it follows from the Cauchy theorem that

$$I_1 + I_2 - I_3 + I_4 = 0 \qquad (13\text{-}94)$$

Consider the integrals I_2 and I_4. It follows, using Eq. (13-30), that I_2 is given approximately, for large values of β, by

$$I_2 = \frac{a\Gamma(\nu+1)}{j(2\pi)^{3\!/\!2}P_i{}^{1\!/\!2}} \int_\epsilon^0 \frac{\exp[uP_i + \sigma_i{}^2/2(u^2 - \beta^2)]\,\exp[j(\beta P_i + \sigma_i{}^2 u\beta)]}{(u+j\beta)^{\nu+3\!/\!2-k}}\,du$$

Hence

$$|I_2| \leq \frac{a\Gamma(\nu+1)\epsilon}{(2\pi)^{3\!/\!2}P_i{}^{1\!/\!2}}\exp\!\left(\epsilon P_i + \frac{\sigma_i{}^2\epsilon^2}{2}\right)\frac{\exp(-\sigma_i{}^2\beta^2/2)}{\beta^{\nu+3\!/\!2-k}} \to 0 \qquad \text{as } \beta \to \infty$$

for any m, k, ν since $\exp(-\sigma_i{}^2\beta^2/2)$ vanishes faster than $1/\beta^{\nu+3\!/\!2-k}$ can

† Cf. Eq. (13-37) in Art. 13-2.
‡ Cf. Watson (I, Art. 13.3), or Middleton (II, Appendix III).

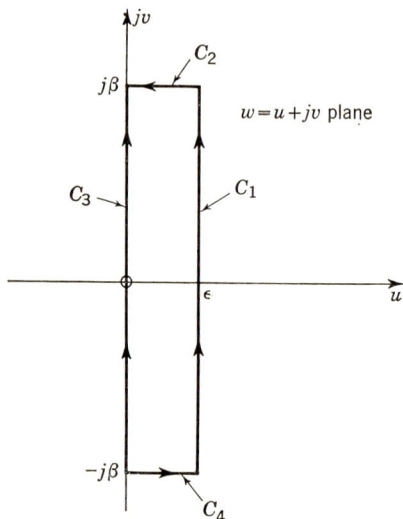

FIG. 13-7. Contour of integration.

increase as $\beta \to \infty$. It therefore follows that $I_2 \to 0$, and similarly $I_4 \to 0$, as $\beta \to \infty$. Hence, using Eq. (13-94),

$$
\begin{aligned}
h_{mk}(t_i) &= \frac{a\Gamma(\nu + 1)}{2\pi j} \int_{-j\infty}^{+j\infty} w^{k-\nu-1} I_m(wP_i) \exp\left(\frac{\sigma_i{}^2 w^2}{2}\right) dw \\
&= a\Gamma(\nu + 1) \frac{\sin(k + m - \nu)\dfrac{\pi}{2}}{\pi} \int_0^\infty v^{k-\nu-1} J_m(vP_i) \exp\left(-\frac{\sigma_i{}^2 v^2}{2}\right) dv
\end{aligned}
$$

when $m + k - \nu - 1 > 0$. The integral in this equation, Hankel's exponential integral,† is (when $m + k - \nu > 0$)

$$
\int_0^\infty v^{k-\nu-1} J_m(vP_i) \exp\left(-\frac{\sigma_i{}^2 v^2}{2}\right) dv =
$$

$$
\frac{(P_i{}^2/2\sigma_i{}^2)^{\frac{m}{2}} \Gamma[(k + m - \nu)/2]}{2m!(\sigma_i{}^2/2)^{(k-\nu)/2}} \, {}_1F_1\left(\frac{k + m - \nu}{2}; m + 1; -\frac{P_i{}^2}{2\sigma_i{}^2}\right) \quad (13\text{-}95)
$$

where ${}_1F_1(a;c;z)$ is the confluent hypergeometric function defined by the series‡

$$
{}_1F_1(a;c;z) = \sum_{r=0}^\infty \frac{(a)_r z^r}{(c)_r r!} = 1 + \frac{a}{c}\frac{z}{1!} + \frac{a(a + 1)}{c(c + 1)}\frac{z^2}{2!} + \cdots \quad (13\text{-}96)
$$

It therefore follows, using Eq. (13-35), that

$$
h_{mk}(t_i) = \frac{a\Gamma(\nu + 1)(P_i{}^2/2\sigma_i{}^2)^{\frac{m}{2}}}{2m!\Gamma[1 - (m + k - \nu)/2](\sigma_i{}^2/2)^{(k-\nu)/2}}
$$

$$
{}_1F_1\left(\frac{m + k - \nu}{2}; m + 1; -\frac{P_i{}^2}{2\sigma_i{}^2}\right) \quad (13\text{-}97)
$$

when $m + k - \nu - 1 > 0$. However, for fixed m and k, the function $h_{mk}(t_i)/\Gamma(\nu + 1)$ as defined by Eq. (13-92) can be shown to be a single-valued analytic function of ν for all ν in any bounded region. Since the right side of Eq. (13-97) divided by $\Gamma(\nu + 1)$ is also a single-valued analytic function of ν for all ν in any bounded region, the theory of analytic continuation permits us to extend $h_{mk}(t_i)$ as given by Eq. (13-97) to all values of ν and so remove the restriction $m + k - \nu - 1 > 0$.

It should be noted that, since $|\Gamma(-n)| = \infty$ for $n = 0, 1, 2, \ldots,$ the coefficient $h_{mk}(t_i)$ vanishes whenever $m + k - \nu$ is an even integer greater than zero.

---

† Watson (I, Art. 13.3) or Magnus and Oberhettinger (I, Chap. 3, Art. 7).
‡ See Whittaker and Watson (I, Chap. XVI), or Magnus and Oberhettinger (I, Chap. VI, Art. I). Convenient tables and plots for the confluent hypergeometric function are given by Middleton and Johnson (I). Some plots, using the symbol $M(\alpha,\gamma,x)$, are also given in Jahnke and Emde (I, Chap. X).

Output Signal and Noise. We shall assume in the remainder of this article that the input to the νth-law detector is the sum of an amplitude-modulated sine-wave signal and a stationary narrow-band gaussian noise whose spectral density is concentrated about the carrier frequency of the signal.

If we further assume, for the moment, that the input sine wave is unmodulated, the mean value m_o of the detector output is, from Eq. (13-56),

$$m_o = [h_{00}(t_i)]_{P_i=P} \tag{13-98}$$

where, from Eq. (13-97) for the half-wave ν-th law detector,

$$h_{00}(t_i) = \frac{a\Gamma(\nu + 1)\sigma^\nu}{\Gamma(\nu/2 + 1)2^{(\nu+2)/2}} \, _1F_1\left(-\frac{\nu}{2};1;-\frac{P_i^2}{2\sigma^2}\right) \tag{13-99}$$

If we now vary slowly the amplitude of the input sine wave, the output mean value will also vary slowly. We shall define the output signal to be the variation of the output mean with respect to its zero signal value;† i.e.,

$$s_o(t_i) = h_{00}(t_i) - [h_{00}(t_i)]_{P_i=0} \tag{13-100}$$

which becomes, for the half-wave νth-law detector,

$$s_o(t_i) = \frac{a\Gamma(\nu + 1)\sigma^\nu}{\Gamma(\nu/2 + 1)2^{(\nu+2)/2}} \left[_1F_1\left(-\frac{\nu}{2};1;-\frac{P_i^2}{2\sigma^2}\right) - 1\right] \tag{13-101}$$

The output signal power is defined to be the mean square of the output signal:

$$S_o(t_i) = E(\{h_{00}(t_i) - [h_{00}(t_i)]_{P_i=0}\}^2) \tag{13-102}$$

which becomes a constant when the input amplitude modulation is stationary. The output signal power from a half-wave νth-law detector then is

$$S_o(t_i) = \frac{a^2\Gamma^2(\nu + 1)\sigma^{2\nu}}{\Gamma^2(\nu/2 + 1)2^{\nu+2}} E\left\{\left[_1F_1\left(-\frac{\nu}{2};1;-\frac{P_i^2}{2\sigma^2}\right) - 1\right]^2\right\} \tag{13-103}$$

The (NXN) portion of the detector output autocorrelation function is, from Eq. (13-88),

$$R_{0(NXN)}(t_1,t_2) = \sum_{\substack{k=2 \\ (k\ even)}}^{\infty} \frac{1}{\left(\dfrac{k}{2}!\right)^2 2^k} E[h_{0k}(t_1)h_{0k}(t_2)]R_\nu^k(\tau) \tag{13-104}$$

where, from Eq. (13-97) for the half-wave νth-law detector,

$$h_{0k}(t_i) = \frac{a\Gamma(\nu + 1)\,_1F_1[(k - \nu)/2;1; -P_i^2/2\sigma^2]}{2\Gamma[1 - (k - \nu)/2](\sigma^2/2)^{(k-\nu)/2}} \tag{13-105}$$

† Just how *output signal* is to be defined depends strongly on the particular application. This definition is useful for the following analyses.

The corresponding noise power output is, on setting $t_2 = t_1$ in Eq. (13-104),

$$N_{o(NXN)}(t_i) = \frac{a^2\Gamma^2(\nu + 1)\sigma^{2\nu}}{2^{\nu+2}} \sum_{\substack{k=2 \\ (k \text{ even})}}^{\infty} \frac{E\{{}_1F_1{}^2[(k - \nu)/2;1;- P_i{}^2/2\sigma^2]\}}{(k/2!)^2\Gamma^2[1 - (k - \nu)/2]}$$

(13-106)

since $R_\nu(0) = R_n(0) = \sigma^2$. The terms $k > \nu$ in Eq. (13-104) and (13-106) vanish when ν is an even integer because of the factor $\Gamma[1 - (k - \nu)/2]$.

The (SXN) part of the detector output autocorrelation function is, from Eq. (13-89),

$$R_{o(SXN)}(t_1,t_2) = \sum_{k=1}^{\infty} \sum_{\substack{m=1 \\ (m+k \text{ even})}}^{k} \frac{E[h_{mk}(t_1)h_{mk}(t_2)]R_\nu{}^k(\tau)}{\left(\dfrac{k + m}{2}\right)! \left(\dfrac{k - m}{2}\right)! 2^{k-1}}$$

(13-107)

where $h_{mk}(t_i)$ is given by Eq. (13-97) for the half-wave νth-law detector. The corresponding noise power output is

$$N_{o(SXN)}(t_i) = \frac{a^2\Gamma^2(\nu + 1)\sigma^{2\nu}}{2^{\nu+1}}$$

$$\sum_{k=1}^{\infty} \sum_{\substack{m=1 \\ (m+k \text{ even})}}^{k} \frac{E\left[\left(\dfrac{P_i{}^2}{2\sigma^2}\right)^m {}_1F_1{}^2\left(\dfrac{m + k - \nu}{2};m + 1;- \dfrac{P_i{}^2}{2\sigma^2}\right)\right]}{(m!)^2 \left(\dfrac{k + m}{2}\right)! \left(\dfrac{k - m}{2}\right)! \Gamma^2\left(1 - \dfrac{m + k - \nu}{2}\right)}$$

(13-108)

The terms $m + k > \nu$ in Eqs. (13-107) and (13-108) vanish when ν is an even integer because of the factor $\Gamma[1 - (m + k - \nu)/2]$.

The total noise power output from the detector is given by the sum of Eqs. (13-106) and (13-108). It should be noted from these equations that the total detector output noise power is independent of the shape of the input noise spectrum.† This fact is, of course, a result of the assumptions of a narrow-band input noise and a zonal output filter.

Small Input Signal-to-noise Ratios. These expressions for output signal and noise powers are rather complicated functions of the input signal and noise powers. Considerably simpler results may be obtained for either very small or very large values of the input signal-to-noise power ratio

$$\frac{S}{N}(t_i)_I = \frac{E(P_i{}^2)}{2\sigma^2}$$

(13-109)

When the input signal-to-noise power ratio is very small, it is con-

† Middleton (II, Sec. 3).

venient to expand the confluent hypergeometric function in series form by Eq. (13-96). In this way we get, from Eq. (13-103),

$$S_o(t_i) = \frac{a^2\Gamma^2(\nu + 1)\sigma^{2\nu}}{2^{\nu+2}\Gamma^2(\nu/2)} \frac{E(P_i^4)}{E^2(P_i^2)} \left[\frac{S}{N}(t_i)_I \right]^2 \tag{13-110}$$

The output signal power of a νth-law detector hence varies as the square of the input signal-to-noise power ratio for small values of the latter. The (NXN) part of the output noise power becomes, from Eq. (13-106),

$$N_{o(NXN)}(t_i) = \frac{a^2\Gamma^2(\nu + 1)\sigma^{2\nu}}{2^{\nu+2}} \sum_{\substack{k=2 \\ (k\ even)}}^{\infty} \frac{1}{\left(\frac{k}{2}!\right)^2 \Gamma^2\left(1 - \frac{k-\nu}{2}\right)} \tag{13-111}$$

which is independent of the input signal-to-noise power ratio. The (SXN) part of the output noise power becomes, on using Eq. (13-96) in Eq. (13-108),

$$N_{o(SXN)}(t_i) = \frac{a^2\Gamma^2(\nu + 1)\sigma^{2\nu}}{2^{\nu+1}} \sum_{\substack{k=1 \\ (m+k\ even)}}^{\infty} \sum_{m=1}^{k} \frac{\frac{E(P_i^{2m})}{E^m(P_i^2)}\left[\frac{S}{N}(t_i)_I\right]^m}{\left(\frac{k+m}{2}\right)!\left(\frac{k-m}{2}\right)!2^{k-1}}$$

For small values of the input signal-to-noise ratio, the dominant terms in this equation are those for which $m = 1$. Hence, approximately,

$$N_{o(SXN)}(t_i) = \frac{a^2\Gamma^2(\nu + 1)\sigma^{2\nu}}{2^{\nu+1}}\left[\frac{S}{N}(t_i)_I\right] \sum_{\substack{k=1 \\ (k\ odd)}}^{\infty} \frac{1}{\left(\frac{k+1}{2}\right)!\left(\frac{k-1}{2}\right)!2^{k-1}}$$

$$\tag{13-112}$$

A comparison of Eqs. (13-111) and (13-112) shows that the output noise is primarily of the (NXN) type when the input signal-to-noise ratio is very small.

On combining the above expressions, the output signal-to-noise power ratio is found to be

$$\frac{S}{N}(t_i)_o = C[\nu,p(P_i)]\left[\frac{S}{N}(t_i)_I\right]^2 \qquad \text{when } \frac{S}{N}(t_i)_I << 1 \tag{13-113}$$

where the constant

$$C[\nu,p(P_i)] = \frac{1}{\Gamma^2(\nu/2)}\frac{E(P_i^4)}{E^2(P_i^2)}\frac{1}{\displaystyle\sum_{\substack{k=2 \\ (k\ even)}}^{\infty}\frac{1}{\left(\frac{k}{2}!\right)^2\Gamma^2\left(1-\frac{k-\nu}{2}\right)}} \tag{13-114}$$

is a function only of ν and the probability distribution of the input amplitude modulation. Since the same ratio is obtained for the full-wave νth-law detector, the output signal-to-noise power ratio for a νth-law detector is proportional to the *square* of the input signal-to-noise power ratio for *small* values of the latter and *all* values of ν.   This is the "small signal" suppression effect.

Large Input Signal-to-Noise Ratios. When the input signal-to-noise power ratio is very large, it is most convenient to expand the confluent hypergeometric function in the asymptotic series†

$$
\begin{aligned}
{}_1F_1(a;c;-z) &= \frac{\Gamma(c)}{\Gamma(c-a)z^a} \sum_{r=0}^{\infty} \frac{(a)_r(a-c+1)_r}{r!z^r} \\
&= \frac{\Gamma(c)}{\Gamma(c-a)z^a} \left[1 + \frac{a(a-c+1)}{z} \right. \\
&\quad \left. + \frac{a(a+1)(a-c+1)(a-c+2)}{2z^2} + \cdots \right]
\end{aligned}
\tag{13-115}
$$

On substituting Eq. (13-115) into Eq. (13-103), we get

$$
S_o(t_i) = \frac{a^2\Gamma^2(\nu+1)\sigma^{2\nu}}{\nu^4\Gamma^4(\nu/2)2^{\nu-2}} \frac{E(P_i^{2\nu})}{E^\nu(P_i^2)} \left[\frac{S}{N}(t_i)_I \right]^\nu
\tag{13-116}
$$

The (NXN) part of the output noise power becomes, approximately, on substituting Eq. (13-115) into Eq. (13-106) and picking out the dominant term,

$$
N_{o(NXN)}(t_i) = \frac{a^2\Gamma^2(\nu+1)\sigma^{2\nu}}{\Gamma^4(\nu/2)2^{\nu+2}} \frac{E[P_i^{2(\nu-2)}]}{E^{(\nu-2)}(P_i^2)} \left[\frac{S}{N}(t_i)_I \right]^{\nu-2}
\tag{13-117}
$$

and the (SXN) part of the output noise power becomes, approximately, on substituting Eq. (13-115) into Eq. (13-108) and keeping the dominant term,

$$
N_{o(SXN)}(t_i) = \frac{a^2\Gamma^2(\nu+1)\sigma^{2\nu}}{\nu^2\Gamma^4(\nu/2)2^{\nu-1}} \frac{E[P_i^{2(\nu-1)}]}{E^{(\nu-1)}(P_i^2)} \left[\frac{S}{N}(t_i)_I \right]^{\nu-1}
\tag{13-118}
$$

A comparison of Eqs. (13-117) and (13-118) shows that the output noise is primarily of the (SXN) type when the input signal-to-noise ratio is very large.

On combining the above expressions, the output signal-to-noise power ratio is found to be

$$
\frac{S}{N}(t_i)_o = K[\nu,p(P_i)] \left[\frac{S}{N}(t_i)_I \right] \qquad \text{when } \frac{S}{N}(t_i)_I >> 1 \tag{13-119}
$$

where the constant

$$
K[\nu,p(P_i)] = \frac{2}{\nu^2} \frac{E(P_i^{2\nu})}{E(P_i^2)E(P_i^{2\nu-2})}
\tag{13-120}
$$

† Magnus and Oberhettinger (I, Chap VI, Art. 1).

is a function only of ν and the probability distribution of the input amplitude modulation. Thus the output signal-to-noise power ratio for a νth-law detector is *directly* proportional to the input signal-to-noise power ratio for *large* values of the latter and *all* values of ν. Hence *all* νth-law detectors behave in essentially the same way as the full-wave square-law detector in so far as ratios of signal-to-noise power *ratios* are concerned.

13-7. Problems

1. Show by direct evaluation of the contour integral† that

$$\frac{1}{2\pi j} \int_{\epsilon-j\infty}^{\epsilon+j\infty} \left[\frac{a\Gamma(\nu + 1)}{w^{\nu+1}} \right] e^{xw} \, dw = \begin{cases} ax^\nu & \text{when } x > 0 \\ 0 & \text{when } x \le 0 \end{cases} \qquad (13\text{-}121)$$

where $\epsilon > 0$, a is real, and ν is real and nonnegative.

2. Let the input to a nonlinear device be

$$x(t) = s(t) + n(t) \qquad (13\text{-}122)$$

where $s(t)$ and $n(t)$ are sample functions of independent real gaussian processes with zero means and variances $\sigma^2(s_t)$ and $\sigma^2(n_t)$, respectively. Show that the autocorrelation function of the output of the nonlinear device can be expressed as

$$R_y(t_1,t_2) = \sum_{k=0}^{\infty} \sum_{m=0}^{\infty} \frac{E[h_{km}(t_1)h_{km}(t_2)]}{k!m!} R_s{}^k(t_1,t_2) R_n{}^m(t_1,t_2) \qquad (13\text{-}123)$$

where the coefficients $h_{km}(t_i)$ are

$$h_{km}(t_i) = \frac{1}{2\pi j} \int_C f(w) w^{k+m} \exp\left(\frac{\sigma_i{}^2 w^2}{2}\right) dw \qquad (13\text{-}124)$$

where $f(w)$ is the transfer function of the nonlinear device and $\sigma_i{}^2 = \sigma^2(s_i) + \sigma^2(n_i)$.

3. Let the nonlinear device of Prob. 2 be an ideal limiter with the transfer characteristic

$$g(x) = \begin{cases} a & \text{for } x > 0 \\ 0 & \text{for } x = 0 \\ -a & \text{for } x < 0 \end{cases} \qquad (13\text{-}125)$$

Show that the coefficients in Prob. 2 become

$$h_{km}(t_i) = \begin{cases} (-1)^{(k+m-1)/2} \dfrac{2^{(k+m)/2}\Gamma[(k+m)/2]}{\pi\sigma_i{}^{k+m}} & \text{for } k + m \text{ odd} \\ 0 & \text{for } k + m \text{ even} \end{cases} \qquad (13\text{-}126)$$

4. Let the input to a nonlinear device be

$$x(t) = \cos(pt + \theta) + A \cos(qt + \phi) \qquad (13\text{-}127)$$

† Cf. Whittaker and Watson (I, Art. 12.22).

where $|q - p| << p$, where θ and ϕ are independent random variables each uniformly distributed over the interval $(0, 2\pi)$, and where A is a constant. Show that the autocorrelation function of the output of the nonlinear device can be expressed as

$$R_y(\tau) = \sum_{m=0}^{\infty} \sum_{k=0}^{\infty} \varepsilon_m \varepsilon_k h^2_{mk} \cos m p\tau \cos kq\tau \qquad (13\text{-}128)$$

where ε_m and ε_k are Neumann numbers and the coefficient h_{mk} is given by

$$h_{mk} = \frac{1}{2\pi j} \int_C f(w) I_m(w) I_k(wA) \, dw \qquad (13\text{-}129)$$

where $f(w)$ is the transfer function of the nonlinear device and where $I_m(z)$ is a modified Bessel function of the first kind.

5. Let the nonlinear device of Prob. 4 be an ideal limiter with the transfer characteristic given by Eq. (13-125). Show that the coefficient h_{mk} in Prob. 4 becomes

$$
h_{mk} =
\begin{cases}
(-1)^{\frac{m+k-1}{2}} \dfrac{a\Gamma\left(\dfrac{m+k}{2}\right) {}_2F_1\left(\dfrac{m+k}{2}, \dfrac{m-k}{2}; m+1; \dfrac{1}{A^2}\right)}{\pi A^m \Gamma\left(1 + \dfrac{k-m}{2}\right) m!} \\
\qquad\qquad\qquad\qquad\qquad\qquad \text{for } m+k \text{ odd and } A > 1 \\[2ex]
(-1)^{\frac{m+k-1}{2}} \dfrac{a}{\pi \left(\dfrac{k+m}{2}\right) \Gamma\left(1 + \dfrac{k-m}{2}\right) \Gamma\left(1 + \dfrac{m-k}{2}\right)} \\
\qquad\qquad\qquad\qquad\qquad\qquad \text{for } m+k \text{ odd and } A = 1 \\[2ex]
(-1)^{\frac{m+k-1}{2}} \dfrac{aA^k \Gamma\left(\dfrac{m+k}{2}\right) {}_2F_1\left(\dfrac{m+k}{2}, \dfrac{k-m}{2}; k+1; A^2\right)}{\pi \Gamma\left(1 + \dfrac{m-k}{2}\right) k!} \\
\qquad\qquad\qquad\qquad\qquad\qquad \text{for } m+k \text{ odd and } A < 1 \\[1ex]
0 \qquad\qquad\qquad\qquad\qquad\qquad \text{for } m+k \text{ even}
\end{cases}
\qquad (13\text{-}130)
$$

where ${}_2F_1(a,b;c;z)$ is a hypergeometric function.†

6. Let the nonlinear device of Probs. 4 and 5 be followed by a band-pass zonal filter, centered about $p/2\pi$, to form an ideal band-pass limiter. Show that the filter output $z(t)$ has the autocorrelation function

$$R_z(\tau) = \frac{2a^2}{\pi^2}\left[\frac{1}{A^2}\cos p\tau + 4\cos q\tau + \frac{1}{A^2}\cos (2q - p)\tau\right] \qquad (13\text{-}131)$$

when $A >> 1$, and

$$R_z(\tau) = \frac{2a^2}{\pi^2}\left[4\cos p\tau + A^2\cos q\tau + A^2\cos (2p - q)\tau\right] \qquad (13\text{-}132)$$

when $A << 1$.

7. Derive Eq. (13-90) and (13-91).

† Cf. Magnus and Oberhettinger (I, Chap. II) or Whittaker and Watson (I, Chap. XIV).

8. Let the input to a half-wave linear detector be an unmodulated sine wave plus a stationary narrow-band real gaussian noise. Show that the mean value of the detector output is, approximately,

$$m_o = \begin{cases} \dfrac{a\sigma_n}{\sqrt{2\pi}} & \text{when } \dfrac{P^2}{2} << \sigma_n^2 \\[3mm] \dfrac{aP}{\pi} & \text{when } \dfrac{P^2}{2} >> \sigma_n^2 \end{cases} \qquad (13\text{-}133)$$

where a is the detector scaling constant, P is the sine-wave amplitude, and σ_n^2 is the noise variance.

9. For the detector, and input, of Prob. 8, show that the detector noise output power can, to a reasonable approximation, be expressed as[†]

$$N_o = \frac{a^2 \sigma_n^2}{8\pi} \left\{ {}_1F_1^2 \left[\frac{1}{2}; 1; -\frac{P^2}{2\sigma_n^2} \right] + \frac{P^2}{\sigma_n^2} {}_1F_1^2 \left[\frac{1}{2}; 2; -\frac{P^2}{2\sigma_n^2} \right] \right\} \qquad (13\text{-}134)$$

10. Let the input to a full-wave (odd)[‡] νth-law device be the sum of an amplitude-modulated sine wave and a stationary narrow-band real gaussian noise. Show that the autocorrelation function of the device output is

$$R_y(t_1, t_2) = \sum_{\substack{m=0 \\ (m+k \text{ odd})}}^{\infty} \sum_{k=0}^{\infty} \frac{4\varepsilon_m}{k!} E[h_{mk}(t_1) h_{mk}(t_2)] R_n^k(\tau) \cos m\omega_c \tau \qquad (13\text{-}135)$$

where ε_m is the Neumann factor, $R_n(\tau)$ is the autocorrelation function of the input noise, $f_c = \omega_c/2\pi$ is the carrier frequency of the signal, and the coefficients $h_{mk}(t_i)$ are given by Eq. (13-92) and hence by Eq. (13-97).

11. Let the full-wave (odd) νth-law device of Prob. 10 be followed by a zonal band-pass filter, centered about f_c, to form a νth-law nonlinear amplifier. Show that[§]

$$\frac{S}{N}(t_i)_o = C'(\nu) \left[\frac{S}{N}(t_i)_I \right] \qquad \text{when } \frac{S}{N}(t_i)_I << 1 \qquad (13\text{-}136)$$

where the constant $C'(\nu)$ is

$$C'(\nu) = \cfrac{1}{\Gamma^2 \left(\dfrac{1+\nu}{2} \right) \left[\displaystyle\sum_{\substack{k=1 \\ (k \text{ odd})}}^{\infty} \cfrac{1}{\Gamma^2 \left(1 - \dfrac{k-\nu}{2} \right) \left(\dfrac{k+1}{2} \right)! \left(\dfrac{k-1}{2} \right)!} \right]} \qquad (13\text{-}137)$$

and hence that the output signal-to-noise power ratio for a νth-law nonlinear amplifier is *directly* proportional to the input signal-to-noise power ratio for *small* values of the latter and *all* values of ν.

12. For the νth-law nonlinear amplifier of Prob. 11, show that

$$\frac{S}{N}(t_i)_o = K'[\nu, p(P_i)] \left[\frac{S}{N}(t_i)_I \right] \qquad \text{when } \frac{S}{N}(t_i)_I >> 1 \qquad (13\text{-}138)$$

[†] Cf. Rice (I, Art. 4.10).
[‡] Cf. Art. 13-2.
[§] Cf. Art. 13-6.

where the constant $K'[\nu, p(P_i)]$ is

$$K'[\nu, p(P_i)] = \frac{2}{1 + \nu^2} \frac{E(P_i^{2\nu})}{E(P_i^2)E(P_i^{2\nu-2})} \tag{13-139}$$

and hence that the output signal-to-noise power ratio for a νth-law nonlinear amplifier is *directly* proportional to the input signal-to-noise power ratio for *large* values of the latter and *all* values of ν.

13. The νth-law nonlinear amplifier of Probs. 11 and 12 becomes an ideal band-pass limiter when $\nu = 0$. Let the input to an ideal band-pass limiter be the sum of an unmodulated sine wave and a stationary narrow-band real gaussian noise. Show that†

$$\left(\frac{S}{N}\right)_o = \frac{\pi}{4}\left(\frac{S}{N}\right)_I, \qquad \text{when } \left(\frac{S}{N}\right)_I << 1 \tag{13-140}$$

and that

$$\left(\frac{S}{N}\right)_o = 2\left(\frac{S}{N}\right)_I, \qquad \text{when } \left(\frac{S}{N}\right)_I >> 1 \tag{13-141}$$

where $(S/N)_o$ and $(S/N)_I$ are, respectively, the output and input signal-to-noise power ratios.

14. Show that a cascade of N ideal band-pass limiters is equivalent to a single ideal band-pass limiter.

† Cf. Davenport (I).

CHAPTER 14

STATISTICAL DETECTION OF SIGNALS

In radio communications and in radar, a signal meant to carry intelligence to a user is always partly masked and distorted in transmission before it is made available at the receiving end. Some of this distortion is due to natural causes which cannot be removed. For example, thermal noise cannot be avoided in any receiver; sea-clutter return cannot be avoided for a radar looking at objects on the ocean; distortion of signal waveforms by multipath propagation cannot be avoided under certain conditions in long-distance radio communication. Thus, even after careful engineering of all parts of a communications or radar system to minimize the disturbing influences, there remains some signal distortion, which can deprive the user of at least part of the intelligence carried by the signal.

The question naturally arises, then, as to how a received signal can be processed to recover as much information from it as possible. If a signal is perturbed in an unknown way by an agent which behaves with some statistical regularity, as those mentioned above do, it is appropriate to apply a statistical analysis to the problem of how the user should process the received signal. This chapter contains a preliminary treatment of the subject of statistical analysis of received signals. Certain statistical procedures are introduced and then applied to a few typical examples from radio and radar. No attempt is made to discuss the practical engineering aspects of these radio and radar problems, and no attempt is made to catalogue those problems for which a satisfactory solution exists. There is a connection between optimizing a receiver and optimizing the form of the signals to be used, but we shall not discuss this second problem at all. There is also a strong tie-in between the optimization procedures discussed here and those discussed in Chap. 11. We shall need certain concepts and results from two parts of the theory of statistical inference: the testing of hypotheses and the estimation of parameters. These are discussed, necessarily very briefly, in Arts. 14-2, 14-3, and 14-4.

14-1. Application of Statistical Notions to Radio and Radar

Two examples will illustrate the applicability of statistics to radio- and radar-receiver design. First, suppose an FSK (frequency-shift keying) radio teletype operates at a frequency between 5 and 15 megacycles over a long distance, say, two to four thousand miles. The basic teletype alphabet is a two-symbol alphabet, with symbols called *mark* (M) and *space* (S). Each letter of the English alphabet is coded as a sequence of five marks and spaces, e.g., $A = MMSSS$. In teletype operation, each mark or space has a fixed time duration of T seconds. The FSK teletype is a system in which a mark is transmitted as a carrier at a given frequency f_0 for a duration of T seconds, and a space is transmitted as a carrier at a slightly different frequency f_1 for a duration of T seconds. Thus, a letter of the English alphabet is broadcast as a signal $s(t)$ given by

$$s(t) = A \cos \left[\omega(t)t + \phi(t) \right] \qquad t > 0 \qquad (14\text{-}1)$$

where
$$\omega(t) = 2\pi f^{(k)} \qquad \text{for } (k-1)T < t \leq kT, \, k = 1, 2, 3, 4, 5$$
$$\phi(t) = \phi^{(k)} \qquad \text{for } (k-1)T < t \leq kT$$

and where $f^{(k)}$ is either f_0 or f_1, depending on the message, and $\phi^{(k)}$ is a fixed phase shift ϕ_0 if $f^{(k)} = f_0$ and ϕ_1 if $f^{(k)} = f_1$.

At the frequencies and ranges indicated, there will be times when there are multiple paths for the radiated waves to take between transmitter and receiver. The different paths will introduce different delays and different attenuations, and these delays and attenuations will change with time. At the receiver there will be added, necessarily, antenna noise and thermal and shot noise. Thus, if a transmitted signal is given by $s(t)$, if we assume that the path delays do not change appreciably over the duration of a message, and if we neglect other disturbing influences, such as frequency dependence of the path parameters and the small doppler shifts of frequency, the received signal is given by a function

$$y(t) = a_1(t)s(t - \tau_1) + a_2(t)s(t - \tau_2) + \cdots + a_N(t)s(t - \tau_N) + n(t)$$
$$(14\text{-}2)$$

where $n(t)$ is stationary gaussian noise. The path delays $\tau_1, \ldots, \tau_N$ and the path attenuations $a_1(t), \ldots, a_N(t)$ are not known exactly; there may or may not be some information about them. Even the number of paths N may not be known.

What the receiver must do is decide for each time interval $(k-1)$ $T < t \leq kT$ whether the transmitter was sending a mark or a space, i.e., whether during that time interval the transmitter frequency was f_0 or f_1. This is a statistical decision problem; one of two alternative hypotheses must be true: that the frequency is f_0 or that it is f_1. A decision between these alternatives must be based on imperfect observations, with only statistical information available about $n(t)$ and perhaps not even that about the $a_i(t)$ and the τ_i.

Even without analyzing the problem in detail, one can make the following observations. If the largest difference between the delays τ_i is small compared to T and the $a_i(t)$ do not change much in time T, then except for a little interference from the previous symbol at the beginning of each new symbol, the effect of the various delays is just to add sine waves of the same frequency but different phases. Since the receiver needs only to measure frequency, the addition of the different sine waves does not distort the signal form in a significant way and the additive noise is essentially the only source of error. However, the different sine waves may combine for certain values of the $a_i(t)$ in such a way as to cancel; consequently there is very little signal energy at the receiver and the noise $n(t)$ completely masks the signal. In this situation, optimizing the receiver can accomplish little; if there is no signal present at the input to the receiver, no amount of sophistication of design in the receiver can recover it. On the other hand, if T is decreased (resulting in an increase

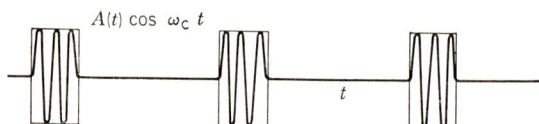

Fig. 14-1. Transmitted signal from a pulsed radar.

in the rate at which messages can be sent) to a point where the differences between the τ_i are comparable with or even greater than T, then in addition to the possible cancellation effect, the signal is smeared out over an observation interval. There will then usually be present at the receiver components of both mark and space frequencies. In this situation, if some information is available about the $a_i(t)$ and the τ_i, a receiver which is optimum according to appropriate statistical criteria may work a good deal better *on the average* than a receiver which is not optimum.

Although not much can be done at a single receiver about fading of signal level due to cancellation in multipath propagation, this effect can be countered to some extent. One thing that can be done is to operate on two different frequency channels simultaneously, i.e., to use so-called *frequency diversity*. Then a combination of path lengths which causes cancellation at one wave length probably will not at the other. Thus there will be two receivers providing information from which to make a decision. This leads to a statistical problem of how to weight the information from each receiver in arriving at a decision.

As a second example, let us consider a narrow-beam scanning pulsed radar used to locate ships at sea. The carrier frequency is, say, several thousand megacycles, and the pulse length is a few microseconds. If $A(t)$ is the modulation—a periodic sequence of pulses, as shown in Fig. 14-1—the signal emitted from the transmitter can be represented by

$$s(t) = A(t) \cos \omega_c t$$

The signal radiated in a fixed direction increases to a maximum as the scanning antenna sweeps by that direction and then falls off essentially to zero. This imposes a second and relatively slowly varying modulation on the signal. Letting $G(t)$ be the antenna (voltage) gain in a fixed direction as the antenna rotates, the signal radiated in one direction is given by

$$G(t)A(t)\cos\omega_c t$$

If the radiated energy is reflected from a fixed object at a distance r from the transmitter and received on the same antenna, the echo signal received at the radar is given by

$$\frac{1}{r^2}\,\alpha\left(t-\frac{r}{c}\right)G(t)G\left(t-\frac{2r}{c}\right)A\left(t-\frac{2r}{c}\right)\cos\omega_c\left(t-\frac{2r}{c}\right)$$

where c is velocity of propagation and $\alpha(t)$ is proportional to the ratio of incident signal field strength to the field strength of the signal reflected back to the radar at time t by the object. Now if the object is moving in a radial direction with constant velocity v, there is a doppler shift of the frequency spectrum of the signal, which, since the modulation is narrow band, can be accounted for approximately by replacing ω_c by $\omega_c\left(1-\dfrac{2v}{c}\right)$. Then, adding the noise $n(t)$ at the receiver, the total received signal is approximately

$$y(t) = \frac{1}{r^2}\,\alpha\left(t-\frac{\tau}{2}\right)G(t)G(t-\tau)A(t-\tau)\cos\left[(\omega_c+\omega_d)t-\tau\omega_c\right]+n(t)$$

$$(14\text{-}3)$$

where
$$\tau = \frac{2r}{c}$$

$$\omega_d = -\frac{2v}{c}\,\omega_c$$

If there are other objects or sea clutter illuminated at the same time, then during the time interval that the return from the first object as given by Eq. (14-3) is not zero, there will be other received signals of the same form as the first term in Eq. (14-3) added in. Sea clutter will cause the superposition of a great many such terms, each of relatively small amplitude.

The usual function of the radar is to detect the presence of an object and determine its position in range and azimuth. The data available at the receiver are signals which have been corrupted by random noise and perhaps by sea-clutter return, which is also random. Thus a decision has to be made from statistical data; and if it is decided an object is present, the parameter's range and azimuth have to be estimated from statistical data. The problem is how to use any information available about the statistical nature of the noise and sea clutter, together with the knowledge of signal structure, to make the detection and location of the object as sure and precise as possible.

Of course, many different situations are encountered in radio and radar reception with respect both to the nature of the signal and to random disturbances of the signal. Common to all, however, is the presence of additive gaussian noise caused by thermal effects. Since often the additive gaussian noise is the only random disturbance of any importance, considerable emphasis has been placed on signal detection and extraction problems in which the received signal is the same as the transmitted signal (or a multiple of it) plus gaussian noise. Generally speaking, any random signal perturbations other than additive gaussian noise complicate the receiver problem considerably.

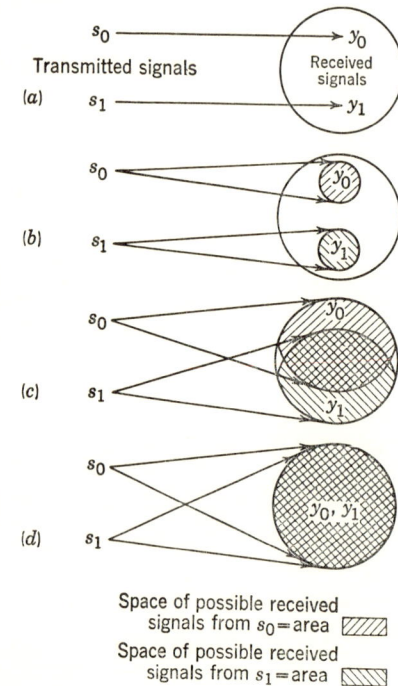

However, it is possible to have a signal present at the receiver which has been "hashed up," by multipath propagation, for example, but which is strong enough completely to override the thermal noise. In such a case the reception problem may or may not be really a statistical problem. For example, in FSK radioteletype communication, if we idealize the situation by supposing the additive gaussian noise to be zero, it may be that the mode of propagation is such that a received signal, although distorted, could have come from only one of the two possible transmitted signals. Then there is no ambiguity at the receiver in deciding whether a mark or space was sent.

Symbolically, for the simple case of point-to-point communication using a two-letter alphabet, the possibilities for ambiguity at the receiver are as shown in Fig. 14-2. The diagram of Fig. 14-2a is meant to indicate that the transmitted signal s_0 is transformed into one received signal y_0, and s_1 into y_1. Thus, although y_0 and y_1 differ from s_0 and s_1, they are fixed and distinct; consequently, if sufficient knowledge about the transmission medium—the channel—is available at the receiver, there need never be any ambiguity in reception. In general, if there is a fixed 1:1

FIG. 14-2. Decision situations at the receiver.

transformation carrying transmitted signals into received signals, the user can tell exactly what was transmitted. In Fig. 14-2b, both possible transmitted signals are perturbed randomly so that there is a whole set of possible received signals that might have come from s_0 and a whole set that might have come from s_1, but these sets have no elements in common. Again, a sure decision can be made at the receiver as to which signal was sent. The statistical problem of deciding between s_0 and s_1 on the basis of received data is said to be *singular* in these cases where a correct decision can always be made. In Fig. 14-2c there is a set of possible received signals which could arise either from s_0 or s_1, and in Fig. 14-2d any received signal could have been caused by either s_0 or s_1. Always, in this last case, and sometimes in the previous one, a nontrivial decision has to be made, after the received signal is observed, as to whether s_0 or s_1 was sent.

14-2. Testing of Statistical Hypotheses†

This article and the two that follow contain some basic statistical principles and procedures presented with no more than passing reference to communications or radar. These principles and procedures cannot be presented in detail here, of course.‡ Let us start by discussing the problem of testing between two alternate statistical hypotheses, the simplest case of the theory of testing of hypotheses.§

The situation is this: An event is observed which could have been produced by either of two possible mutually exclusive causes, or which is a concomitant of one of two possible mutually exclusive states of nature. The hypothesis that a particular one of the two possible causes was responsible for the event is designated by H_0, and the hypothesis that the other possible cause was responsible, by H_1. It is desired to establish a reasonable rule for deciding between H_0 and H_1, that is, for deciding which possible cause of the observed event is, in some sense, most likely. The decision rule is to be laid down before the event is observed. Some knowledge about the causes and their connections with the events that can occur is obviously necessary before a useful decision rule can be formulated. In statistical decision theory this knowledge is taken to be in the form of given probability distributions.

† The statistical treatment of signal detection and extraction problems can be unified and generalized by using the statistical-decision theory associated with the name of Wald. Cf. Middleton and Van Meter (I).

‡ The reader is referred to Mood (I) for general statistical background. The chapters on hypothesis testing and estimation are particularly pertinent.

§ A comprehensive treatment of the subject of hypothesis testing on a more advanced level than in Mood is given in Lehman (I). There is a large body of published material on this subject; some of the early papers of Neyman and Pearson, who are responsible for much of the development of the subject, may be particularly interesting, e.g., Neyman and Pearson (I) and (II).

To be more precise, we postulate a space Y of all possible values of some observed variable (or variables) which depends on two mutually exclusive causes C_0 and C_1, one of which must be in effect. We assume that if C_0 is in effect, there is a *known* probability law P_0 induced on Y which determines the probability that the observed variable takes on any specified set of values. Similarly, we assume that if C_1 is in effect, there is a different known probability law P_1 induced on Y. We shall suppose that these probabilities have probability density functions $p_0(y)$ and $p_1(y)$, respectively, for points y in Y. In addition we *sometimes* assume that there are known probabilities π_0 and π_1, where $\pi_0 + \pi_1 = 1$, for the existence of the causes C_0 and C_1. The problem is to establish a decision rule for choosing H_0 or H_1. This amounts exactly to partitioning the space Y into two parts, so that for any observation y we can say that we believe either C_0 or C_1 to be the true cause.

Bayes' Solution—A Posteriori Probability. Consider the problem formulated above with known probabilities π_0 and π_1 for C_0 and C_1. These probabilities are usually called *a priori* probabilities. If M is a set of values of y, by Bayes' rule the conditional probabilities of C_0 and C_1, given M, are

$$P(C_0|M) = \frac{P(M|C_0)P(C_0)}{P(M)} \qquad (14\text{-}4a)$$

$$P(C_1|M) = \frac{P(M|C_1)P(C_1)}{P(M)} \qquad (14\text{-}4b)$$

Letting M be the set of values $y \leq \eta < y + dy$ and using the notation introduced above, the ratio of these conditional probabilities is

$$\frac{P(C_0|y)}{P(C_1|y)} = \frac{p_0(y)\pi_0}{p_1(y)\pi_1} \qquad (14\text{-}5)$$

The probabilities $P(C_0|y)$ and $P(C_1|y)$ are sometimes called *a posteriori* probabilities of C_0 and C_1. An eminently reasonable rule for making a decision in many examples is: once an observation y has been made, choose hypothesis H_0 if the conditional probability $P(C_0|y)$ exceeds $P(C_1|y)$; choose H_1 if $P(C_1|y)$ exceeds $P(C_0|y)$; and choose either one, say H_1, if they are equal. That is,

accept hypothesis H_0 if $\dfrac{p_0(y)}{p_1(y)} > \dfrac{\pi_1}{\pi_0}$ $\qquad (14\text{-}6)$

accept hypothesis H_1 if $\dfrac{p_0(y)}{p_1(y)} \leq \dfrac{\pi_1}{\pi_0}$, or if $p_0(y) = p_1(y) = 0$

If there is no difference in importance between the two possible kinds of mistakes, i.e., choosing H_0 when H_1 is in fact true, and choosing H_1 when H_0 is in fact true, then by adopting this decision rule an observer will guarantee himself a minimum penalty due to incorrect inferences over a long sequence of repeated independent observations.

If the two kinds of error are not of equal importance, then it seems reasonable to prejudice the test one way or the other by multiplying the threshold value π_1/π_0 by some constant different from one. Exactly how to do this shows up more clearly when we look at the same problem from a different point of view in the following paragraph.

After a decision has been made, one of four possible situations obtains: (1) H_0 may be true, and the decision is that H_0 is true; (2) H_0 true, decision is H_1; (3) H_1 true, decision is H_0; (4) H_1 true, decision is H_1. Each one of these situations represents an addition of true or false information to the observer's state of knowledge. Let us suppose that we can assign a numerical gain or loss to each of these four possibilities. We may as well assign a loss to each with the understanding that a negative loss is a gain. Thus we assign loss values, which are numbers designated by L_{00}, L_{01}, L_{10}, L_{11} to each of the four situations listed above, in the order listed. The expected loss to the observer is the average value of the four loss values, each weighted according to the probability of its occurrence. That is,

$$L = E[\text{loss}] = \sum_{\substack{j=0,\,1 \\ k=0,\,1}} L_{jk} P(H_j \text{ is true and } H_k \text{ is chosen}) \qquad (14\text{-}7)$$

The probabilities involved depend on the decision rule used. Thus, if we denote by Y_0 that part of Y such that if the observation y belongs to Y_0, we choose H_0; and if correspondingly we denote by Y_1 the remainder of Y, where we choose H_1, Eq. (14-7) becomes

$$L = E[\text{loss}] = \sum_{\substack{j=0,\,1 \\ k=0,\,1}} L_{jk} \pi_j P_j(Y_k) \qquad (14\text{-}8)$$

where $P_j(Y_k)$ is the probability that y falls in Y_k if H_j is actually true. It is desired to choose Y_0 (and hence $Y_1 = Y - Y_0$) so as to minimize the loss. In order to do this we impose one highly reasonable constraint on the loss values L_{jk}; we assume that whichever hypothesis is actually true, a false decision carries a greater loss than a true decision. Thus

$$\begin{aligned} L_{01} &> L_{00} \\ L_{10} &> L_{11} \end{aligned} \qquad (14\text{-}9)$$

Now, using the facts that

$$P_j(Y_k) = \int_{Y_k} p_j(y)\,dy$$

and

$$P_j(Y_1) = 1 - P_j(Y_0)$$

we have, from Eq. (14-8),

$$L = L_{01}\pi_0 + L_{11}\pi_1$$
$$+ \int_{Y_0} [-\pi_0(L_{01} - L_{00})p_0(y) + \pi_1(L_{10} - L_{11})p_1(y)]\,dy \qquad (14\text{-}10)$$

The first two terms in Eq. (14-10) are constants and hence not affected by the choice of Y_0, so all that has to be done to minimize L is to choose Y_0 so as to minimize the integral in Eq. (14-10). This is easy, because since the first term in the integrand is everywhere less than or equal to zero and the second term is everywhere greater than or equal to zero, it is sufficient to choose for Y_0 that set of points y where the first term in the integrand is greater in absolute value than the second. That is, Y_0 is the set of points y where

$$\frac{p_0(y)}{p_1(y)} > \frac{\pi_1(L_{10} - L_{11})}{\pi_0(L_{01} - L_{00})} \tag{14-11}$$

Y_1 is all the rest of Y, including those points y for which $p_0(y) = p_1(y) = 0$. This criterion for a decision reduces to that of the inequality (14-6) if $L_{10} - L_{11} = L_{01} - L_{00}$. But if $L_{10} - L_{11} = L_{01} - L_{00}$ the minimization which led to the inequality (14-11) determines Y_0 so as to minimize the over-all probability of error. Thus, minimizing the probability of error gives a decision rule which is identical to the following rule: Choose that hypothesis H_0 or H_1 with the greater a posteriori probability.

The solution to the problem of testing between two hypotheses given by the inequality (14-11), which we shall call the *Bayes' solution*, seems to be satisfactory when it can be applied. However, there are often practical difficulties in assigning loss values, and there are both semantic and practical difficulties in assigning a priori probabilities. Since only the differences $L_{10} - L_{11}$ and $L_{01} - L_{00}$ enter into the Bayes' solution, one can without sacrificing any freedom of choice always set $L_{11} = L_{00} = 0$. Then it is necessary to give values to L_{10} and L_{01} which reflect the relative loss to the observer of the two kinds of error. Obviously, this can be hard to do in some circumstances, because all the implications of the two kinds of wrong decision may not be known. For example, if the problem is to test whether a radar echo indicates the presence of an object at sea or not, it is sometimes a much more serious error to decide there is no object present when one actually is than to decide an object is present when there is none. But rarely can one say with assurance that the first kind of error is a thousand times more serious than the second.

The meaning of a priori probabilities is a delicate question. When one assigns a probability π_0, as we have used the term probability, to a hypothesis H_0, he is taking H_0 to be an "event," i.e., a subset of a sample space. Whether or not H_0 turns out to be true should then depend on the outcome of a random experiment. But actually, the truth or falsity of H_0 is often determined in a nonrandom way, perhaps by known natural laws, or by the will of an individual; and the only reason there is a statistical test at all is that the observer does not know the true state of things. In such a circumstance, no random experiment is performed which has the hypothesis H_0 as a possible outcome. When one assigns a priori probabilities, he is introducing an apparent randomness to account for his ignorance. Thus, in the example of reception of a radio-teletype message, whether the symbol mark or space is actually transmitted is the result of a deliberate rational choice made by the sender. One could perhaps justify attaching a priori probabilities to the symbols in this case on the ground that the fact that a particular message is sent at a particular time is the result of what may be regarded as a random set of circumstances.

This brief remark indicates that the notion of an a priori probability may not always fit with the usual formulation of probability as a weighting of subsets of a sample space. On the other hand, it can be maintained that subjective probabilities, which represent degrees of belief, however arrived at, can legitimately be used for a priori probabilities.

But even if one concedes that it is meaningful to postulate a priori probabilities for hypotheses, he will most likely agree that in some situations it is awkward to give actual values for π_0 and π_1. In the radio-teletype example mentioned above, there is no practical difficulty; if H_0 is the hypothesis that the transmitted signal is a space, then it is natural to take for π_0 the relative frequency with which spaces are known to occur in coded English text. But in a radar detection problem, if H_0 is the hypothesis that a ship lies, say, in a one-mile-square area located at the maximum range of the radar and there is no recent auxiliary information about ship movements, then a believable value for π_0 is obviously hard to determine.

To summarize, the Bayes' test, as given by the inequality (14-11), guarantees minimum expected loss to the observer. There is often difficulty in applying it, first because of difficulty in assigning losses, second because of difficulty in assigning a priori probabilities. If the losses for the two kinds of errors are set equal, the test reduces to the simpler Bayes' test of (14-6), which minimizes the total probability of error, or maximizes the a posteriori probability of the possible hypotheses. Let us emphasize this last point. A reasonable mode of behavior for the observer is to minimize his expected loss, i.e., to use the test specified by the inequality (14-11) if he has sufficient information to do so. If he feels he cannot justify assigning particular loss values, then it is reasonable (and common practice) for him to minimize probability of error, i.e., to use the test specified by the inequality (14-6). But this means his decision rule is exactly the same as if he had set the loss values equal in the first place, even though he set up his test without considering loss.

14-3. Likelihood Tests

Simple Hypotheses. If we are not willing to introduce a priori probabilities into a hypothesis-testing problem, we cannot calculate over-all expected loss or probability of error. The decision rule must then be based on some different idea. A more or less obvious basis for a decision rule is given by what we call the *likelihood principle*, by which we infer that the cause (or state of nature) exists which would more probably yield the observed value. That is, if $p_0(y) > p_1(y)$, one chooses H_0; if $p_1(y) \geq p_0(y)$, one chooses H_1 (where again the decision for the case $p_1(y) = p_0(y)$ is arbitrary). This can be restated as follows: take the set Y_0 to be that set of values y for which $p_0(y) \neq 0$ and

$$\frac{p_0(y)}{p_1(y)} > 1 \qquad (14\text{-}12)$$

Here a test is given which is not founded on considerations involving a priori probability or loss and yet which gives the same decision rule as the Bayes' test with $\pi_1(L_{10} - L_{11}) = \pi_0(L_{01} - L_{00})$.

Every decision criterion discussed to this point has involved the ratio $p_0(y)/p_1(y)$. This function of y, which we shall denote by $l(y)$, is called the *likelihood ratio*. Since $p_0(y)/p_1(y)$ is not defined when both $p_0(y) = 0$ and $p_1(y) = 0$, a special definition for $l(y)$ is required for these points; we take $l(y) = 0$ when both $p_0(y)$ and $p_1(y)$ vanish. The likelihood ratio has a central role in statistical hypothesis testing; in fact, it will transpire that all the hypothesis tests we consider are likelihood ratio tests, sometimes with a more general form of the likelihood ratio. Since $l(y)$ is a function defined on the sample space Y (the space of observations), it is a random variable. It has two different probability distributions. If H_0 is true, there is a probability law effective on Y given by $p_0(y)$, so the probability distribution of l is

$$P_0(l < a) = \int_{\frac{p_0(y)}{p_1(y)} < a} p_0(y) \, dy \qquad (14\text{-}13)$$

If H_1 is true, the probability distribution of l is

$$P_1(l < a) = \int_{\frac{p_0(y)}{p_1(y)} < a} p_1(y) \, dy \qquad (14\text{-}14)$$

Before going further, it is convenient to introduce here a few conventional statistical terms that apply to hypothesis testing. Thus far we have considered only the testing of one hypothesis against one other hypothesis. These individual hypotheses are described as *simple* in order to distinguish this situation from one we shall discuss later, where so-called *composite hypotheses,* which are groups of simple hypotheses, are considered. As has been made clear by now, any decision rule for testing between two simple hypotheses is equivalent to partitioning the space of observations Y into two parts Y_0 and Y_1. If an observation y occurs which belongs to Y_0, the rule is to choose H_0; if y belongs to Y_1, the rule is to choose H_1. A rule is determined when either of the sets Y_0 or Y_1 is specified, because the other set is the remaining part of Y. It is common practice to discuss hypothesis tests in terms of one hypothesis H_0. Thus, in terms of H_0, Y_0 is called the *acceptance region* and Y_1 is called the *rejection region,* or more commonly, the *critical region.* If the observation y falls in Y_1, so that we reject H_0, when in fact H_0 is true, we say we have made an error of the *first kind;* if y falls in Y_0 when H_1 is true, we say we have made an error of the *second kind.* The probability of rejecting H_0 when it is true, $P_0(Y_1)$, is called the *level* or

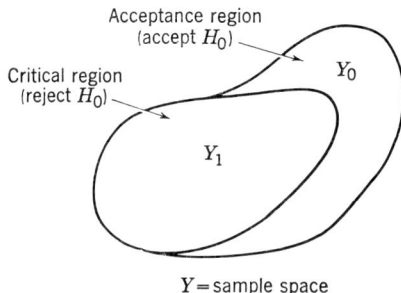

Fig. 14-3. Test between two hypotheses.

the *size* of the test. The probability of rejecting H_0 when it is false, $P_1(Y_1)$, is called the *power* of the test. These definitions are repeated in Fig. 14-3.

As mentioned above, when there are no a priori probabilities given, one cannot determine expected loss or over-all probability of error and hence one cannot establish a test to minimize either of these quantities. The likelihood principle was pointed out as being one criterion on which a test can then be based. Another criterion, which applies in different circumstances but is perhaps more convincing, is to keep the probability of error of the first kind less than or equal to a prescribed value and minimize the probability of error of the second kind. A famous theorem of Neyman and Pearson† states that a likelihood ratio test will satisfy this criterion. Precisely, if η is a real nonnegative number, the critical region $Y_1(\eta)$ which consists of all y for which $p_1(y)/p_0(y) \geq \eta$ gives a test of H_0 against H_1 of maximum power from among all tests of level $\leq P_0[Y_1(\eta)]$.

The proof of the theorem is as follows. Let $\alpha = P_0(Y_1) =$ the level of the test. Let T_1 be any region in Y for which $P_0(T_1) \leq \alpha$. We shall show that the power of the test which has critical region Y_1 is greater than or equal to the power of the test which has critical region T_1. Let U be the subset of Y which contains those points y in both T_1 and Y_1 (U may be empty), and let the notation $Y_1 - U$ denote the set of points

† See Lehman (I, Art. 2.1) or Neyman and Pearson (I, Art. III).

y in Y_1 which does not belong to U. First, for every point y in Y_1, and thus in $Y_1 - U$, $p_1(y) \geq \eta p_0(y)$; hence

$$P_1(Y_1 - U) = \int_{Y_1 - U} p_1(y)\, dy \geq \eta \int_{Y_1 - U} p_0(y)\, dy = \eta P_0(Y_1 - U) \quad (14\text{-}15)$$

Therefore

$$P_1(Y_1) = P_1(Y_1 - U) + P_1(U) \geq \eta P_0(Y_1 - U) + P_1(U) \quad (14\text{-}16)$$

But

$$\eta P_0(Y_1 - U) + P_1(U) = \eta P_0(Y_1) - \eta P_0(U) + P_1(U) \quad (14\text{-}17)$$

and since, by hypothesis

$$P_0(T_1) \leq \alpha = P_0(Y_1)$$

it follows that

$$\eta P_0(Y_1) - \eta P_0(U) + P_1(U) \geq \eta P_0(T_1) - \eta P_0(U) + P_1(U) \quad (14\text{-}18)$$

Now

$$\eta P_0(T_1) - \eta P_0(U) + P_1(U) = \eta P_0(T_1 - U) + P_1(U) \quad (14\text{-}19)$$

Since points in $T_1 - U$ do not belong to Y_1, $p_1(y) \leq \eta p_0(y)$ for all y in $T_1 - U$. Hence

$$\eta P_0(T_1 - U) + P_1(U) \geq P_1(T_1 - U) + P_1(U) \quad (14\text{-}20)$$

Finally,

$$P_1(T_1 - U) + P_1(U) = P_1(T_1) \quad (14\text{-}21)$$

The relations (14-16) through (14-21) combine to give

$$P_1(Y_1) \geq P_1(T_1)$$

which is what we wished to show.

One observes that as η increases the size of the set Y_1 decreases and the level of the test, $\alpha = P_0(Y_1)$, decreases (or, strictly speaking, does not increase). Other than this one cannot say anything in general about the functional dependence of α on η. In particular, α need not take on all values between zero and one; that is, α as a function of η may have jump discontinuities. It may also be observed that the set Y_1 may be taken to be those values of y for which $p_1(y)/p_0(y) > \eta$ (instead of $\geq$) and the statement and proof of the theorem remain unchanged. The difference is that the level α corresponding to η may be different. If Y_1 is chosen to be the set of y's for which $p_1(y)/p_0(y) \geq \eta$, then Y_0 is the set where $p_1(y)/p_0(y) < \eta$, which is equivalent to $p_0(y)/p_1(y) > 1/\eta$. By the preceding remark, this fact implies that Y_0 is a critical region of maximum power for a test of H_1 against H_0 at level $P_1(Y_0)$.

Example 14-3.1. Let the space of observations Y be the real numbers. Let H_0 be the hypothesis that the observed values are distributed according to a gaussian distribution with mean zero and variance one. Let H_1 be the hypothesis that the observed

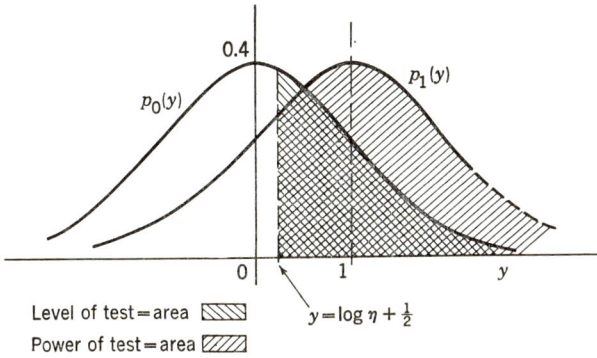

Level of test=area ▨▨ $y = \log \eta + \frac{1}{2}$
Power of test=area ▨▨

FIG. 14-4. Test for the mean of a gaussian distribution.

values are distributed according to a gaussian distribution with mean one and variance one.   (See Fig. 14-4.)   We test H_0 against H_1.   Then

$$p_0(y) = \frac{1}{\sqrt{2\pi}} \exp \frac{-y^2}{2}$$

$$p_1(y) = \frac{1}{\sqrt{2\pi}} \exp \left[-\frac{(y-1)^2}{2} \right]$$

and

$$\frac{p_1(y)}{p_0(y)} = \exp \left[\frac{1}{2}(2y - 1) \right]$$

Since the logarithm is a real-valued strictly increasing function when its argument is positive, we can replace the condition

$$\frac{p_1(y)}{p_0(y)} = \exp \left(y - \frac{1}{2} \right) \geq \eta$$

by the equivalent condition

$$\log \frac{p_1(y)}{p_0(y)} = y - \frac{1}{2} \geq \log \eta = \eta'$$

If $\eta' = -1/2$, Y_1 is then the set of all positive real numbers and Y_0 is the set of all negative real numbers and zero.   The level of the test is

$$\alpha = P_0(Y_1) = \frac{1}{2}$$

The power of the test is

$$P_1(Y_1) = \frac{1}{\sqrt{2\pi}} \int_0^\infty \exp \left[-\frac{(y-1)^2}{2} \right] dy = \frac{1}{\sqrt{2\pi}} \int_{-1}^\infty \exp \left(-\frac{y^2}{2} \right) dy$$

$$= 0.841$$

In general, Y_1 is the set of numbers $y \geq \eta' + 1/2$; the level is

$$\alpha = \frac{1}{\sqrt{2\pi}} \int_{\eta' + \frac{1}{2}}^\infty \exp \left(-\frac{y^2}{2} \right) dy$$

and the power is

$$\frac{1}{\sqrt{2\pi}} \int_{\eta' - \frac{1}{2}}^\infty \exp \left(\frac{-y^2}{2} \right) dy$$

Example 14-3.2. Let Y be the set of all real numbers from zero to three. Let $p_0(y)$ and $p_1(y)$ be given by

$$p_0(y) = \tfrac{1}{2} \quad \text{for } 0 \leq y \leq 2$$
$$= 0 \quad \text{otherwise}$$
$$p_1(y) = \tfrac{1}{2} \quad \text{for } 1 \leq y \leq 3$$
$$= 0 \quad \text{otherwise}$$

Then

$$\frac{p_1(y)}{p_0(y)} = 0 \quad \text{for } 0 \leq y < 1$$
$$= 1 \quad \text{for } 1 \leq y \leq 2$$
$$= +\infty \quad \text{for } 2 < y \leq 3$$

If $\eta > 1$, Y_1 is the set of numbers $2 < y \leq 3$. The level of the test is zero and the power is $\tfrac{1}{2}$. If $\eta \leq 1$, Y_1 is the set of numbers $1 \leq y \leq 3$; the level is $\tfrac{1}{2}$ and the power is 1. This is a very simple example of what is called a *singular* hypothesis-testing problem, where there is some subset of Y which has non-zero probability under one hypothesis and zero probability under the other.

Example 14-3.3. The problem is to test between two N-variate gaussian distributions with the same covariance matrices but different means. This is a test between two simple hypotheses. The space of observations Y is N-dimensional vector space; i.e., each observation y is a vector $y = (y_1, \ldots, y_N)$. We suppose the coordinates are chosen so that $y_1, \ldots, y_N$ are uncorrelated random variables. The hypotheses are

$$H_0: p_0(y_1, \ldots, y_N) = \frac{1}{(2\pi)^{N/2}\sigma_1 \cdots \sigma_N} \exp\left[-\frac{1}{2} \sum_{k=1}^{N} \frac{(y_k - a_k)^2}{\sigma_k^2} \right] \quad (14\text{-}22a)$$

$$H_1: p_1(y_1, \ldots, y_N) = \frac{1}{(2\pi)^{N/2}\sigma_1 \cdots \sigma_N} \exp\left[-\frac{1}{2} \sum_{k=1}^{N} \frac{(y_k - b_k)^2}{\sigma_k^2} \right] \quad (14\text{-}22b)$$

thus $a = (a_1, \ldots, a_N)$ is the mean of p_0 and $b = (b_1, \ldots, b_N)$ is the mean of p_1. The likelihood ratio is

$$l(y_1, \ldots, y_N) = \frac{p_0(y_1, \ldots, y_N)}{p_1(y_1, \ldots, y_N)} = \exp\left[\frac{1}{2} \sum_{k=1}^{N} \frac{(y_k - b_k)^2}{\sigma_k^2} - \frac{1}{2} \sum_{k=1}^{N} \frac{(y_k - a_k)^2}{\sigma_k^2} \right]$$

$$= \exp\left[\frac{1}{2} \sum_{k=1}^{N} \frac{b_k^2 - a_k^2}{\sigma_k^2} + \sum_{k=1}^{N} y_k \frac{(a_k - b_k)}{\sigma_k^2} \right] \quad (14\text{-}23)$$

A likelihood-ratio test is determined by choosing Y_0 to contain those $y = (y_1, \ldots, y_N)$ for which

$$l(y_1, \ldots, y_N) > \eta \qquad \eta \geq 0$$

We may as well consider the logarithm of $l(y_1, \ldots, y_N)$, and it is convenient to do so. Then the test becomes

$$\log l(y_1, \ldots, y_N) > \log \eta = \eta'$$

This gives

$$\sum_{k=1}^{N} y_k \frac{(a_k - b_k)}{\sigma_k^2} > \eta' - \frac{1}{2} \sum_{k=1}^{N} \frac{b_k^2 - a_k^2}{\sigma_k^2} = \text{const.} = c \quad (14\text{-}24)$$

as the inequality determining Y_0. Geometrically this inequality means that Y_0 is the region on one side (if c is positive, it is the far side looking from the origin) of a hyperplane in N-dimensional space. The hyperplane lies perpendicular to a vector from the origin with components

$$\left[\frac{a_1 - b_1}{\sigma_1^2}, \frac{a_2 - b_2}{\sigma_2^2}, \cdots, \frac{a_N - b_N}{\sigma_N^2} \right]$$

and is at a distance from the origin of

$$\frac{c}{\left[\sum_{k=1}^{N} \frac{(a_k - b_k)^2}{\sigma_k^4} \right]^{1/2}}$$

Since the y_k are independent gaussian random variables,

$$\xi = \sum_{k=1}^{N} y_k \frac{(a_k - b_k)}{\sigma_k^2} \tag{14-25}$$

is also a gaussian random variable. If H_0 is true,

$$E_0(\xi) = \sum_{k=1}^{N} \frac{a_k(a_k - b_k)}{\sigma_k^2} \tag{14-26a}$$

$$\sigma_0^2(\xi) = \sum_{k=1}^{N} \frac{(a_k - b_k)^2}{\sigma_k^2} \tag{14-26b}$$

If H_1 is true,

$$E_1(\xi) = \sum_{k=1}^{N} \frac{b_k(a_k - b_k)}{\sigma_k^2} \tag{14-27a}$$

$$\sigma_1^2(\xi) = \sum_{k=1}^{N} \frac{(a_k - b_k)^2}{\sigma_k^2} \tag{14-27b}$$

The error probabilities then are found by substituting the appropriate parameters in a gaussian distribution. For example,

$$\text{level} = P_0(Y_1) = \frac{1}{\sqrt{2\pi}\,\sigma_0(\xi)} \int_{-\infty}^{c} \exp\left\{ -\frac{1}{2} \frac{[\xi - E_0(\xi)]^2}{\sigma_0^2(\xi)} \right\} d\xi \tag{14-28}$$

where $E_0(\xi)$ and $\sigma_0^2(\xi)$ are given by Eqs. (14-26a and b).

Composite Alternatives. Thus far we have considered hypothesis tests in which each of the two hypotheses relates to a single possible cause of the observations. That is, each hypothesis is simple. We now want to consider hypothesis tests in which one or both of the hypotheses relate to a whole set of possible causes, that is, in which one or both hypotheses are composite. Let us denote by Ω a set of labels or indices of all the possible simple hypotheses, so that a simple hypothesis is represented by

H_ω where ω is an element of Ω. Let Ω_0 be a subset of Ω and Ω_1 be the set of all points of Ω not contained in Ω_0, $\Omega_1 = \Omega - \Omega_0$. Corresponding to each ω, there is a probability law defined on Y with a density $p_\omega(y)$. The *composite hypothesis* H_0 is that the probability density actually governing the observations y is a $p_\omega(y)$ with ω an element of Ω_0; the *composite hypothesis* H_1 is that a density $p_{\omega'}(y)$ is in effect with ω' an element of Ω_1. This formulation includes the case of testing between simple hypotheses; for if Ω_0 and Ω_1 each contain only one element, the hypotheses are simple. If Ω_0 contains only one element, but Ω_1 contains many, then the problem is to test a simple hypothesis H_0 against a composite alternative, a fairly commonly occurring situation in statistics.

In the problems in which we are interested, composite hypotheses usually arise because of the existence of unknown parameters which affect the observations. For example, in the radar problem alluded to in Art. 14-1, H_0 might be taken to be the hypothesis that there is no echo returned from objects in a certain area. Then H_1 is the hypothesis that there is a returned echo. But because of the unknown size of the reflecting object and the variability of propagation, the amplitude of the returned signal is unknown. Since any observation made on the returned signal is influenced not only by the presence or absence of an object but also by the strength of the return if there is an object, H_1 is a composite hypothesis. The set Ω is composed of a set Ω_0 containing only one point ω_0, which gives the simple hypothesis H_0, and a set Ω_1 containing an infinity of points (corresponding to the infinity of possible amplitudes), which gives the hypothesis H_1.

If a probability law is given for Ω, that is, if a set of a priori probabilities is given for the possible causes, then a minimum expected loss or a minimum probability of error test (a Bayes' solution) can be obtained in essentially the same way as in the case of simple hypotheses. In fact, if $\pi(\omega)$ is the a priori probability density on Ω, $\pi_0 P_0(Y_k)$ is replaced in Eq. (14-8) by

$$\int_{\Omega_0} P_\omega(Y_k)\pi(\omega)\,d\omega \qquad k = 0, 1$$

and $\pi_1 P_1(Y_k)$ is replaced by

$$\int_{\Omega_1} P_\omega(Y_k)\pi(\omega)\,d\omega \qquad k = 0, 1$$

The region Y_0 is then the set of points y where

$$\frac{\displaystyle\int_{\Omega_0} p_\omega(y)\pi(\omega)\,d\omega}{\displaystyle\int_{\Omega_1} p_\omega(y)\pi(\omega)\,d\omega} > \frac{L_{10} - L_{11}}{L_{01} - L_{00}} \tag{14-29}$$

which is analogous to the condition (14-11).

When no a priori probability distribution for Ω is given and one or both of the hypotheses are composite, it is often hard to find a satisfactory decision rule. We shall discuss briefly two possible approaches to the problem of testing hypotheses when there are no a priori probabilities and the hypotheses are composite: a direct application of the maximum-likelihood principle, and maximization of the power of a test when the level is held fixed. These two ideas were used in the preceding section in the discussion of simple alternatives.

The maximum-likelihood principle, which is closely related to maximum-likelihood estimation, is that the observer should choose that ω from Ω which renders the observation y most likely; that is, given an observation y, he should choose ω so as to maximize $p_\omega(y)$. Then if ω belongs to Ω_0, the observer decides on H_0; if ω belongs to Ω_1, he decides on H_1. This criterion can be stated in terms of a likelihood-ratio test. We define the likelihood ratio for the general case to be

$$l(y) = \frac{\max\limits_{\omega_0} p_{\omega_0}(y)}{\max\limits_{\omega_1} p_{\omega_1}(y)} \qquad (14\text{-}30)$$

where ω_0 ranges over the set Ω_0 and ω_1 ranges over the set Ω_1.† A *generalized likelihood-ratio test* is a test in which Y_0 is the set of points y for which

$$l(y) > \eta$$

where η is some predetermined nonnegative number.

The likelihood principle can be untrustworthy in certain applications. Obviously, if the observer uses a likelihood-ratio test when there is actually a probability distribution on Ω of which he is unaware, he may get a test which differs radically from the test for minimum probability of error.

In considering tests of maximum power at a given level, let us, for the sake of simplicity, consider only the special case that the null hypothesis H_0 is simple and the alternative H_1 is composite. Then for each ω in Ω_1 one could test H_0 at level $\leq \alpha$ against the simple hypothesis that corre-

---

† The usual definition for the likelihood ratio is

$$l(y) = \frac{\max\limits_{\omega_0} p_{\omega_0}(y)}{\max\limits_{\omega} p_\omega(y)}$$

where ω ranges over all of Ω. It is more convenient in this chapter to use the definition of Eq. (14-30), and the two are essentially equivalent.

sponds to ω. A maximum power test to do this would have a critical region Y_ω consisting of the set of x's for which

$$\frac{p_\omega(x)}{p_0(x)} \geq \eta_\omega$$

for some number η_ω. If the critical regions Y_ω are identical for all ω in Ω_1, then the test with critical region $Y_1 = Y_\omega$, ω in Ω_1, is said to be a *uniformly most powerful* test of H_0 against H_1 at level α. When this fortuitous situation occurs, one is in just as good a position to test H_0 against H_1 as if H_1 were simple. The ambiguity introduced by the fact that H_1 is composite turns out in this case to be of no importance. In some useful examples, as we shall see later, there exists a uniformly most powerful test. When there does not, there is no simple criterion as to what is the best test of level α. The situation is clarified perhaps by reference to the power function. The *power function* of a test with critical region Y_1 is the probability of the critical region Y_1 as a function of ω. Suppose that Ω is an interval on the real line so that every element ω is a real number, say, between a and b. Then typical power-function curves are as shown in Fig. 14-5, in which ω_0 is the point in Ω corresponding to H_0, so that the value of the power function at ω_0 is the level of the test. Each curve in Fig. 14-5 is the power-function graph of a different test; all have level α. If there is a test with power function lying above all the rest which have the same value at ω_0, like curve (a) in Fig. 14-5, it is a uniformly most powerful test at its level. Usually every power function will for some values of ω_1 lie under the power function of another test, like curves (b) and (c) in Fig. 14-5.

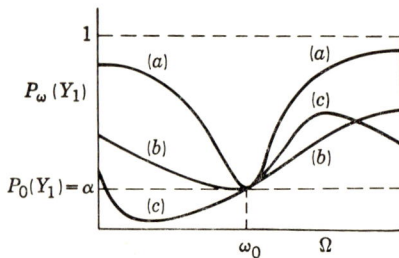

FIG. 14-5. Power-function curves.

When there is no uniformly most powerful test, some further criterion must be used to find a "best" test. There are various possibilities: the class of tests considered may be made smaller by considering only those tests which have some particular desired property. There may then be a uniformly most powerful test with respect to this smaller class. One such class is that of *unbiased tests*, whose power functions have a minimum at ω_0; curves (a) and (b) in Fig. 14-5 are power functions of unbiased tests, but curve (c) is not. Another possibility is to judge the goodness of tests at the same level according to some over-all property of the power function.†

† See Lehman (I, Chap. 6).

Example 14-3.4. Let us modify Example 14-3.3 by setting $a_k = 0$, $k = 1, \ldots, N$, and replacing b_k by βb_k where β is an unknown real positive number. H_0 is then the simple hypothesis

$$H_0: p_0(y_1, \ldots, y_N) = \frac{1}{(2\pi)^{N/2}\sigma_1 \cdots \sigma_N} \exp\left[-\frac{1}{2}\sum_{k=1}^{N}\frac{y_k^2}{\sigma_k^2}\right] \qquad (14\text{-}31a)$$

and H_1 is the composite hypothesis

$$H_1: p_\beta(y_1, \ldots, y_N) = \frac{1}{(2\pi)^{N/2}\sigma_1 \cdots \sigma_N} \exp\left[-\frac{1}{2}\sum_{k=1}^{N}\frac{(y_k - \beta b_k)^2}{\sigma_k^2}\right] \qquad \beta > 0 \tag{14-31b}$$

We now show for this example that a likelihood ratio test of H_0 against H_1 is a uniformly most powerful test at its level. First, we observe that with nonsingular gaussian probability densities, as in this example, the level α, for any fixed β, is a continuous as well as monotonic function of η which runs from zero to one. Hence at least one $\eta = \eta(\beta)$ can be found to yield any specified level α. From Example 14-3.3, for a fixed value of β we have that the critical region Y_β for a most powerful test of H_0 against H_β at level α consists of those $y = (y_1, \ldots, y_N)$ for which

$$\sum_{k=1}^{N}\frac{y_k\beta b_k}{\sigma_k^2} \geq c(\beta)$$

or

$$\sum_{k=1}^{N}\frac{y_k b_k}{\sigma_k^2} \geq \frac{c(\beta)}{\beta} = k(\beta) \tag{14-32}$$

Now if for two different values of β, β_1 and β_2, $k(\beta_1)$ is strictly greater than $k(\beta_2)$, then it follows from Eq. (14-32) that Y_{β_2} contains at least all the points $y = (y_1, \ldots, y_N)$ contained in Y_{β_1}. But

$$P_0(Y_{\beta_2}) = \alpha = P_0(Y_{\beta_1})$$

Hence $P_0(Y_{\beta_2} - Y_{\beta_1}) = 0$. But then, by our convention regarding sets of y's where $p_0(y) = 0$ in the definition of the critical region for a maximum power test, $Y_{\beta_2} = Y_{\beta_1}$. Hence all the Y_β are equal, and we can take $Y_1 = Y_\beta$. That is, the maximum-power test at level α of H_0 against each of the simple hypotheses comprising H_1 is the same as each other one; so this test is a uniformly most powerful test of H_0 against H_1.

Example 14-3.5. We now further modify the preceding examples to get an example in which both hypotheses are composite. Let β be an unknown real number. It is desired to test between the hypotheses

$$H_0: p_0(y_1, \ldots, y_N; \beta) = \frac{1}{(2\pi)^{N/2}\sigma_1 \cdots \sigma_N} \exp\left[-\frac{1}{2}\sum_{k=1}^{N}\frac{(y_k - \beta a_k)^2}{\sigma_k^2}\right] \qquad (14\text{-}33a)$$

$$H_1: p_1(y_1, \ldots, y_N; \beta) = \frac{1}{(2\pi)^{N/2}\sigma_1 \cdots \sigma_N} \exp\left[-\frac{1}{2}\sum_{k=1}^{N}\frac{(y_k - \beta b_k)^2}{\sigma_k^2}\right] \qquad (14\text{-}33b)$$

We shall determine the general likelihood ratio, as defined in Eq. (14-30),

$$l(y) = \frac{\max_\beta p_0(y_1, \ldots, y_N; \beta)}{\max_\beta p_1(y_1, \ldots, y_N; \beta)}$$

The maximum of $p_0(y_1, \ldots ,y_N;\beta)$ occurs for the same value of β as the maximum of $\log p_0(y_1, \ldots ,y_N;\beta)$. Since $\log p_0(y_1, \ldots ,y_N;\beta)$ is a second-degree polynomial in β, we can find its stationary point by setting its derivative equal to zero. Thus

$$\log p_0(y_1, \ldots ,y_N;\beta) = \text{const.} - \frac{1}{2} \sum_{k=1}^{N} \frac{(y_k - \beta a_k)^2}{\sigma_k^2}$$

and

$$\frac{\partial}{\partial \beta} \log p_0 = \sum_{k=1}^{N} \frac{(y_k - \beta a_k)a_k}{\sigma_k^2} = 0$$

$$\beta \sum_{k=1}^{N} \frac{a_k^2}{\sigma_k^2} = \sum_{k=1}^{N} \frac{y_k a_k}{\sigma_k^2} \tag{14-34}$$

$$\beta = \frac{\displaystyle\sum_{k=1}^{N} \frac{y_k a_k}{\sigma_k^2}}{\displaystyle\sum_{k=1}^{N} \frac{a_k^2}{\sigma_k^2}}$$

From the form of $\log p_0$ it is obvious that this value of β in fact gives a maximum for $\log p_0$ and hence for p_0. The value of β to maximize $p_1(y_1, \ldots ,y_N;\beta)$ is similarly

$$\beta = \frac{\displaystyle\sum_{k=1}^{N} \frac{y_k b_k}{\sigma_k^2}}{\displaystyle\sum_{k=1}^{N} \frac{b_k^2}{\sigma_k^2}} \tag{14-35}$$

Substituting these values of β in the expressions for p_0 and p_1, respectively, and performing a slight reduction yields

$$l(y) = \frac{\exp\left\{\dfrac{-1}{2\displaystyle\sum_{k=1}^{N} \frac{a_k^2}{\sigma_k^2}}\left[\sum_{k=1}^{N} \frac{y_k^2}{\sigma_k^2}\sum_{k=1}^{N}\frac{a_k^2}{\sigma_k^2} - \left(\sum_{k=1}^{N}\frac{y_k a_k}{\sigma_k^2}\right)^2\right]\right\}}{\exp\left\{\dfrac{-1}{2\displaystyle\sum_{k=1}^{N} \frac{b_k^2}{\sigma_k^2}}\left[\sum_{k=1}^{N} \frac{y_k^2}{\sigma_k^2}\sum_{k=1}^{N}\frac{b_k^2}{\sigma_k^2} - \left(\sum_{k=1}^{N}\frac{y_k b_k}{\sigma_k^2}\right)^2\right]\right\}} \tag{14-36}$$

Let

$$c_N^2 = \sum_{k=1}^{N} \frac{a_k^2}{\sigma_k^2} \qquad d_N^2 = \sum_{k=1}^{N} \frac{b_k^2}{\sigma_k^2} \tag{14-37}$$

Then

$$\log l(y_1, \ldots ,y_N) = \frac{1}{2c_N^2}\left(\sum_{k=1}^{N}\frac{y_k a_k}{\sigma_k^2}\right)^2 - \frac{1}{2d_N^2}\left(\sum_{k=1}^{N}\frac{y_k b_k}{\sigma_k^2}\right)^2 \tag{14-38}$$

and Y_0 is the region where this expression is greater than $\log \eta$.

14-4. Statistical Estimation

Statistical estimation can be described briefly as follows: An event is observed which could have been produced by any one of a family of mutually exclusive possible causes (or is related to any one of a family of possible states of nature). Each of these possible causes is indexed by a parameter or by a set of parameters. It is desired to infer from the observation which cause could most reasonably be supposed to exist, that is, which parameter value or sets of values obtained. Thus, as for the problem of testing statistical hypotheses, we postulate a space Y of all possible values of some observed variable (or variables) and a set of known probabilities P_ω on Y where ω belongs to a set Ω. But now Ω is partitioned into a collection of mutually exclusive subsets Ω_a where a is a number or a vector labeling the subset. The sets Ω_a may consist of only single elements ω from Ω. An *estimator* of a is a function $\hat{a}(y)$ defined on Y—which is intended, of course, to yield a reasonable guess of the true value a. Thus an estimator is a random variable on the space of observations.

This abstract formulation is perhaps made clearer in terms of a common example. Suppose it is known that some quantity y is determined by a gaussian distribution of unknown mean m and variance σ^2 and an estimate of m is wanted. A set of N independent measurements of y is made. We designate these $y_1, y_2, \ldots, y_N$. Then the space of observations Y is N-dimensional vector space, and a point y in Y has coordinates $(y_1, \ldots, y_N)$. The set Ω is the upper half of the plane, with the abscissa of a point ω representing a mean m and the ordinate (which is positive) representing a variance σ^2. The parameter a to be estimated is m, and the sets Ω_a are lines drawn vertically up from the horizontal axis. An example of an estimator for m is the sample mean†

$$\hat{m} = \frac{y_1 + y_2 + \cdots + y_N}{N}$$

which is a function of the point y in Y. Under the same conditions, it might be desired to estimate m and σ^2. Then Y and Ω are the same as before, but now a is a vector with coordinates (m, σ^2) and Ω_a is a single point in Ω.

According to the definition of an estimator, any function of y is one, whether it gives a good, fair, or ridiculous guess of the true value of the parameter. There are several criteria for judging the reasonableness and quality of estimators. First, if the distributions on Y are completely determined by a so that any one of the family of probability densities

† See Art. 5-2.

on Y can be written as $p_a(y)$, we say an estimator $\hat{a}$ of a is unbiased†
if it gives the correct value on the average; i.e., $\hat{a}$ is unbiased if for all a

$$E_a(\hat{a}) = a \qquad (14\text{-}39)$$

where $E_a(\hat{a})$ is the expectation of $\hat{a}$ when a is true. This condition can
also be written

$$\int_Y \hat{a}(y)p_a(y)\,dy = a$$

If the probability distributions on Y are not completely determined by a
single parameter, a condition corresponding to Eq. (14-39) may still be
written, but it represents a more severe restriction. Thus, for example,
if the family of distributions are determined by two parameters a and b
we may say that an estimator $\hat{a}$ of a is unbiased if for every a

$$E_{a,b}(\hat{a}) = a \qquad \text{for all } b \qquad (14\text{-}40)$$

The property of being unbiased does not constitute a sufficient guarantee
of the goodness of an estimator. An estimator $\hat{a}(y)$ can be unbiased and
yet in repeated identical independent trials yield values which fluctuate
wildly around the true value. This situation reflects the fact that for
some set of densities $p_a(y)$, $\hat{a}(y)$ considered as a random variable has a
distribution which does not concentrate the probability weighting closely
around the mean. One measure of the fluctuations of $\hat{a}(y)$ about its
mean is given by its variance

$$\sigma_a{}^2(\hat{a}) = E_a\{[\hat{a} - E_a(\hat{a})]^2\} \qquad (14\text{-}41)$$

If $\hat{a}(y)$ is an unbiased estimator, this reduces to

$$\sigma_a{}^2(\hat{a}) = E_a[(\hat{a} - a)^2] \qquad (14\text{-}42)$$

A condition of optimality of an estimator which we shall use is that it
be an estimator of minimum variance‡ from some class of unbiased
estimators.

As with hypothesis testing, we are often interested in how the statistical inference can be improved as the number of independent observations
becomes large. Let a sequence $\hat{a}_N(y)$ of estimators be given, where $\hat{a}_N(y)$
is an estimator on the joint sample space of N independent observations.
If this sequence has the property that for any true value of the parameter a, $\hat{a}_N(y)$ converges in probability§ to a, the sequence of estimators is
said to be *consistent.*¶

† See Cramér (I, Art. 32.3).
‡ See Cramér (I, Art. 32.3).
§ See Art. 4-6.
¶ See Art. 5-3.

Maximum-likelihood Estimation. The likelihood principle mentioned in a preceding section applies also to estimation. The intuitive idea is that if any one of a family of probability densities $p_a(y)$ on Y might be operative when a particular observation y is made, it is reasonable to suppose the true value of a is one which establishes a probability distribution on Y so as to make y most likely, i.e., that the parameter a is such as to maximize $p_a(y)$, which as a function of a and y is called the *likelihood function* and sometimes written $\lambda(y;a)$. Since the logarithm is a real increasing function for positive arguments, we may consider $\log \lambda(y;a)$ instead of $\lambda(y;a)$ when searching for values of a which yield a maximum. It turns out usually to be convenient to do this. If $\log \lambda(y;a)$ has a continuous derivative with respect to a, then we may often expect to find the value or values of a which maximize the likelihood among the solutions of the equation

$$\frac{\partial \log \lambda(y;a)}{\partial a} = 0 \qquad (14\text{-}43)$$

This equation is called the likelihood equation. A solution for a is a function $\hat{a}(y)$, which is an estimator of a. Solutions can occur of the form $a \equiv$ constant;† these are not reasonable estimators, since they do not depend on y, and we drop them from consideration. The remaining solutions of Eq. (14-43) are called *maximum-likelihood estimators.* Maximum-likelihood estimators are known to have properties which help justify their use in some circumstances beyond the rough intuitive justification given above. Under certain conditions, maximum-likelihood estimators have minimum variance in a wide class of unbiased estimators.‡ Under more general conditions, maximum-likelihood estimators are consistent and have asymptotically minimum variance.§

If there are n different parameters $a_1, \ldots, a_n$ to be estimated, then the maximum-likelihood method consists of solving the n equations

$$\frac{\partial}{\partial a_i} \log \lambda(y;a_1, \ldots, a_n) = 0 \qquad i = 1, \ldots, n \qquad (14\text{-}44)$$

for $a_1, \ldots, a_n$. If an a priori probability distribution is known for the parameter values, with density function $\pi(a)$, then there is a Bayes' estimate for a obtained by maximizing $\pi(a)p_a(y)$ for each y. This is closely analogous to the situation in hypothesis testing with given a priori probabilities.

† For example, if $\lambda(y;a) = \dfrac{1}{\sqrt{2\pi}} \exp \dfrac{-(y - a^2)^2}{2}$, $a \equiv 0$ is a solution of the likelihood equation.

‡ Cramér (I, Art. 33.2).

§ Cramér (I, Art. 33.3).

Example 14-4.1. Let Y be n-dimensional vector space. There is given a set of gaussian N-variate probability density functions,

$$p_\beta(y) = p_\beta(y_1, \ldots, y_N) = \frac{1}{(2\pi)^{N/2}\sigma_1 \cdots \sigma_N} \exp\left[-\frac{1}{2} \sum_{k=1}^{N} \frac{(y_k - \beta a_k)^2}{\sigma_k^2} \right] \quad (14\text{-}45)$$

and it is desired to estimate β. A maximum-likelihood estimator has already been obtained for this problem in Example 14-3.5. By Eq. (14.34),

$$\hat\beta(y) = \frac{\displaystyle\sum_{k=1}^{N} y_k a_k/\sigma_k^2}{\displaystyle\sum_{k=1}^{N} a_k^2/\sigma_k^2} \quad (14\text{-}46)$$

This estimator is unbiased, for

$$E_\beta[\hat\beta(y)] = \frac{\displaystyle\sum_{k=1}^{N} E_\beta(y_k) a_k/\sigma_k^2}{\displaystyle\sum_{k=1}^{N} a_k^2/\sigma_k^2} = \beta \quad (14\text{-}47)$$

The variance of $\hat\beta(y)$ is, since the y_k are independent,

$$\sigma_\beta^2[\hat\beta(y)] = \sum_{k=1}^{N} \frac{\sigma_k^2 a_k^2}{\sigma_k^4} \left(\frac{1}{\displaystyle\sum_{k=1}^{n} a_k^2/\sigma_k^2} \right)^2 = \frac{1}{\displaystyle\sum_{k=1}^{N} a_k^2/\sigma_k^2} \quad (14\text{-}48)$$

which does not depend on β. The variance of the estimator does not depend on β because the parameter β influences only the mean of the distribution on Y.

The estimator $\hat\beta(y)$ is linear, i.e., if $y = (y_1, \ldots, y_N)$ and $z = (z_1, \ldots, z_N)$ are two observations and c_1 and c_2 are any constants,

$$\hat\beta(c_1 y + c_2 z) = c_1 \hat\beta(y) + c_2 \hat\beta(z)$$

We can now show that $\hat\beta$ has minimum variance in the class of all unbiased, linear estimators of β. Any linear estimator $\beta'(y)$ can be written in the form

$$\beta'(y) = \sum_{k=1}^{N} d_k y_k \quad (14\text{-}49)$$

with some set of numbers d_k. In order for β' to be unbiased

$$E_\beta[\beta'(y)] = \sum_{k=1}^{N} d_k E_\beta(y_k) = \beta \sum_{k=1}^{N} d_k a_k = \beta$$

hence the following relation must be satisfied by the d_k,

$$\sum_{k=1}^{N} a_k d_k = 1 \qquad (14\text{-}50)$$

The variance of $\beta'(y)$ is

$$\sigma_\beta^2[\beta'(y)] = \sum_{k=1}^{N} d_k^2 \sigma_k^2 \qquad (14\text{-}51)$$

Now, from Eq. (14-50) and the Schwartz inequality,

$$1 = \left(\sum_{k=1}^{N} a_k d_k\right)^2 = \left[\sum_{k=1}^{N} (d_k \sigma_k)\left(\frac{a_k}{\sigma_k}\right)\right]^2 \leq \sum_{k=1}^{N} d_k^2 \sigma_k^2 \sum_{k=1}^{N} \frac{a_k^2}{\sigma_k^2}$$

Hence,

$$\sigma_\beta^2[\beta'(y)] = \sum_{k=1}^{N} d_k^2 \sigma_k^2 \geq \frac{1}{\sum_{k=1}^{N} a_k^2/\sigma_k^2} = \sigma_\beta^2[\hat\beta(y)]$$

Observations and Statistics. Thus far, in the discussions of both the hypothesis-testing and estimation problems, the space of observations Y, or the sample space, has been assumed given. In many classical statistical testing procedures, some number N of prescribed measurements or observations are made, and these provide the statistical data. These N measurements then provide a set of N real numbers $y_1, \ldots, y_N$ which can be thought of as the coordinates of a point $y = (y_1, \ldots, y_N)$ in an N-dimensional vector space Y, the sample space. Once the method of making the measurements is decided, Y is fixed. Y contains the raw data of the test from which a statistical inference is to be drawn. In some types of problem, however—particularly in the applications we consider here—there is too much raw data to be handled directly, and the analysis really proceeds in two stages. First, the original data are treated so as to yield a smaller set of data, and then statistical procedures are applied to these reduced data. In this situation the sample space Y is the space of possible values of the reduced data.

For example, suppose a pulsed radar is used to determine whether there is an object in a certain direction at a distance of between 20 and 21 miles. Ideally the data available at the radar receiver from which a decision can be made consist of a sequence of sections, one for each pulse, of a continuous record of voltage against time of duration 2/186,000 seconds (the time for a radio wave to travel two miles). A continuous record of this sort involves an uncountable infinity of values. Various things may be done to reduce the quantity of data. One is to sample each returned pulse once; i.e., the voltage amplitude received is measured once during each time interval corresponding to the range interval from 20 to 21 miles. Suppose the radar is trained in the required direction long enough for K pulses to be returned. Then a sample point has K coordinates and Y is K-dimensional space.

The reduction of data from the original space of observations to the sample space can be regarded as a mapping or transformation. If the original space of observations is denoted by M, with points m, then for each observation m a point y in the sample space is determined. Thus a mapping $y(m)$ is defined from M to Y; such a mapping is called a *statistic*. In general, any mapping from a space of observations to a sample space or from one sample space to another is called a statistic. Thus, in the situation described in the preceding paragraph, the mapping which carries a received waveform into a point in K-dimensional space is a statistic. If, further, an average is taken of the coordinates $y_1, \ldots,$ y_K of the point in Y, it constitutes a mapping of Y into a new one-dimensional sample space and is another statistic, the *sample mean*.

It is clear that the choice of a statistic is part of the over-all statistical problem and that it must be made with some care. Usually, when the original data are reduced, some information pertinent to the decision to be made is lost, but not always. Thus, for example, it can be shown[†] that, if a set of numbers is known to be distributed according to a gaussian probability law with unknown mean and variance, and if n independent samples are taken, the sample mean and the sample variance contain just as much information about the distribution as do the n sample values. Here the statistic with the two coordinates, sample mean and sample variance, maps an n-dimensional sample space onto a two-dimensional one.

Grenander has introduced the term "observable coordinates" for the initial statistics used in making statistical inferences on random processes. This term seems apt for the situations we are concerned with, and we shall use it in the articles that follow.

14-5. Communication with Fixed Signals in Gaussian Noise[‡]

The first statistical signal-reception problem we shall consider is the following: One of two known possible signals $s_0(t)$ and $s_1(t)$ is transmitted for a fixed interval of time $0 \leq t < T$. The transmitted signal is corrupted by the addition of stationary gaussian noise with a known auto-

---

[†] Cramér (I, p. 494, Example 1). The sample mean and variance are "sufficient estimators," which implies the statement above.

[‡] The material of this article is an immediate application of parts of a theory developed in Grenander (I, particularly Arts. 3 and 4).

correlation function, so the received signal $y(t)$ is given by the equation

$$y(t) = s_i(t) + n(t) \qquad 0 \le t \le T, \, i = 0, 1 \qquad (14\text{-}52)$$

where $n(t)$ is the noise. The decision to be made at the receiver is whether $s_0(t)$ or $s_1(t)$ was actually transmitted. If $s_0(t)$ and $s_1(t)$ are sine waves of different frequencies, then this problem becomes an idealized per-symbol analysis of the FSK radio teletype discussed in Art. 14-1. The $s_i(t)$ may be quite arbitrary, however, so the application is more general. Application may also be made to radar, as we shall see in Art. 14-7. The chief idealization of the problem represented by Eq. (14-52) with respect to almost any practical application is that no parameters have been introduced to account for ambiguous amplitude and phase of the signal $s_i(t)$. Modifications in which unknown signal amplitude or phase is introduced will be discussed later.

This is a problem in the testing of statistical hypotheses. The hypothesis H_0 is that $s_0(t)$ was actually sent; the hypothesis H_1 is that $s_1(t)$ was actually sent. The observation is a real-valued function $y(t)$ on the fixed interval $0 \le t < T$; the observation space is the set of all such functions. We are going to derive a likelihood-ratio test to choose between the two hypotheses. The significance of the likelihood-ratio test and the establishing of a threshold depend on what particular application one has in mind. For example, in radio-teletype transmission, it is usually reasonable to assign a priori probabilities $\pi_0 = \frac{1}{2}$, $\pi_1 = \frac{1}{2}$ and equal losses to each kind of error. Then the threshold for the likelihood-ratio test is one.

The first step is to choose a statistic, or set of observable coordinates, and thereby specify a sample space on which to calculate the likelihood ratio. Following Grenander† we take as observable coordinates a set of weighted averages of $y(t)$ as follows: From Art. 6-4, we know that the noise $n(t)$ can be written

$$n(t) = \sum_k z_k \phi_k(t) \qquad 0 \le t \le T \qquad (14\text{-}53)$$

where

$$z_k = \int_0^T n(t) \phi_k^*(t) \, dt \qquad (14\text{-}54)$$

$$E(z_k) = 0$$
$$E(z_k z_m^*) = 0 \qquad \text{if } k \ne m$$
$$= \sigma_k{}^2 \qquad \text{if } k = m$$

and where the $\phi_k(t)$ are a set of orthonormal functions satisfying

$$\int_0^T R_n(s - t) \phi_k(t) \, dt = \sigma_k{}^2 \phi_k(s) \qquad 0 \le s \le T \qquad (14\text{-}55)$$

† Grenander (I), Art. 3.

Since $R_n(t)$ is real, the $\phi_k(t)$ may be taken to be real,† and we shall assume that this is done. We take for observable coordinates

$$y_k = \int_0^T y(t)\phi_k(t)\, dt \qquad k = 1, 2, \ldots \qquad (14\text{-}56)$$

If we define a_k and b_k, $k = 1, 2, \ldots$, by

$$
\begin{aligned}
a_k &= \int_0^T s_0(t)\phi_k(t)\, dt \\
b_k &= \int_0^T s_1(t)\phi_k(t)\, dt
\end{aligned}
\qquad (14\text{-}57)
$$

then from Eqs. (14-56), (14-57), and (14-54) we have

$$
\begin{aligned}
y_k &= a_k + z_k \qquad \text{if } i = 0 \\
&= b_k + z_k \qquad \text{if } i = 1
\end{aligned}
\qquad (14\text{-}58)
$$

From Eq. (14-54) it follows, according to a principle presented in Art. 8-4, that any finite collection of z_k's has a joint gaussian distribution. Each z_k has mean zero, and any two different z_k's are uncorrelated; hence the z_k's are mutually independent (gaussian) random variables. The a_k's and b_k's are numbers, not random variables. Hence, if $i = 0$, the y_k's are mutually independent gaussian random variables with mean value a_k and variance σ_k^2. Similarly, if $i = 1$, the y_k's are mutually independent gaussian random variables with mean value b_k and variance σ_k^2.

The reason for choosing the observable coordinates in this way should now be clear. The coordinates y_k are mutually independent, and hence it is straightforward to write joint probability densities for $(y_1, \ldots, y_N)$, where N is arbitrary. An approximate likelihood ratio can be written that involves only $y_1, \ldots, y_N$, and then a limit taken as $N \to \infty$. The important point is that this choice of observable coordinates leads to independent random variables. This decomposition in terms of the orthogonal expansion of $n(t)$ is the infinite-dimensional analogue of the diagonalization of the covariance matrix of a finite set of random variables.

The orthonormal functions $\phi_k(t)$ may or may not form a complete‡ set if no restrictions are put on $R_n(t)$ except that it be a correlation function.

† Since $R_n(t)$ is real, if $\phi_k(t)$ is a characteristic function, $\Re[\phi_k(t)]$ and $\Im[\phi_k(t)]$ are characteristic functions. It can be shown that if $\phi_1(t), \ldots, \phi_K(t)$ is a set of linearly independent complex-valued functions, some set of K of their real and imaginary parts is linearly independent. Hence, given a set of K linearly independent complex-valued characteristic functions for a characteristic value λ, a set of K linearly independent *real-valued* characteristic functions for λ can be found, and from this a set of K real-valued orthonormal characteristic functions.

‡ See Appendix 2, Art. A2-1.

If the functions $\phi_k(t)$ are not complete, then there are functions $\psi(t)$ which are orthogonal to all the $\phi_k(t)$, i.e., with the property that

$$\int_0^T \psi(t)\phi_k(t)\, dt = 0 \qquad (14\text{-}59)$$

for all k. If in addition some function $\psi(t)$ which satisfies Eq. (14-59) for all k also satisfies

$$\int_0^T \psi(t)s_0(t)\, dt \neq \int_0^T \psi(t)s_1(t)\, dt \qquad (14\text{-}60)$$

then a correct decision may be made with probability one. This can be shown as follows: let $\psi(t)$ be a function which satisfies Eqs. (14-59) and (14-60) for all k; then

$$\int_0^T \psi(t)y(t)\, dt = \int_0^T \psi(t)s_0(t)\, dt + \int_0^T \psi(t)n(t)\, dt \qquad \text{if } i = 0$$

$$= \int_0^T \psi(t)s_1(t)\, dt + \int_0^T \psi(t)n(t)\, dt \qquad \text{if } i = 1$$

But the second integral in each of the expressions on the right is equal to zero because, using the representation of Eq. (14-53), since $\psi(t)$ is orthogonal to all the $\phi_k(t)$, it is orthogonal to $n(t)$. Hence, if $s_0(t)$ was sent

$$\int_0^T \psi(t)y(t)\, dt = \int_0^T \psi(t)s_0(t)\, dt = c_0$$

whereas if $s_1(t)$ was sent,

$$\int_0^T \psi(t)y(t)\, dt = \int_0^T \psi(t)s_1(t)\, dt = c_1$$

and c_0 and c_1 are different. The situation described in this paragraph is called the *extreme singular case* by Grenander. It does not occur in conventional problems, because, if $n(t)$ is filtered white noise, the $\phi_k(t)$ are a complete set of orthonormal functions.† Intuitively, this extreme singular case is that in which the noise can be completely "tuned out" (see Prob. 14-5 for a simple example).

We now turn to the usual situation in which there is no function $\psi(t)$ orthogonal to all the $\phi_k(t)$ and satisfying the condition (14-60). This includes the case that the $\phi_k(t)$ are a complete set. Since $(y_1, \ldots, y_N)$ is a set of mutually independent gaussian random variables with variances $(\sigma_1^2, \ldots, \sigma_N^2)$ and with mean values $(a_1, \ldots, a_N)$ under hypothesis H_0 and $(b_1, \ldots, b_N)$ under hypothesis H_1, the natural logarithm of the likelihood ratio for the first N observable coordinates is, from Eq. (14-25),

$$\log l_N(y_1, \ldots, y_N) = \frac{1}{2} \sum_{k=1}^N \frac{b_k^2 - a_k^2}{\sigma_k^2} + \sum_{k=1}^N y_k \frac{(a_k - b_k)}{\sigma_k^2} \qquad (14\text{-}61)$$

† See Appendix 2, Art. A2-3.

A likelihood-ratio test based on just these N observable coordinates is then, from Eq. (14-24), to choose H_0 if

$$\sum_{k=1}^{N} y_k \frac{(a_k - b_k)}{\sigma_k^2} > \log \eta - \frac{1}{2} \sum_{k=1}^{N} \frac{b_k^2 - a_k^2}{\sigma_k^2} = c \qquad (14\text{-}62)$$

and to choose H_1 otherwise. Before discussing the limiting behavior as $N \to \infty$, it is convenient to put the likelihood ratio in a different form. Let

$$f_N(t) = \sum_{k=1}^{N} \frac{(a_k - b_k)}{\sigma_k^2} \phi_k(t) \qquad (14\text{-}63)$$

Then

$$\log l_N(y_1, \ldots, y_N) = -\frac{1}{2} \sum_{k=1}^{N} (a_k + b_k) \frac{(a_k - b_k)}{\sigma_k^2} + \frac{1}{2} \sum_{k=1}^{N} y_k \frac{(a_k - b_k)}{\sigma_k^2}$$

$$= -\frac{1}{2} \sum_{k=1}^{N} \int_0^T s_0(t) \phi_k(t) \frac{(a_k - b_k)}{\sigma_k^2} \, dt - \frac{1}{2} \sum_{k=1}^{N} \int_0^T s_1(t) \phi_k(t) \frac{(a_k - b_k)}{\sigma_k^2} \, dt$$

$$+ \sum_{k=1}^{N} \int_0^T y(t) \phi_k(t) \frac{(a_k - b_k)}{\sigma_k^2} \, dt = \int_0^T f_N(t) \left[y(t) - \frac{s_0(t) + s_1(t)}{2} \right] dt$$

$$(14\text{-}64)$$

We note also, for future use, that by Eq. (14-63),

$$\int_0^T R_n(u - t) f_N(t) \, dt = \sum_{k=1}^{N} (a_k - b_k) \phi_k(u) \qquad (14\text{-}65)$$

It can be shown that $\log l_N(y_1, \ldots, y_N)$ converges as $N \to \infty$ † (if either hypothesis is true). Thus the limiting form of the likelihood ratio test is given by

$$\log l(y) = \lim_{N \to \infty} \int_0^T f_N(t) \left[y(t) - \frac{s_0(t) + s_1(t)}{2} \right] dt > \log \eta \qquad (14\text{-}66)$$

If

$$\sum_{k=1}^{\infty} \frac{(a_k - b_k)^2}{\sigma_k^2} = +\infty \qquad (14\text{-}67)$$

† Grenander (I, Art. 4).

it can be shown by a relatively straightforward calculation using the Chebyshev inequality that

$$\log l_N(y_1, \ldots , y_N) \to + \infty \text{ in probability if } H_0 \text{ is true}$$
and
$$\log l_N(y_1, \ldots , y_N) \to - \infty \text{ in probability if } H_1 \text{ is true}$$

This is also a singular case in which perfect detection is possible in the limit. If the infinite series of Eq. (14-67) converges to a finite limit (the only other alternative, since the series has positive terms) the limiting likelihood ratio is not singular, and we call this the *regular case*. In some types of systems, natural constraints on signals and noise guarantee that the series of Eq. (14-67) converges. For example, if the noise is considered to be introduced as white noise at the input to the receiver, then the singular case cannot occur. A proof of this is outlined in Prob. 14-6. Let us now discuss further the regular case.

Fixed Signal in Gaussian Noise—Regular Case. The form of the likelihood-ratio test given by Eq. (14-66) is inconvenient in that the likelihood ratio is obtained as a limit and the test function $f_N(t)$ is defined by a series which becomes an infinite series as $N \to \infty$. Formally one can see that if there is no trouble connected with passing to the limit and the limit of $f_N(t)$ is denoted by $f(t)$, then Eqs. (14-64) and (14-65) give

$$\log l(y) = \int_0^T f(t) \left[y(t) - \frac{s_0(t) + s_1(t)}{2} \right] dt \qquad (14\text{-}68)$$

where $f(t)$ is the solution of the integral equation

$$\int_0^T R_n(u - t)f(t) \, dt = \sum_{k=1}^{\infty} a_k \phi_k(u) - \sum_{k=1}^{\infty} b_k \phi_k(u)$$
$$= s_0(u) - s_1(u) \qquad 0 \le u \le T \qquad (14\text{-}69)$$

It can be shown† rigorously that if $f_N(t)$ converges in mean square to a function $f(t)$ of integrable square, then the logarithm of the likelihood ratio is given by Eq. (14-68) and $f(t)$ satisfies the integral equation (14-69). Conversely, if a function $f(t)$ of integrable square is a solution of Eq. (14-69), then it can be used in Eq. (14-68) to give the likelihood ratio. If the characteristic functions of $R_n(t)$ are not a complete set, $f(t)$ will not be unique.

We shall assume for the rest of this section that the likelihood ratio is specified by Eqs. (14-68) and (14-69). First, let us point out the very close connection between this likelihood-ratio test and the maximum

† Grenander (I, Art. 4.6).

signal-to-noise ratio filter discussed in Art. 11-7. Let us suppose for convenience that $s_1(t) \equiv 0$. Define the function $g(t)$ by

$$g(T - t) = f(t) \qquad 0 \le t \le T$$

Then Eq. (14-69) becomes

$$s_0(u) = \int_0^T R_n(u - t)g(T - t)\, dt$$

$$= \int_0^T R_n(u + v - T)g(v)\, dv$$

or $\qquad\qquad s_0(T - t) = \int_0^T R_n(v - t)g(v)\, dv \qquad\qquad (14\text{-}70)$

and the test inequality becomes

$$\int_0^T g(T - t)y(t)\, dt > \tfrac{1}{2} \int_0^T g(T - t)s_0(t)\, dt + \log \eta \qquad (14\text{-}71)$$

From Eq. (14-70) it follows that $g(t)$ is the weighting function of a filter which will maximize signal-to-noise ratio at time T when the input signal is $s_0(t)$, $0 \le t \le T$, and the noise has autocorrelation $R_n(t)$. Hence the test given by Eq. (14-71) can be interpreted as putting the received signal $y(t)$ into this maximum signal-to-noise ratio filter and comparing the output with what is obtained when $s_0(t)$ is put into the same filter.

The probabilities of the two kinds of error can be calculated straightforwardly because $\log l(y)$ is a gaussian random variable under either hypothesis. If $s_1(t)$ was actually sent, using Eq. (14-68),

$$m_{l_1} = E[\log l(y)] = \tfrac{1}{2} \int_0^T f(t)[s_1(t) - s_0(t)]\, dt \qquad (14\text{-}72)$$

$$\sigma^2(l) = \sigma^2[\log l(y)] = E\left[\int_0^T \int_0^T f(t)n(t)f(u)n(u)\, dt\, du \right]$$

$$= \int_0^T \int_0^T R_n(t - u)f(t)f(u)\, dt\, du \qquad (14\text{-}73)$$

From Eq. (14-69) it follows that m_{l_1} can be written in the form

$$m_{l_1} = -\tfrac{1}{2} \int_0^T \int_0^T R_n(t - u)f(t)f(u)\, dt\, du \qquad (14\text{-}74)$$

Then the probability that the received signal is identified as $s_0(t)$ when it is actually $s_1(t)$ is

$$\frac{1}{\sigma(l)\sqrt{2\pi}} \int_{\log \eta}^{\infty} \exp\left[-\frac{1}{2} \frac{(x - m_{l_1})^2}{\sigma^2(l)} \right] dx$$

or $\qquad\qquad \dfrac{1}{\sqrt{2\pi}} \displaystyle\int_{\log \eta/\sigma(l) + \frac{1}{2}\sigma(l)}^{\infty} \exp\left(-\frac{1}{2} u^2 \right) du \qquad (14\text{-}75)$

The probability that the received signal is identified as $s_1(t)$ when it is actually $s_0(t)$ is

$$\frac{1}{\sqrt{2\pi}} \int_{-\infty}^{\log \eta/\sigma(l) - \frac{1}{2}\sigma(l)} \exp\left(-\frac{1}{2}u^2\right) du \qquad (14\text{-}76)$$

If the noise is white noise so that the correlation function is an impulse function $R_n(t) = N\delta(t)$, then Eq. (14-69) reduces to

$$Nf(u) = s_0(u) - s_1(u)$$

where N is the noise power. Substituting this in Eq. (14-68) yields

$$\log l(y) = \frac{1}{N} \int_0^T s_0(t)y(t)\, dt - \frac{1}{N} \int_0^T s_1(t)y(t)\, dt$$

$$- \frac{1}{2N} \int_0^T s_0{}^2(t)\, dt + \frac{1}{2N} \int_0^T s_1{}^2(t)\, dt \qquad (14\text{-}77)$$

If now the two signals $s_0(t)$ and $s_1(t)$ have equal energy, the last two terms in Eq. (14-77) cancel, and if $\eta = 1$, the likelihood test becomes simply: choose H_0 if

$$\int_0^T s_0(t)y(t)\, dt > \int_0^T s_1(t)y(t)\, dt \qquad (14\text{-}78)$$

and H_1 otherwise. A detector which instruments the criterion of (14-78) is called a *correlation detector*. The correlation function cannot be strictly an impulse function, of course, but it can be an approximation to one as far as the integral equation (14-69) is concerned if the spectral density of the noise is fairly flat over a frequency range considerably wider than the frequency band of the signals. This approximation is still reasonable in the usual case of narrow-band signals and white narrow-band noise; the requirements are that the center frequency ω_0 be very much greater than the noise bandwidth and the noise bandwidth be greater than the signal bandwidth.

14-6. Signals with Unknown Parameters in Gaussian Noise

A more realistic model of a received radio signal than that given by Eq. (14-52) can be had by introducing parameters to account for unknown amplitude, phase, additional terms due to multipath transmission, etc. The simplest model of a signal in gaussian noise which includes such a parameter is

$$y(t) = \beta s_i(t) + n(t) \qquad 0 \le t \le T, i = 0, 1 \qquad (14\text{-}79)$$

Let us discuss detection of a signal of the form given by Eq. (14-79) under two different statistical assumptions: that β is a random variable with known probability distribution, and that β is completely unknown.

Using the notations of the preceding article, we have, instead of Eq. (14-58), the equations

$$y_k = \beta a_k + z_k \qquad \text{if } i = 0, \, k = 1, 2, \ldots \qquad (14\text{-}80)$$
$$= \beta b_k + z_k \qquad \text{if } i = 1$$

Suppose now that β is a random variable independent of the noise $n(t)$ and with a probability density function $p(\beta)$. Then on the sample space of the first N of the y_k's there are the probability densities

$$p_0(y_1, \ldots, y_N) = \int_{-\infty}^{\infty} \frac{1}{(2\pi)^{N/2}\sigma_1 \cdots \sigma_N}$$
$$\exp\left[-\frac{1}{2}\sum_{k=1}^{N} \frac{(y_k - \beta a_k)^2}{\sigma_k{}^2}\right] p(\beta) \, d\beta \qquad (14\text{-}81)$$

$$p_1(y_1, \ldots, y_N) = \int_{-\infty}^{\infty} \frac{1}{(2\pi)^{N/2}\sigma_1 \cdots \sigma_N}$$
$$\exp\left[-\frac{1}{2}\sum_{k=1}^{N} \frac{(y_k - \beta b_k)^2}{\sigma_k{}^2}\right] p(\beta) \, d\beta \qquad (14\text{-}82)$$

if $s_0(t)$ was actually sent or $s_1(t)$ was actually sent, respectively. The likelihood ratio $l_N(y_1, \ldots, y_N)$ is then

$$l_N(y_1, \ldots, y_N) = \frac{\displaystyle\int_{-\infty}^{\infty} \exp\left[-\frac{1}{2}\sum_{k=1}^{N}(y_k - \beta a_k)^2/\sigma_k{}^2\right] p(\beta) \, d\beta}{\displaystyle\int_{-\infty}^{\infty} \exp\left[-\frac{1}{2}\sum_{k=1}^{N}(y_k - \beta b_k)^2/\sigma_k{}^2\right] p(\beta) \, d\beta} \qquad (14\text{-}83)$$

The limit of l_N as $N \to \infty$ is the likelihood ratio on the sample space of all the y_k.

If there is actually very little known about β, it may be reasonable to assign it a uniform distribution between two limits β_1 and β_2. Then

$$p(\beta) = \frac{1}{\beta_2 - \beta_1} \qquad \beta_1 \le \beta \le \beta_2 \qquad (14\text{-}84)$$
$$= 0 \text{ otherwise}$$

After canceling common factors, the likelihood ratio is

$$l_N(y_1, \ldots, y_N) = \frac{\displaystyle\int_{\beta_1}^{\beta_2} \exp\left(\beta \sum_{k=1}^{N} a_k y_k / \sigma_k{}^2 - \beta^2/2 \sum_{k=1}^{N} a_k{}^2/\sigma_k{}^2\right) d\beta}{\displaystyle\int_{\beta_1}^{\beta_2} \exp\left(\beta \sum_{k=1}^{N} b_k y_k / \sigma_k{}^2 - \beta^2/2 \sum_{k=1}^{N} b_k{}^2/\sigma_k{}^2\right) d\beta}$$

This cannot be expressed in a very simple form.   If we let

$$Y_0 = \sum_{k=1}^{N} \frac{a_k y_k}{\sigma_k^2} \qquad A^2 = \sum_{k=1}^{N} \frac{a_k^2}{\sigma_k^2}$$

$$Y_1 = \sum_{k=1}^{N} \frac{b_k y_k}{\sigma_k^2} \qquad B^2 = \sum_{k=1}^{N} \frac{b_k^2}{\sigma_k^2}$$

then

$$l_N(y_1, \ldots, y_N) = \frac{B}{A}$$

$$\exp\left(\frac{Y_0^2}{2A^2} - \frac{Y_1^2}{2B^2}\right) \frac{F(\beta_2 A - Y_0/A) - F(\beta_1 A - Y_0/A)}{F(\beta_2 B - Y_1/B) - F(\beta_1 B - Y_1/B)} \qquad (14\text{-}85)$$

where F is the cumulative normal distribution

$$F(X) = \frac{1}{\sqrt{2\pi}} \int_{-\infty}^{X} \exp\left(-\frac{x^2}{2}\right) dx$$

It is not difficult in principle to introduce parameters with known probability densities, but as indicated even by this relatively simple example, the calculations become very unwieldy.

If the amplitude parameter β is unknown and there is no a priori information about it, then deciding at the receiver whether $s_0(t)$ or $s_1(t)$ was sent is a problem of testing between two composite hypotheses.   The generalized likelihood ratio on the space of the first N observable coordinates has already been calculated in Example 14-3.5 and is given by Eq. (14-38).   The limit of this as $N \to \infty$ is the likelihood ratio on the space of all the y_k.

This likelihood test can also be interpreted in terms of filters which maximize signal-to-noise ratio.   Using the notations c_N and d_N as in Eq. (14-37), we shall suppose that

$$c^2 = \lim_{N \to \infty} c_N^2 \qquad d^2 = \lim_{N \to \infty} d_N^2 \qquad (14\text{-}86)$$

are finite and that the integral equations

$$\int_0^T f_0(t) R_n(u - t)\, dt = s_0(u) \qquad 0 \le u \le T \qquad (14\text{-}87)$$

$$\int_0^T f_1(t) R_n(u - t)\, dt = s_1(u) \qquad 0 \le u \le T \qquad (14\text{-}88)$$

have solutions $f_0(t)$ and $f_1(t)$ which are of integrable square. Then, by a development like that which produced Eq. (14-69), we have

$$f_0(t) = \sum_{k=1}^{\infty} \frac{a_k \phi_k(t)}{\sigma_k^2} \qquad 0 \le t \le T \tag{14-89a}$$

$$f_1(t) = \sum_{k=1}^{\infty} \frac{b_k \phi_k(t)}{\sigma_k^2} \qquad 0 \le t \le T \tag{14-89b}$$

From Eq. (14-38) this gives, for the logarithm of the likelihood ratio on the space of all the y_k,

$$\log l(y) = \frac{1}{2c^2} \left[\int_0^T f_0(t) y(t)\, dt \right]^2 - \frac{1}{2d^2} \left[\int_0^T f_1(t) y(t)\, dt \right]^2$$

If now the functions $g_0(t)$ and $g_1(t)$ are defined by

$$g_0(T - t) = f_0(t) \qquad 0 \le t \le T \tag{14-90a}$$
$$g_1(T - t) = f_1(t) \qquad 0 \le t \le T \tag{14-90b}$$

we have, as in Eq. (14-71),

$$\log l(y) = \frac{1}{2c^2} \left[\int_0^T g_0(T - t) y(t)\, dt \right]^2 - \frac{1}{2d^2} \left[\int_0^T g_1(T - t) y(t)\, dt \right]^2$$
$$\tag{14-91}$$

where $g_0(t)$ and $g_1(t)$ satisfy

$$\int_0^T R_n(v - t) g_0(t)\, dt = s_0(T - v) \qquad 0 \le t \le T \tag{14-92a}$$

$$\int_0^T R_n(v - t) g_1(t)\, dt = s_1(T - v) \qquad 0 \le t \le T \tag{14-92b}$$

Thus the terms in parentheses in Eq. (14-91) can be regarded as outputs of filters which maximize the signal-to-noise ratio for signals $s_0(t)$ and $s_1(t)$ respectively. If the signals are normalized so that $c^2 = d^2$ and the likelihood threshold $\eta = 1$, the likelihood test may be implemented by putting the received signal through each of these two "matched filters" and choosing $s_0(t)$ or $s_1(t)$ according to which of the filters has the greater output at time T. (See Fig. 14-6.)

The generalized maximum-likelihood test which we have just used is clearly equivalent to the following statistical procedure: Assuming H_0 is true, a maximum-likelihood estimate is made for the unknown parameter (or parameters); this estimated value (or values) is then used in place of the actual parameter value in determining a probability density on the

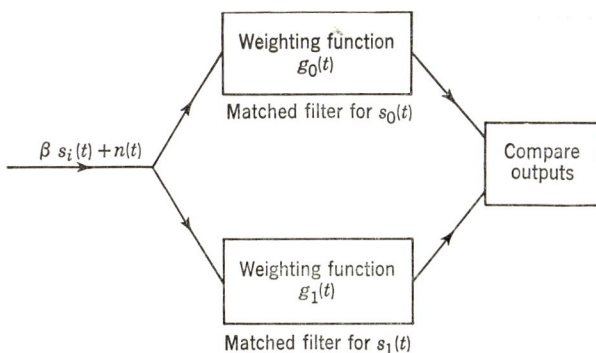

FIG. 14-6. Optimum receiver for binary signals in gaussian noise.

sample space under hypothesis H_0. Then another estimate (or estimates) is made, assuming H_1 is true, and this estimated value is used in determining a probability density under hypothesis H_1. The ratio of these two densities is the generalized likelihood ratio. Thus, parameter estimation is an integral part of the hypothesis-testing procedure.

Let us consider very briefly the problem of estimating β and τ in

$$y(t) = \beta s(t - \tau) + n(t) \qquad 0 \leq t \leq T \qquad (14\text{-}93)$$

where now $s(t)$ is presumed to be defined outside the interval $0 \leq t \leq T$ as well as inside it. Let

$$s_k(\beta, \tau) = \int_0^T \beta s(t - \tau) \phi_k(t) \, dt \qquad (14\text{-}94)$$

Then the probability density for the first N observable coordinates for fixed β and τ is

$$p(y_1, \ldots, y_N; \beta, \tau) = \frac{1}{(2\pi)^{N/2} \sigma_1 \cdots \sigma_N} \exp\left\{ -\frac{1}{2} \sum_{k=1}^{N} \frac{[y_k - s_k(\beta, \tau)]^2}{\sigma_k^2} \right\}$$

The likelihood equations are

$$\frac{\partial}{\partial \beta} \log p(y_1, \ldots, y_N; \beta, \tau) = \sum_{k=1}^{N} \frac{[y_k - s_k(\beta, \tau)]}{\sigma_k^2} \frac{\partial s_k(\beta, \tau)}{\partial \beta} = 0 \quad (14\text{-}95)$$

$$\frac{\partial}{\partial \tau} \log p(y_1, \ldots, y_N; \beta, \tau) = \sum_{k=1}^{N} \frac{[y_k - s_k(\beta, \tau)]}{\sigma_k^2} \frac{\partial s_k(\beta, \tau)}{\partial \tau} = 0 \quad (14\text{-}96)$$

Solutions of these equations for β and τ as functions of $y_1, \ldots, y_N$ are maximum-likelihood estimates. It is hard to say much about these solutions until the form of $s(t)$ is specified. As an example, let us take $s(t)$

to be a sine wave with an integral number of periods in the time interval $0 \leq t \leq T$ and estimate τ.

$$s(t) = \cos m\omega_0 t \qquad \omega_0 = \frac{2\pi}{T} \tag{14-97}$$

Then, from Eq. (14-94),

$$s_k(\beta,\tau) = \beta \int_0^T \cos m\omega_0(t - \tau)\phi_k(t)\,dt \tag{14-98}$$

$$= \beta c_k \cos m\omega_0\tau + \beta d_k \sin m\omega_0\tau$$

where

$$c_k = \int_0^T \phi_k(t)\cos m\omega_0 t\,dt \tag{14-99a}$$

$$d_k = \int_0^T \phi_k(t)\sin m\omega_0 t\,dt \tag{14-99b}$$

Substituting $s_k(\beta,\tau)$ from Eq. (14-98) and its derivatives into Eqs. (14-95) and (14-96) and eliminating β yields

$$\tan m\omega_0\tau = -\frac{\left(\sum\limits_{k=1}^N \dfrac{y_k d_k}{\sigma_k{}^2}\right)\left(\sum\limits_{k=1}^N \dfrac{c_k{}^2}{\sigma_k{}^2}\right) - \left(\sum\limits_{k=1}^N \dfrac{y_k c_k}{\sigma_k{}^2}\right)\left(\sum\limits_{k=1}^N \dfrac{d_k c_k}{\sigma_k{}^2}\right)}{\left(\sum\limits_{k=1}^N \dfrac{y_k d_k}{\sigma_k{}^2}\right)\left(\sum\limits_{k=1}^N \dfrac{d_k c_k}{\sigma_k{}^2}\right) - \left(\sum\limits_{k=1}^N \dfrac{y_k c_k}{\sigma_k{}^2}\right)\left(\sum\limits_{k=1}^N \dfrac{d_k{}^2}{\sigma_k{}^2}\right)} \tag{14-100}$$

If the limits exist, this becomes

$$\tan m\omega_0\tau = \frac{E\int_0^T y(t)g(t)\,dt - C\int_0^T y(t)h(t)\,dt}{E\int_0^T y(t)h(t)\,dt - D\int_0^T y(t)g(t)\,dt} \tag{14-101}$$

where

$$C = \sum_{k=1}^\infty \frac{c_k{}^2}{\sigma_k{}^2} \qquad D = \sum_{k=1}^\infty \frac{d_k{}^2}{\sigma_k{}^2} \qquad E = \sum_{k=1}^\infty \frac{c_k d_k}{\sigma_k{}^2}$$

$$\int_0^T R_n(u - t)h(t)\,dt = \sin m\omega_0 u \qquad 0 \leq u \leq T$$

$$\int_0^T R_n(u - t)g(t)\,dt = \cos m\omega_0 u \qquad 0 \leq u \leq T$$

If the noise is white noise, Eq. (14-101) reduces to

$$\tan m\omega_0\tau = \frac{\int_0^T y(t)\sin m\omega_0 t\,dt}{\int_0^T y(t)\cos m\omega_0 t\,dt} \tag{14-102}$$

as one would anticipate. A principal value for τ from Eqs. (14-100), (14-101), and (14-102) is then an estimate of the phase of $s(t)$.

General Remarks. Although the choice of observable coordinates which we have made is in a sense the most natural one for problems involving gaussian noise and finite observation times, it is of course not the only possible one. Another natural way to make observations is to sample the received signal at discrete intervals. If the sampling intervals are appreciably longer than the "correlation time," the samples are approximately independent gaussian random variables; hence it is relatively easy to write approximately the joint probability density functions for the samples in order to calculate the likelihood ratio. Some information available in the received signal is lost when this is done. If the sampling intervals are taken to be short, the joint probability densities of the samples require the inversion of the correlation matrix for the samples. The detection of a sine wave in noise has been treated using sampling at intervals.† In the limit as the intervals between samples approach zero, the matrix inversion becomes formally the solution of the integral equation (14-69).

We have restricted ourselves in Arts. 14-5 and 14-6 to what might be termed synchronous communication, in which every symbol is represented by a function of the same time duration. Radio teletype often meets this condition; Morse code does not. However, one can think of Morse-code symbols as being further divided into symbols on basic intervals which do meet this condition.‡ Thus if the dot, dash, symbol space, and letter space have relative time durations of, say, 1, 3, 1, 3, then the length of a dot may be taken as the basic time interval and each of the Morse-code symbols may be expressed in terms of marks and spaces on this basic interval. In general, of course, any code can be broken up into a more elementary one in which each symbol has the same duration, if the symbol durations in the original code are rationally related.

Also, we have concerned ourselves only with a per-symbol analysis of the received signal. In a more complete study, one should analyze the interrelations among symbols. The parameters which we have treated as unknown constants might actually be constants, or might vary with time, but slowly enough so that for the duration of one symbol they are nearly constant. In either case, information is obtained implicitly with the reception of each symbol which should be useful in estimating parameter values which exist with succeeding symbols.

One of the more difficult practical communication problems to which statistical theory can be fruitfully applied is that of long-distance multipath propagation, referred to in Art. 14-1. Rather complicated statistical models of the received signals are necessary to account for the distortion caused by the fluctuating multiple paths. Some results have been obtained in this area, for which the reader is referred to the literature.§

---

† See, for example, Reich and Swerling (I).

‡ This is strictly true only if the keying is done by a machine rather than by hand.

§ See Price (I), Price (III), Turin (I), and Root and Pitcher (II).

An additional restriction on the material of Arts. 14-5 and 14-6 is that it applies only to two-symbol alphabets. The use of more than two symbols does not change the problem much, however, as long as the number of symbols is finite. If a priori probabilities exist for the symbols, then a decision may be made that will minimize the probability of error; if no a priori probabilities are given, the likelihood may be maximized. In either case the calculations are essentially the same as those above.

If the communication is not carried out with a discrete set of symbols, as with ordinary voice radio, for example, the reception problem is to recover the form of the transmitted signal as accurately as possible. That is, the reception problem is a smoothing problem such as those discussed in Chap. 11. The maximum-likelihood method has also been applied to the extraction of continuously-modulated signals from noise.†

14-7. Radar Signals in Gaussian Noise

Most of the early applications of statistical methods to radio-wave detection were to radar.‡ The initial problem was to find an optimum receiver for detecting target-reflected signals in noise using a pulsed transmitter, and we shall confine the discussion here to this question. Statistical methods are appropriate in the treatment of many other radar problems, however, such as detecting a target in the presence of sea clutter, measuring accurately range and radial velocity with either pulsed or modulated continuous-wave radars, and measuring accurately azimuth and elevation, particularly with scanning radars.

We shall consider a "conventional" radar emitting a periodic sequence of rectangular pulses of radio waves. The signal is as shown in Fig. 14-1. For simplicity we shall assume the target to be stationary and to have a fixed reflecting surface, and we shall assume the radar to be trained permanently in one direction. The expression for the received signal given by Eq. (14-3) for a target at range r then becomes simply

$$y(t) = A(t - \tau) \cos \omega_0(t - \tau) + n(t) \qquad (14\text{-}103)$$

Finally, we shall assume that the radar does what is known as *range gating*. This means that the time between pulses is divided into equal intervals, which correspond to intervals of range, and a decision is required for each interval as to whether a reflected signal is present or not. The pulse length T is taken to be equal to the length of these intervals. (See Fig. 14-7.)

If we look at any one of these range intervals, the problem is to make a choice between two alternative hypotheses: H_0, there is no target present;

† Youla (I).

‡ For example, Lawson and Uhlenbeck (I), Woodward (I), Marcum (I), Hanse (I), Schwartz (I), Middleton (IV).

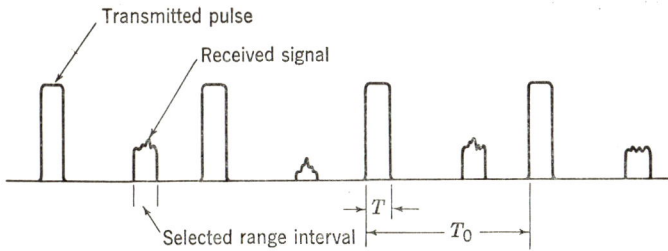

FIG. 14-7. Range-gated pulsed radar return.

and H_1, there is a target present. It will be assumed that if a target is present, it is at the beginning of the range interval. For a single pulse, then, H_0 is the hypothesis that a particular section of received signal of T seconds duration is noise alone, and H_1 is the hypothesis that it is sine-wave-plus noise. This is a special case of the hypothesis-testing problem discussed in Art. 14-5. As in Eqs. (14-70) and (14-71) we see that the outcome of a maximum-likelihood test is to choose H_1, i.e., to decide a target is present, if

$$\int_0^T g(T - t)y(t)\, dt > \text{const.} \tag{14-104}$$

where $g(t)$ is determined by

$$\int_0^T R_n(v - t)g(v)\, dv = \cos \omega_0(T - t) \qquad 0 \le t \le T \tag{14-105}$$

and the constant is determined by the level of the test. In the usual radar terminology the *false-alarm rate* is the average ratio of false target sightings to the total number of decisions when no target is present. The false-alarm rate is therefore equal to the probability of rejecting H_0 when H_0 is true; that is, the false-alarm rate is equal to the level of the test. It was shown in Example 14-3.1 that the maximum-likelihood test on $y_1, \ldots, y_N$ is a uniformly most powerful test of the null hypothesis H_0 against a composite hypothesis which includes all positive amplitudes of the signal. The argument given there carries over immediately to the limiting case as $N \to \infty$, hence the test described above is uniformly most powerful at its level, or false-alarm rate, against the hypothesis that there is a target echo of any positive amplitude in the received signal.

Usual radar practice is to have a sequence of returned pulses to examine instead of only one. Suppose the radar is trained in one direction long enough for K pulses to be returned, again considering only one range interval. The portion of the received waveform which is of interest is as shown in Fig. 14-7. It will be assumed, quite reasonably, that the noise bandwidth is great enough so that the noise is completely uncorrelated

with itself one pulse period removed.   Thus the noise in each section of
the received waveform, as shown in Fig. 14-7, will be independent of
that in each other section.

Take the pulse period to be T_0 and let t be measured from the leading edge of the range interval in question after each pulse.   Let
$y^{(1)}(t) = y(t)$, $0 \leq t \leq T$ be the first section of received waveform,
$y^{(2)}(t) = y(t + T_0)$, $0 \leq t \leq T$ be the second section of received waveform, and so on.   Suppose that the modulating pulses are phase-locked
to the rf carrier so that the signal $s_1(t)$ is identical for each section of the
received waveform.   Let

$$y_m^{(1)} = \int_0^T y^{(1)}(t)\,\phi_m(t)\ dt$$

$$y_m^{(2)} = \int_0^T y^{(2)}(t)\,\phi_m(t)\ dt \qquad (14\text{-}106)$$

$$y_m^{(K)} = \int_0^T y^{(K)}(t)\,\phi_m(t)\ dt$$

Then, since $y_m^{(p)}$ is independent of $y_m^{(q)}$, $p \neq q$, we have the probability
densities

$$p_0[y_1^{(1)}, \ldots, y_N^{(1)}; y_1^{(2)}, \ldots, y_N^{(2)}; \ldots ; y_1^{(K)}, \ldots, y_N^{(K)}]$$

$$= \frac{1}{(2\pi)^{NK/2}\sigma_1^K \cdots \sigma_N^K} \exp\left\{ -\frac{1}{2}\sum_{m=1}^{N} \frac{[y_m^{(1)}]^2}{\sigma_m^2} - \frac{1}{2}\sum_{m=1}^{N}\frac{[y_m^{(2)}]^2}{\sigma_m^2} \right.$$

$$\left. - \cdots - \frac{1}{2}\sum_{m=1}^{N}\frac{[y_m^{(K)}]^2}{\sigma_m^2} \right\} \qquad (14\text{-}107)$$

if there is no target present, and

$$p_1[y_1^{(1)}, \ldots, y_N^{(1)}; y_1^{(2)}, \ldots, y_N^{(2)}; \ldots ; y_1^{(K)}, \ldots, y_N^{(K)}]$$

$$= \frac{1}{(2\pi)^{NK/2}\sigma_1^K \cdots \sigma_N^K} \exp\left\{ -\frac{1}{2}\sum_{m=1}^{N}\frac{[y_m^{(1)} - b_m]^2}{\sigma_m^2} \right.$$

$$\left. - \cdots - \frac{1}{2}\sum_{m=1}^{N}\frac{[y_m^{(K)} - b_m]^2}{\sigma_m^2} \right\} \qquad (14\text{-}108)$$

if there is a target present.   The logarithm of the likelihood ratio is

$$\log l_N[y_1^{(1)}, \ldots, y_N^{(K)}] = \frac{K}{2}\sum_{m=1}^{N}\frac{b_m^2}{\sigma_m^2} - \sum_{m=1}^{N}\frac{y_m^{(1)}b_m}{\sigma_m^2}$$

$$- \cdots - \sum_{m=1}^{N}\frac{y_m^{(K)}b_m}{\sigma_m^2} \qquad (14\text{-}109)$$

Passing to the limit and introducing the test function $f(t)$ defined by Eq. (14-69) gives

$$\log l(y) = \int_0^T f(t) \left[y^{(1)}(t) + y^{(2)}(t) + \cdots + y^{(K)}(t) - \frac{K s_1(t)}{2} \right] dt \tag{14-110}$$

From this formula for $\log l(y)$ one can interpret the likelihood-ratio test in terms of signal-to-noise ratio maximizing filters as before. The result is as follows: The received waveform for the selected time interval after each pulse is put into such a filter. The output from the filter at the end of each interval is stored, and after K pulses, these outputs are added. If their sum exceeds an established threshold, the decision is that a target is present. Again, this is a uniformly most powerful test at its level for any positive signal amplitude. As given by Eq. (14-110), $\log l(y)$ is a gaussian random variable, so the probability of each kind of error can be determined from the means and variances. With the notations used in Eqs. (14-72) and (14-73), the probability that a target is missed is

$$\frac{1}{\sqrt{2\pi K}\, \sigma(l)} \int_{K \log \eta}^{\infty} \exp\left[-\frac{1}{2} \frac{(x - K m_{l_1})^2}{K \sigma^2(l)} \right] dx$$

or
$$\frac{1}{\sqrt{2\pi}} \int_{\sqrt{K}\left[\frac{\log \eta}{\sigma(l)} + \frac{1}{2}\sigma(l) \right]}^{\infty} \exp\left(-\frac{1}{2} u^2 \right) du \tag{14-111}$$

and the probability of a false alarm is

$$\frac{1}{\sqrt{2\pi}} \int_{-\infty}^{\sqrt{K}\left[\frac{\log \eta}{\sigma(l)} - \frac{1}{2}\sigma(l) \right]} \exp\left(-\frac{1}{2} u^2 \right) du \tag{14-112}$$

These error probabilities reduce of course to the expressions given by (14-75) and (14-76) when $K = 1$.

The probability-of-error formulas (14-111) and (14-112) are the same as would be obtained if instead of K pulses only one pulse were received but with magnitude $\sqrt{K}$ times greater. This can be easily verified by multiplying $s_1(t)$ by $\sqrt{K}$ and substituting in Eqs. (14-69), (14-72), and (14-73). Thus the effective voltage signal-to-noise ratio at the receiver (for the detector being considered) is proportional to the square root of the number of independent pulses received from one target.

Envelope Detector. In the preceding section, the "received" signal, here written $z(t)$, was taken to be

$$z(t) = s_i(t) + n(t) \qquad i = 0, 1, \quad 0 \leq t \leq T \qquad (14\text{-}113)$$
$$s_0(t) \equiv 0$$

with $n(t)$ gaussian noise. The sort of matched-filter techniques which have been developed for optimum reception of such signals are sometimes difficult to implement practically because they are sensitive to RF phase. For this reason, it is often desirable to take the envelope of the signal first, i.e., to rectify the signal and put it through a low-pass filter in the conventional way so that only the modulating frequencies are left, before doing any special signal processing. It should be realized that doing this involves throwing away some information of possible value.

When dealing with a signal $z(t)$ as given by Eq. (14-113), it is expedient to expand $z(t)$ in the orthogonal series using the characteristic functions of the correlation function of the noise, because the random coefficients of the expansion are gaussian and statistically independent. However, nothing is gained by expanding the envelope of $z(t)$ in such a series; the coefficients are no longer either gaussian or statistically independent. Consequently, a different set of observable coordinates will be chosen. The usual procedure, which we shall follow here, is to use for observable coordinates samples of the envelope waveform taken at regular intervals and far enough apart so that it is a reasonable approximation to suppose them statistically independent. We suppose, as before, that the time between pulses is divided into intervals of equal length and that it is desired to test for the presence of a target echo in each one of these intervals. The simplifying assumptions made in the previous section are in effect.

We also suppose the radar is trained in a fixed direction long enough for K pulses to be returned from a target, and we look at a particular range interval. The observable coordinates $y_1, \ldots, y_K$ are chosen to be a sequence of values of the envelope, one from the time interval in question after each pulse.

The signal input to the second detector (demodulator) of the radar receiver is

$$z(t) = s_i(t) + n(t) \qquad 0 \leq t \leq T, \quad i = 0, 1$$
$$s_0(t) \equiv 0$$

as before. This can be written in the form of a narrow-band signal. From Eq. (8-82) we have

$$z(t) = A_i \cos \omega_c t + x_c(t) \cos \omega_c t - x_s(t) \sin \omega_c t \qquad i = 0, 1 \qquad (14\text{-}114)$$
$$A_0 = 0, \quad A_1 = \text{const.}$$

where x_{ct} and x_{st} are gaussian random variables with the properties stated in Art. 8-5. The demodulator is assumed to be a perfect envelope detector, so its output $y(t)$ is, from Eqs. (8-79) and (8-85a),

$$y(t) = \{[A_i + x_c(t)]^2 + x_s^2(t)\}^{\frac{1}{2}} \qquad i = 0, 1 \qquad (14\text{-}115)$$

From Eq. (8-91), the probability density for the envelope at time t on the hypothesis that the sampled waveform is noise alone is

$$p_0(y_t) = \frac{y_t}{\sigma_n^2} \exp\left(-\frac{y_t^2}{2\sigma_n^2}\right) \qquad y_t \geq 0 \qquad (14\text{-}116)$$
$$= 0 \qquad\qquad\qquad \text{otherwise}$$

where $\sigma_n^2 = R_n(0)$. The probability density for the envelope at time t on the hypothesis that the sampled waveform is sine-wave-plus noise is, from Eq. (8-115),

$$p_1(y_t) = \frac{y_t}{\sigma_n^2} \exp\left(-\frac{y_t^2 + A_1^2}{2\sigma_n^2}\right) I_0\left(\frac{A_1 y_t}{\sigma_n^2}\right) \qquad y_t \geq 0$$
$$= 0 \qquad\qquad\qquad\qquad\qquad\qquad\qquad \text{otherwise}$$

The likelihood ratio for the samples $y_1 = y_{t_1}, \ldots, y_K = y_{t_K}$ is, to a good approximation,

$$l_K(y_1, \ldots, y_K) = \frac{\displaystyle\prod_{k=1}^{K} p_0(y_k)}{\displaystyle\prod_{k=1}^{K} p_1(y_k)} = \prod_{k=1}^{K} \exp\left(\frac{A_1^2}{2\sigma_n^2}\right) \Big/ I_0\left(\frac{y_k A_1}{\sigma_n^2}\right) \qquad (14\text{-}117)$$

since the y_k are nearly independent. Thus a likelihood test is to choose hypothesis H_1 (target present) if

$$\sum_{k=1}^{K} \log I_0\left(\frac{y_k A_1}{\sigma_n^2}\right) > -\log \eta + \frac{K A_1^2}{2\sigma_n^2} = \text{const.} \qquad (14\text{-}118)$$

The Bessel function $I_0(z)$ has the series expansion

$$I_0(z) = \sum_{m=0}^{\infty} \frac{z^{2m}}{2^{2m}(m!)^2}$$

Hence if the target echo has a small amplitude (low signal-to-noise ratio) it is approximately true that

$$I_0\left(\frac{y_k A_1}{\sigma_n^2}\right) = 1 + \left(\frac{y_k A_1}{\sigma_n^2}\right)^2 \frac{1}{2^2(2!)^2} \qquad (14\text{-}119)$$

Using the further approximation, valid for small values of the argument,

$$\log (1 + a) = a \tag{14-120}$$

the test given by (14-118) becomes, for *small signals*,

$$\frac{A_1{}^2}{16\sigma_n{}^2} \sum_{k=1}^{K} y_k{}^2 > \text{const.}$$

or simply

$$\sum_{k=1}^{K} y_k{}^2 > \text{const.} \tag{14-121}$$

The small-signal approximation can often be justified in practice by the idea that it is only necessary to have a near-optimum detector for weak echoes, since strong echoes will be detected even if the receiver is well below optimum. This test is easy to implement, since the $y_k{}^2$ are just the outputs at regularly spaced time intervals of a square-law demodulator. Detection probabilities for the square-law envelope detector have been tabulated by Marcum.†

Phase Detector. As a final example, we shall consider what is called a phase detector, which can be useful for the detection of moving objects. Again we want to test for the presence of a target echo in a fixed range interval. The assumptions made for the above examples are still to hold. The input signal to the demodulator is given by Eq. (14-114), which can be rewritten in the form

$$z(t) = y_i(t) \cos [\omega_c t + \phi_i(t)] \qquad i = 0, 1 \tag{14-122}$$

where

$$y_0(t) = [x_c{}^2(t) + x_s{}^2(t)]^{\frac{1}{2}}$$
$$y_1(t) = \{[A_1 + x_c(t)]^2 + x_s{}^2(t)\}^{\frac{1}{2}}$$

and

$$\phi_0(t) = \tan^{-1} \left[\frac{x_s(t)}{x_c(t)} \right] \tag{14-123}$$

$$\phi_1(t) = \tan^{-1} \left[\frac{x_s(t)}{A_1 + x_c(t)} \right]$$

The observable coordinates are taken to be a sequence of K values of the phase $\phi(t)$, one from the time interval in question after each pulse (these values of phase must themselves be got by what amounts to an estimation procedure; see Art. 14-6). If there is noise alone the probability density function for the phase at any time t is

$$p_0(\phi_t) = \frac{1}{2\pi} \qquad -\pi \le \phi \le \pi$$
$$= 0 \qquad \text{otherwise} \tag{14-124}$$

If there is a sine wave present in the noise, the probability density of

† Marcum (I).

ϕ_t is 2π times $p(\phi_t,\psi)$ as given by Eq. (8-118). For large signal-to-noise ratio, $A_1{}^2/2\sigma_n{}^2$ very much greater than 1, $\phi(t)$ is usually small and the approximate density function $p(\phi_t,\psi)$ given by Eq. (8-114) can be used. Replacing cos ϕ_t by 1 and sin ϕ_t by ϕ_t and multiplying by 2π so as to get the density $p(\phi_t)$ for fixed ψ, one obtains

$$p_1(\phi_t) = \frac{1}{\sqrt{2\pi}} \frac{A_1}{\sigma_n} \exp\left(-\frac{A_1{}^2\phi_t{}^2}{2\sigma_n{}^2}\right) \qquad -\pi \le \phi \le \pi$$
$$= 0 \qquad\qquad\qquad\qquad\qquad\qquad \text{otherwise}$$

Then the likelihood ratio for K independent samples is

$$l_K(\phi_1, \ldots, \phi_K) = \frac{1}{(2\pi)^{K/2}} \frac{\sigma_n{}^K}{A_1{}^K} \exp\left[\frac{A_1{}^2}{2\sigma_n{}^2}(\phi_1{}^2 + \cdots + \phi_K{}^2)\right]$$

and the likelihood test gives the decision criterion: no target present if

$$\sum_{k=1}^{K} \phi_k{}^2 > \text{const.} \tag{14-125}$$

This test can be modified slightly to be made a test to detect targets moving with a known constant velocity toward the radar. The effect of target velocity is to give a doppler shift of frequency. For ordinary pulse lengths, and for targets such as ships or airplanes, this shift in frequency results in an almost imperceptible total phase shift during one pulse length. Thus one can think of a pure echo signal as having constant phase over one pulse length but with this phase advancing linearly from pulse to pulse. The kth returned pulse should then have phase $k\gamma$ where γ is a constant depending on the velocity of the target. The observable coordinates are then taken to be $\phi(t_k) - k\gamma$ instead of $\phi(t_k)$, and the test for no target at the given velocity is

$$\sum_{k=1}^{K} (\phi_k - k\gamma)^2 > \text{const.}$$

14-7. Problems

1. Find a test for deciding between the hypothesis H_0 and H_1 which minimizes the expected loss. H_0 is the hypothesis that an observed real number x is distributed according to the rectangular distribution

$$p_0(x) = \tfrac{1}{4} \qquad 0 \le x \le 4$$
$$= 0 \qquad \text{otherwise}$$

H_1 is the hypothesis that x is distributed according to

$$p_1(x) = \frac{1}{\sqrt{2\pi}} \exp\left(-\frac{x^2}{2}\right)$$

The a priori probabilities are $\pi_0 = \tfrac{1}{4}$, $\pi_1 = \tfrac{3}{4}$. The loss values for each kind of error are equal. Find the total probability of error.

2. a. Show that the likelihood ratio as defined by Eq. (14-30) gives a hypothesis test which is equivalent to a test using either

$$l'(y) = \frac{\max_{\omega_0} p_{\omega_0}(y)}{\max_{\omega} p_{\omega}(y)}$$

or

$$l''(y) = \frac{\max_{\omega_1} p_{\omega_1}(y)}{\max_{\omega} p_{\omega}(y)}$$

b. H_0 is the simple hypothesis that an observed real number x is distributed according to

$$p_0(x) = \frac{1}{\sqrt{2\pi}} e^{-x^2/2}$$

$H_1(b)$ is the simple hypothesis

$$p_{1b}(x) = \frac{b}{2} e^{-b|x|}$$

H_1 is the composite hypothesis containing all the $H_1(b)$ for $b > 0$.   Determine the generalized likelihood ratio for H_0 against H_1.

3. Let $p(y;a)$ be the family of probability density functions

$$
\begin{aligned}
p(y;a) &= 0 & y &< a - 1 \\
&= y + 1 - a & a - 1 &\le y < a \\
&= -y + 1 + a & a &\le y < a + 1 \\
&= 0 & a + 1 &< y
\end{aligned}
$$

Estimate the parameter a by maximizing the likelihood:

a. One observation y_1 of y is made.

$$y_1 = 2$$

b. Three independent observations of y are made.

$$y_1 = 1, \qquad y_2 = 2\tfrac{1}{2}, \qquad y_3 = 2$$

4. N independent samples $y_1, \ldots, y_N$ are drawn from a population which is known to be distributed according to a gaussian law but with unknown mean m and variance σ^2.

a. Show that the maximum likelihood estimator of the mean is the *sample mean* $\bar{y}$, given by

$$\bar{y} = \frac{1}{N} \sum_{k=1}^{N} y_k$$

and the maximum likelihood estimator of the variance is the *sample variance* s^2, given by

$$s^2 = \frac{1}{N} \sum_{k=1}^{N} (y_k - \bar{y})^2$$

b. Show that $\bar{y}$ is an unbiased estimator of m but that s^2 is a biased estimator of σ^2. In particular

$$E_{m,\sigma}(s^2) = \frac{N-1}{N}\sigma^2$$

5. Let a sample waveform be given by

$$y(t) = s_i(t) + n(t) \qquad 0 \leq t \leq 2\pi, \; i = 0, 1$$

where
$$s_0(t) = \cos 3t$$
$$s_1(t) = \cos 2t$$

and
$$n(t) = x_1 \cos t + x_2 \sin t$$

with x_1 and x_2 independent gaussian random variables with mean zero and variance σ^2. Find a function $\psi(t)$ so that with probability one

$$\int_0^{2\pi} \psi(t)y(t)\, dt = a \qquad \text{if } s_0(t)$$
$$= b \qquad \text{if } s_1(t)$$

where $a \neq b$. This is an example of the extreme singular case defined in Art. 14-5. Note that $n(t)$ as defined above is already expanded in the orthogonal series defined in Art. 6-4.

6. Let $s_0(t) \neq 0$ and $s_1(t) \equiv 0$, $0 \leq t \leq T$. In Art. 14-5 it is pointed out, using the notation used there, that a singular case obtains (i.e., perfect detection of $s_0(t)$ in noise is possible) if

$$\sum_{k=1}^{\infty} \frac{a_k^2}{\sigma_k^2} = +\infty$$

Suppose that a radio or radar system can be described as shown in Fig. 14-8. The noise $n(t)$ is assumed to have a spectrum shaped by the input filter to the receiver from

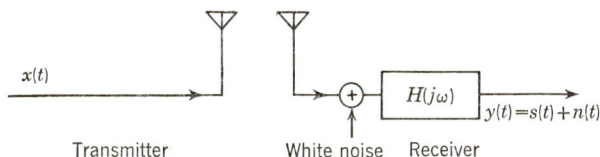

FIG. 14-8. Diagram for Prob. 6.

a white-noise input at the antenna. The signal received at the antenna must also pass through this filter. The signal $x(t)$ at the transmitter is arbitrary except that it is to be of integrable square.

a. From

$$a_k = \int_0^T \phi_k(t)s_0(t)\, dt$$

show that
$$a_k = \int_{-\infty}^{+\infty} X(f)H(j2\pi f)\Phi_k^*(f)\, df$$

where
$$X(f) = \int_{-\infty}^{+\infty} x(t)e^{-j2\pi ft}\, dt$$

and
$$\Phi_k(f) = \int_{-\infty}^{+\infty} \phi_k(t)e^{-j2\pi ft}\, dt = \int_0^T \phi_k(t)e^{-j2\pi ft}\, dt$$

b. Show that

$$\int_{-\infty}^{+\infty} [H^*(j2\pi f)\Phi_m(f)][H(j2\pi f)\Phi_k^*(f)]\, df = \sigma_k^2 \qquad \text{if } m = k$$
$$= 0 \qquad \text{if } m \neq k$$

c. Observe, from **b**, that for some set of numbers c_k

$$X(f) = \sum_k \frac{c_k}{\sigma_k^2} H^*(j2\pi f)\Phi_k(f) + U(f)$$

where

$$\int_{-\infty}^{+\infty} U(f)H(j2\pi f)\Phi_k^*(f)\, df = 0$$

d. Using **a**, **b**, and **c**, show that

$$\sum_k \frac{a_k^2}{\sigma_k^2} \leq \int_{-\infty}^{+\infty} |x^2(t)|\, dt$$

and hence that perfect detection is impossible in this situation.

7. With regard to band-pass and white-noise approximations for optimum filters, consider the integral equation

$$\int_0^T R(t - u)f(u)\, du = s(t) \qquad 0 \leq t \leq T \qquad (14\text{-}126)$$

Suppose $s(t)$ is a narrow-band waveform about a center angular frequency ω_c and $R(t)$ is the correlation function of a narrow-band noise about ω_c. Write

$$s(t) = a(t) \cos \omega_c t + b(t) \sin \omega_c t$$
$$R(t) = R_{lp}(t) \cos \omega_c t$$

a. Show that for large ω_c, $f(u)$ is given approximately by

$$f(u) = g(u) \cos \omega_c u + h(u) \sin \omega_c u$$

where $g(u)$ and $h(u)$ satisfy

$$\int_0^T R_{lp}(t - u)g(u)\, du = 2a(t) \qquad 0 \leq t \leq T$$
$$\int_0^T R_{lp}(t - u)h(u)\, du = 2b(t)$$

b. Let $R_{lp}(t) = \beta e^{-\beta|t|}$ and $a(t) = \cos \omega_M t$, $b(t) \equiv 0$. Show that if the noise bandwidth is much greater than the signal bandwidth, $\beta >> \omega_M$,

$$f(u) = \cos \omega_M u \cos \omega_c u$$

satisfies Eq. (14-126) approximately except at the end points of the interval $0 \leq t \leq T$.

8. Let a received signal be

$$y(t) = s_i(t) + n(t) \qquad 0 \leq t \leq T, i = 0, 1$$

where

$$\int_0^T s_i^2(t)\, dt = 1 \qquad i = 0, 1$$
$$\int_0^T s_1(t)s_0(t)\, dt = 0$$

and $n(t)$ is white noise with autocorrelation function $R(t) = N\delta(t)$. The decision at the receiver is that $s_0(t)$ was sent if

$$\int_0^T y(t)s_0(t)\,dt > \int_0^T y(t)s_1(t)\,dt$$

and $s_1(t)$ otherwise. Both $s_0(t)$ and $s_1(t)$ have a priori probabilities of $\frac{1}{2}$. Find the over-all probability of error for $N = 0.1, 0.5, 1, 2$.

9. Determine the likelihood ratio given by Eq. (14-83) if $p(\beta)$ is a gaussian density function with mean m and variance σ^2.

10. Let

$$y(t) = \beta s(t) + n(t) \qquad 0 \le t \le T$$

where $n(t)$ is stationary gaussian noise, $s(t)$ is a known function, and β is unknown. Find an estimator $\hat{\beta}$ for β which is:

i. of the form

$$\hat{\beta} = \int_0^T f(t)y(t)\,dt \tag{14-127}$$

for some $f(t)$ of integrable square
ii. unbiased, i.e., $E_\beta(\hat{\beta}) = \beta$
iii. of minimum variance among all estimators which satisfy **i** and **ii**.

Find this estimator by directly minimizing the variance of β as given by Eq. (14-127) subject to the constraint **ii**. The result is an estimator which is the limit as $N \to \infty$ of the estimator given by Eq. (14-46) if y_k, a_k, and σ_k are defined as in Eqs. (14-56), (14-57), and (14-54), respectively.

11. The detection criterion for the small-signal approximation to the optimum envelope detector is, from Eq. (14-121),

$$\sum_{k-1}^k y_k^2 > \eta = \text{const.}$$

where the density function for y_k is given by Eq. (14-116) for noise alone, and by Eq. (14-117) for signal plus noise. Show that the false-alarm probability is

$$P(\text{false alarm}) = \int_\eta^\infty \frac{z^{K-1}}{(K-1)!} e^{-z}\,dz$$

and the probability of detecting a target when one is actually present is †

$$P(\text{detection}) = \int_\eta^\infty \left(\frac{z}{K\gamma}\right)^{(K-1)/2} e^{-z-K\gamma} I_{K-1}(2\sqrt{K\gamma z})\,dz$$

where $\gamma = A^2/2\sigma_n^2$. (Hint: Use the Fourier transform

$$\frac{1}{2\pi}\int_{-\infty}^\infty \frac{1}{(1-j\lambda)^n} \exp\left[\frac{jna\lambda}{1-j\lambda}\right] \exp(-j\lambda z)\,d\lambda$$
$$= \left(\frac{z}{n\gamma}\right)^{(n-1)/2} e^{-z-na} I_{n-1}(2\sqrt{naz}) \qquad n = 1, 2, \ldots$$

which can be got by change of variable from a standard integral representation of $J_n(z)$.‡)

† Marcum (I).
‡ Whittaker and Watson (I, Art. 17.1).

APPENDIX 1

THE IMPULSE FUNCTION

A1-1. Definitions

The *unit impulse function* $\delta(x - x_o)$, also called the Dirac delta function, is defined to be infinite when its argument is zero, to be zero when its argument is nonzero, and to have a unit area.† Thus

$$\delta(x - x_o) = \begin{cases} \infty & \text{when } x = x_o \\ 0 & \text{when } x \neq x_o \end{cases} \tag{A1-1}$$

and

$$\int_{-\infty}^{+\infty} \delta(x - x_o) \, dx = 1 \tag{A1-2}$$

Further, it is often desirable to define the impulse function to be an even function of its argument:

$$\delta(x - x_o) = \delta(x_o - x) \tag{A1-3a}$$

In this case,

$$\int_{-\infty}^{x_o} \delta(x - x_o) \, dx = \tfrac{1}{2} = \int_{x_o}^{+\infty} \delta(x - x_o) \, dx \tag{A1-3b}$$

Suppose that the unit impulse function is integrated over the interval $(-\infty, X)$. It then follows that the result of this integral will be equal to zero, one-half, or one according to whether X is less than, equal to, or greater than x_o, respectively; i.e.,

$$\int_{-\infty}^{X} \delta(x - x_o) \, dx = U(X - x_o) \tag{A1-4}$$

where $U(x - x_o)$ is the *unit step function:*

$$U(x - x_o) = \begin{cases} 0 & \text{if } x < x_o \\ \tfrac{1}{2} & \text{if } x = x_o \\ 1 & \text{if } x > x_o \end{cases} \tag{A1-5}$$

The unit step function is thus the integral of the unit impulse function,

† Cf. Guillemin (I, Chap. VII, Arts. 24 and 25) or van der Pol and Bremmer (I, Chap. V).

and we shall therefore consider the unit impulse function to be the derivative of the unit step function. Thus

$$U'(x - x_o) = \delta(x - x_o) \tag{A1-6}$$

The unit impulse function and the unit step function are shown in Fig. A1-1.

Although the impulse function is not quite respectable mathematically, its properties are often useful. For example, we extended the probability

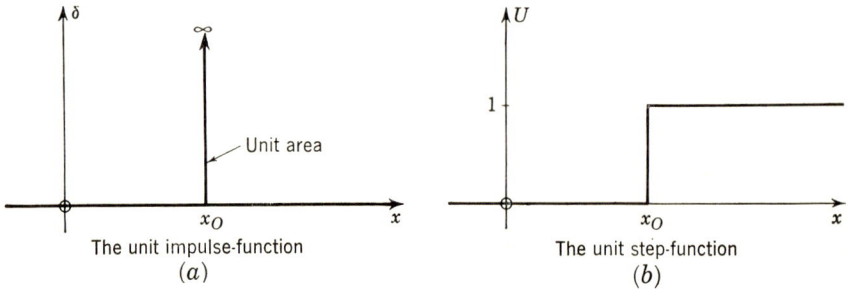

The unit impulse-function
(a)

The unit step-function
(b)

Fig. A1-1. Singularity functions.

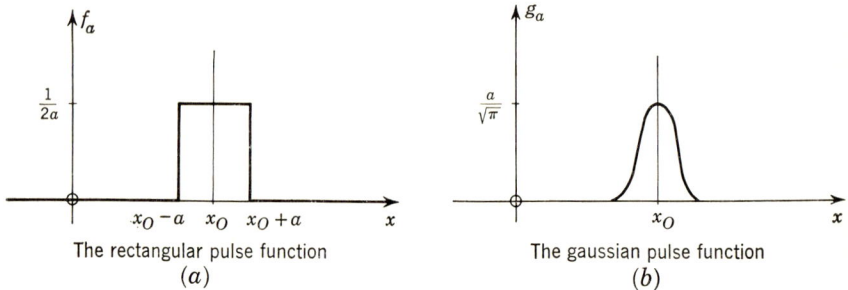

The rectangular pulse function
(a)

The gaussian pulse function
(b)

Fig. A1-2.

density function to cover the case of discrete random variables through the use of the unit impulse-function. To give some validity to the unit impulse function, or rather to the operations we shall perform with it, it is often convenient to consider the unit impulse function to be the limiting member of some infinite sequence of ordinary functions.

Consider the rectangular pulse function

$$f_a(x - x_o) = \begin{cases} \dfrac{1}{2a} & \text{when } x_o - a < x < x_o + a \\ 0 & \text{otherwise} \end{cases} \tag{A1-7}$$

where $a \geq 0$. This function is shown in Fig. A1-2. For this function

$$\int_{-\infty}^{+\infty} f_a(x - x_o)\, dx = 1$$

for all values of $a > 0$. If now we let $a \to 0$, the width of the pulse approaches zero and the height approaches infinity, whereas the area remains constant at unity. The unit impulse function could therefore be considered to be the limiting member of a sequence of rectangular pulse functions:

$$\delta(x - x_o) = \lim_{a \to 0} f_a(x - x_o) \tag{A1-8}$$

Although the rectangular pulse function is a simple and convenient prototype for the impulse function, it is a discontinuous function. In certain problems of interest, it is convenient to use a prototype which possesses derivatives. One such function is the gaussian pulse function

$$g_a(x - x_o) = \frac{a}{\sqrt{\pi}} \exp\left[-a^2(x - x_o)^2\right] \tag{A1-9}$$

where $a > 0$. This function is also shown in Fig. A1-2. Now†

$$\int_{-\infty}^{+\infty} g_a(x - x_o)\,dx = 1$$

for all values of $a > 0$. Further, the height of $g_a(x - x_o)$ increases toward infinity as $a \to \infty$, while the skirts collapse in towards zero. The gaussian pulse function hence satisfies the defining equations for the unit impulse function in the limit $a \to \infty$, and we may set

$$\delta(x - x_o) = \lim_{a \to \infty} g_a(x - x_o) \tag{A1-10}$$

A1-2. The Sifting Integral

Consider the integral

$$I = \int_{-\infty}^{+\infty} f(x)\delta(x - x_o)\,dx$$

where $f(x)$ is continuous at x_o. From the properties of the unit impulse function, the integrand of I is nonzero only at the point $x = x_o$. The only contribution of $f(x)$ to the integral is thus at the point $x = x_o$, and we can write

$$I = f(x_o) \int_{-\infty}^{+\infty} \delta(x - x_o)\,dx$$

Hence, using Eq. (A1-2),

$$\int_{-\infty}^{+\infty} f(x)\delta(x - x_o)\,dx = f(x_o) \tag{A1-11}$$

Thus the effect of integrating the product of some given function and the unit impulse function centered on the point x_o is simply to evaluate the given function at that point. The integral in Eq. (A1-11) is known as the *sifting integral*.

† Cf. Dwight (I, Eq. (861.3)).

A1-3. Fourier Transforms

The Fourier transform $\Delta(u,x_o)$ of the unit impulse function $\delta(x - x_o)$ is

$$\Delta(u,x_o) = \int_{-\infty}^{+\infty} \delta(x - x_o)e^{-jux}\,dx \qquad \text{(A1-12)}$$

Then, from the sifting property of the impulse function,

$$\Delta(u,x_o) = e^{-jux_o} \qquad \text{(A1-13)}$$

and hence
$$\Delta(u,0) = 1 \qquad \text{(A1-14)}$$

Formal application of the Fourier inversion integral then gives

$$\frac{1}{2\pi}\int_{-\infty}^{+\infty} e^{-jux_o}e^{jux}\,du = \delta(x - x_o) \qquad \text{(A1-15)}$$

and
$$\frac{1}{2\pi}\int_{-\infty}^{+\infty} 1 e^{jux}\,du = \delta(x) \qquad \text{(A1-16)}$$

From Eqs. (A1-3) and (A1-14), we see that both the unit impulse function $\delta(x)$ and its Fourier transform are even functions. Therefore

$$\int_{-\infty}^{+\infty} \delta(x) \cos ux\,dx = 1$$

and
$$\frac{1}{2\pi}\int_{-\infty}^{+\infty} 1 \cos ux\,du = \delta(x) \qquad \text{(A1-17)}$$

From Eqs. (A1-12) and (A1-15) and the identity

$$\cos ux_o = \frac{e^{jux_o} + e^{-jux_o}}{2}$$

we obtain the Fourier transform pair

$$\tfrac{1}{2}\int_{-\infty}^{+\infty} [\delta(x - x_o) + \delta(x + x_o)]e^{-jux}\,dx = \cos ux_o$$

and
$$\frac{1}{2\pi}\int_{-\infty}^{+\infty} \cos ux_o e^{jux}\,du = \frac{1}{2}[\delta(x - x_o) + \delta(x + x_o)] \qquad \text{(A1-18)}$$

Since both $\cos ux_o$ and the pair of impulses are even functions, these can also be written in the form

$$\tfrac{1}{2}\int_{-\infty}^{+\infty} [\delta(x - x_o) + \delta(x + x_o)] \cos ux\,dx = \cos ux_o$$

and
$$\frac{1}{2\pi}\int_{-\infty}^{+\infty} \cos ux_o \cos ux\,du = \frac{1}{2}[\delta(x - x_o) + \delta(x + x_o)] \qquad \text{(A1-19)}$$

A1-4. Derivatives of Impulse Functions†

Since, from Eqs. (A1-5) and (A1-7), the rectangular pulse function can be expressed in terms of the unit step function as

$$f_a(x - x_o) = \frac{1}{2a} \{ U[x - (x_o - a)] - U[x - (x_o + a)] \} \quad \text{(A1-20)}$$

the derivative of that function is, from Eq. (A1-6),

$$f_a'(x - x_o) = \frac{1}{2a} \{ \delta[x - (x_o - a)] - \delta[x - (x_o + a)] \} \quad \text{(A1-21)}$$

Suppose now that we integrate the product of this derivative and some function $f(x)$ which has a continuous derivative at $x = x_o$. Using Eq. (A1-11) we get

$$\int_{-\infty}^{+\infty} f(x) f_a'(x - x_o) \, dx$$

$$= \frac{1}{2a} \int_{-\infty}^{+\infty} f(x) \{ \delta[x - (x_o - a)] - \delta[x - (x_o + a)] \} \, dx$$

$$= \frac{f(x_o - a) - f(x_o + a)}{2a}$$

The limit of this as $a \to 0$ is minus the derivative of $f(x)$ evaluated at $x = x_o$; i.e.,

$$\lim_{a \to 0} \int_{-\infty}^{+\infty} f(x) f_a'(x - x_o) \, dx = -f'(x_o) \quad \text{(A1-22)}$$

We shall define the *derivative of the unit impulse function* to be the appropriate limit of the derivative of some one of its prototypes; thus, for example,

$$\delta'(x - x_o) = \lim_{a \to 0} f_a'(x - x_o) \quad \text{(A1-23)}$$

We can then rewrite Eq. (A1-22) as

$$\int_{-\infty}^{+\infty} f(x) \delta'(x - x_o) \, dx = -f'(x_o) \quad \text{(A1-24)}$$

The result of integrating the product of some given function with a continuous derivative at $x = x_o$ and the derivative of the unit impulse function centered on x_o is therefore minus the derivative of the given function at that point.

The nth derivative of the unit impulse function may similarly be defined as a limit of the nth derivative of some one of its prototypes.

† Cf. van der Pol and Bremmer (I, Chap. V, Art. 10).

It can then be shown that, if $f(x)$ has a continuous nth derivative at $x = x_o$,

$$\int_{-\infty}^{+\infty} f(x) \delta^{(n)}(x - x_o) \, dx = (-1)^n f^{(n)}(x_o) \tag{A1-25}$$

The Fourier transform of the nth derivative of the unit impulse function is therefore

$$\begin{aligned} \Delta_n(u,x_o) &= \int_{-\infty}^{+\infty} \delta^{(n)}(x - x_o) e^{-jux} \, dx \\ &= (ju)^n e^{-jux_o} \end{aligned} \tag{A1-26}$$

Hence, using Eq. (A1-13),

$$\Delta_n(u,x_o) = (ju)^n \, \Delta(u,x_o) \tag{A1-27}$$

INTEGRAL EQUATIONS

In Chaps. 6, 9, 11, and 14, we deal with certain linear integral equations. These take either the form

$$\int_a^b R(s,t)\phi(t)\,dt = \lambda\phi(s) \qquad a \le s \le b \qquad \text{(A2-1)}$$

where a, b are constants, $R(s,t)$ is an autocorrelation function, and λ and $\phi(t)$ are unknown, or the form

$$\int_a^b R(s,t)x(t)\,dt = y(s) \qquad a \le s \le b \qquad \text{(A2-2)}$$

where a, b, $R(s,t)$ are as before, $y(s)$ is known, and $x(t)$ is unknown. In either Eq. (A2-1) or (A2-2), $R(s,t)$ is called the *kernel* of the equation. We shall state here some standard results from the theory of integral equations pertaining to Eqs. (A2-1) and (A2-2).

First, however, some definitions are necessary.

A2-1. Definitions†

A real or complex-valued function $f(t)$ of a real variable t is said to be of *integrable square* on the interval $a \le t \le b$ if

$$\int_a^b |f(t)|^2\,dt < \infty \qquad \text{(A2-3)}$$

From the definition, it follows that if a function $f(t)$ is of integrable square, so also are its conjugate $f^*(t)$ and its magnitude $|f(t)|$.

For two functions $f(t)$ and $g(t)$ of integrable square, it can be shown (using the Schwartz inequality) that the integral

$$\int_a^b f(t)g(t)\,dt$$

exists. If

$$\int_a^b f(t)g^*(t)\,dt = 0 \qquad \text{(A2-4)}$$

† See Courant and Hilbert (I, p. 49). They consider mostly the class of piecewise continuous functions which is a subclass of the functions of integrable square. See also Courant and Hilbert (I, p. 110).

the functions $f(t)$ and $g(t)$ are said to be *orthogonal* on the interval $a \leq t \leq b$. If

$$\int_a^b |f^2(t)|\ dt = 1$$

$f(t)$ is said to be *normalized*. A class of functions $f_k(t)$ (containing either finitely or infinitely many member functions) is an *orthogonal class* of functions if every pair of functions in the class is orthogonal. If, in addition, every $f_k(t)$ is normalized, the class of functions is said to be *orthonormal*.

A class of orthogonal functions $f_k(t)$ is *complete* in the class of functions of integrable square on the interval $a \leq t \leq b$ if any function $g(t)$ of integrable square on $a \leq t \leq b$ can be approximated arbitrarily closely in mean square by a linear combination of the $f_k(t)$, i.e., if there are constants a_k so that

$$\lim_{N \to \infty} \int_a^b \left[g(t) - \sum_{k=1}^K a_k f_k(t) \right]^2 dt = 0 \qquad \text{(A2-5)}$$

We write this

$$g(t) = \text{l.i.m.}_{K \to \infty} \sum_k^K a_k f_k(t)$$

where "l.i.m." stands for *limit in the mean*.

Any function which is continuous except at most at a finite number of points in any finite interval, at which points the limits from each side are finite, is called *piecewise continuous*. Such a function is necessarily of integrable square over any finite interval $a \leq t \leq b$. However, there are functions of integrable square which are not piecewise continuous.

A function $K(s,t)$ of two variables which satisfies

$$K(s,t) = K^*(t,s) \qquad \text{(A2-6)}$$

is *symmetric*.† If $K(t,s)$ is real valued, this condition reduces, of course, to

$$K(s,t) = K(t,s)$$

A symmetric function $K(s,t)$ which has the property that

$$\int_a^b \int_a^b K(s,t)g(s)g^*(t)\ ds\ dt \geq 0 \qquad \text{(A2-7)}$$

for any $g(s)$ of integrable square is *non-negative definite*.‡ If the inequality $\geq$ in (A2-7) is replaced by $>$ for any $g(t)$ such that

$$\int_a^b |g(t)|^2\ dt > 0$$

$K(s,t)$ is *positive definite*.‡ As pointed out in Chap. 6, autocorrelation

---

† See Courant and Hilbert (I, p. 122).

‡ The terminology is not standardized. Often, for example, our term "non-negative definite" is replaced by "positive definite" and our "positive definite" by "strictly positive definite."

functions $R(s,t)$ satisfy (A2-6) and (A2-7) and hence are non-negative definite. They may or may not be positive definite.

A2-2. Theorems

We can now state without proof the fundamental theorems† for integral equations of the type (A2-1).

If $R(s,t)$ is symmetric and

$$\int_a^b \int_a^b |R(s,t)|^2 \, ds \, dt < \infty \tag{A2-8}$$

Eq. (A2-1) is satisfied for at least one real number $\lambda \neq 0$ and some function $\phi(t)$ such that

$$0 < \int_a^b |\phi(t)|^2 \, dt < \infty$$

A number λ and a function $\phi(t)$ which satisfy Eq. (A2-1) are called a *characteristic value* and an associated *characteristic function*, respectively, of the integral equation. The following properties all hold:

1. If $\phi_1(t)$ and $\phi_2(t)$ are characteristic functions associated with the characteristic value λ, then $a\phi_1(t) + b\phi_2(t)$ is also a characteristic function associated with λ, for any numbers a and b. Thus in particular to each characteristic value λ there corresponds at least one normalized characteristic function.

2. If λ_k and λ_m are different characteristic values, then any characteristic functions $\phi_k(t)$ and $\phi_m(t)$ associated with λ_k and λ_m, respectively, are orthogonal.

3. There are at most a countably infinite set (or sequence) of characteristic values λ_k and for some constant $A < \infty$, $|\lambda_k| < A$, all k.

4. For each characteristic value λ_k there are at most a finite number N_k of linearly independent characteristic functions associated with λ_k. The integer N_k is called the *multiplicity* of λ_k. Any N_k linearly independent characteristic functions can be transformed into N_k orthonormal characteristic functions by the Gram–Schmidt process.

Thus, counting each λ_k as many times as its multiplicity, there is a sequence λ_1, λ_2, . . . (finite or infinite) of characteristic values and a sequence $\phi_1(t)$, $\phi_2(t)$, . . . of orthonormal characteristic functions such that $\phi_k(t)$ is associated with $\lambda_k(t)$ and such that there are no more characteristic functions orthogonal to all the $\phi_k(t)$.

---

† Courant and Hilbert (I, Chap. III, Arts. 4 and 5). Also Riesz and Nagy (I, p. 242).

5. Every function $g(t)$ of integrable square admits the expansion, convergent in mean-square,

$$g(t) = h(t) + \underset{K \to \infty}{\text{l.i.m.}} \sum_{k=1}^{K} g_k \phi_k(t) \tag{A2-9}$$

where

$$g_k = \int_a^b g(t) \phi_k^*(t) \, dt \tag{A2-10}$$

and where $h(t)$ is some function satisfying

$$\int_a^b R(s,t) h(t) \, dt = 0 \tag{A2-11}$$

6. The kernel $R(s,t)$ can be expanded in the series

$$R(s,t) = \underset{K \to \infty}{\text{l.i.m.}} \sum_{k}^{K} \lambda_k \phi_k(s) \phi_k^*(t) \tag{A2-12}$$

7. If $R(s,t)$ is non-negative definite, all the nonzero characteristic values λ_k are positive real numbers.

8. If in addition $R(s,t)$ is positive definite, the orthonormal set of characteristic functions forms a complete orthonormal set, and the $h(t)$ in Eq. (A2-9) may be taken to be zero. That is, for any function $g(t)$ of integrable square, there is a generalized Fourier-series expansion in terms of the orthonormal characteristic functions:

$$g(t) = \underset{K \to \infty}{\text{l.i.m.}} \sum_{k=1}^{K} g_k \phi_k(t) \tag{A2-13}$$

where the g_k are given by Eq. (A2-10).

In addition to these results, there is Mercer's theorem,† which is superficially similar to property 6 above but is stronger in the case to which it applies: If $R(s,t)$ is non-negative definite, then

$$R(s,t) = \sum_{k=1}^{\infty} \lambda_k \phi_k(s) \phi_k^*(t) \tag{A2-14}$$

where the convergence is uniform for s, t, satisfying $a \leq s \leq b, a \leq t \leq b$.

The integral equation (A2-2) is closely related to Eq. (A2-1). Obviously if $y(s)$ is a characteristic function of Eq. (A2-1) with characteristic value λ, $y(s)/\lambda$ is a solution of Eq. (A2-2). More generally, if

$$y(s) = a_1 \phi_1(s) + \cdots + a_n \phi_n(s)$$

the solution of Eq. (A2-2) is, again obviously,

$$x(s) = \frac{a_1}{\lambda_1} \phi_1(s) + \cdots + \frac{a_n}{\lambda_n} \phi_n(s)$$

† Courant and Hilbert (I, p. 138).   Also Riesz and Nagy (I, p. 245).

This extends, with restrictions, to a $y(s)$, which is an infinite linear combination of characteristic functions. A general theorem, due to Picard,† says in this context:

Equation (A2-2) has a solution $x(t)$ of integrable square if and only if

$$y(t) = \underset{N \to \infty}{\text{l.i.m.}} \sum_{n=1}^{N} y_n \phi_n(t)$$

where

$$y_n = \int_a^b y(t) \phi_n^*(t) \, dt$$

and the series

$$\sum_{n=1}^{\infty} \frac{|y_n|^2}{\lambda_n^2} < \infty \tag{A2-15}$$

A solution, if one exists, is

$$x(t) = \underset{N \to \infty}{\text{l.i.m.}} \sum_{n=1}^{N} \frac{y_n}{\lambda_n} \phi_n(t) \qquad a \le t \le b \tag{A2-16}$$

and this is unique if the $\phi_n(t)$ are complete.

A2-3. Rational Spectra

The existence of solutions to the characteristic-value problem, Eq. (A2-1), and some important properties of them are thus established quite generally. There remains the problem of actually finding the characteristic values and functions. For one special class of autocorrelation functions $R(s,t)$ which is very important in engineering applications, Eq. (A2-1) can always be solved directly for the λ_k and $\phi_k(t)$. The functions $R(s,t)$ referred to are those which are Fourier transforms of rational functions. In this case the kernel $R(s,t)$ is $R(s - t)$ (to be strictly correct, we should introduce a different symbol, since the first R is a function of two arguments and the second of one) where

$$R(s - t) = R(\tau) = \int_{-\infty}^{\infty} e^{j2\pi f \tau} S(f) \, df \tag{A2-17}$$

$S(f)$ being a nonnegative rational integrable even function. The nonnegativeness and integrability of $S(f)$ are necessary in order that $R(\tau)$ be non-negative definite, i.e., that $R(\tau)$ be an autocorrelation function.‡ The evenness of $S(f)$ makes $R(\tau)$ real. Introducing the variable $p = j2\pi f$ and making use of the special properties of $S(f)$,§ we can write

$$R(\tau) = \int_{-\infty}^{\infty} e^{p\tau} \frac{N(p^2)}{D(p^2)} \, df \tag{A2-18}$$

† See Courant and Hilbert (I, p. 160).
‡ See Art. 6-6.
§ See Art. 11-4.

where $N(p^2)$ is a polynomial of degree n in p^2 and $D(p^2)$ is a polynomial of degree d in p^2. $D(p^2)$ can have no real roots and $d > n$. One can see very easily heuristically that for $\phi(t)$ to be a solution of Eq. (A2-1), it must satisfy a linear differential equation with constant coefficients. In fact, substituting from Eq. (A2-18) in Eq. (A2-1), we have

$$\lambda\phi(s) = \int_a^b \phi(t)\, dt \int_{-\infty}^{\infty} e^{p(s-t)}\, \frac{N(p^2)}{D(p^2)}\, df \qquad a \le s \le b \quad \text{(A2-19)}$$

Differentiating the right-hand side with respect to s is equivalent to multiplying the integrand by p. Hence

$$\lambda D\left(\frac{d^2}{ds^2}\right)\phi(s) = N\left(\frac{d^2}{ds^2}\right)\left[\int_a^b \phi(t)\, dt \int_{-\infty}^{\infty} e^{p(s-t)}\, df\right]$$

$$= N\left(\frac{d^2}{ds^2}\right)\int_a^b \delta(s-t)\phi(t)\, dt \qquad \text{(A2-20)}$$

$$= N\left(\frac{d^2}{ds^2}\right)\phi(s) \qquad a < s < b$$

To solve Eq. (A2-1), one first solves the homogeneous differential equation (A2-20). The solution will contain the parameter λ and $2d$ arbitrary constants $c_1, c_2, \ldots, c_{2d}$. This solution is substituted for $\phi(s)$ in the integral equation (A2-1). It will be found that the integral equation can be satisfied only for a discrete set of values for λ, $\lambda = \lambda_k$, $k = 1, 2, \ldots$, and that for each value λ_k the constants $c_1, \ldots, c_{2d}$ must satisfy certain conditions. These λ_k are the characteristic values. Any $\phi(s)$ which satisfies Eq. (A2-20) and the conditions on $c_1, \ldots, c_{2d}$ that arise when $\lambda = \lambda_k$ is a characteristic function $\phi_k(s)$ associated with λ_k. If there are more than one linearly independent functions $\phi(s)$ associated with λ_k, they may be orthogonalized by the Gram–Schmidt process[†] and then normalized. A rigorous demonstration that Eq. (A2-20) imposes a necessary condition on $\phi(s)$, and that solutions of this equation are guaranteed to yield the characteristic values λ_k and the characteristic functions $\phi_k(s)$, is given by Slepian.[‡] An example of this procedure is contained in Art. 6-3.

It is often useful to expand some arbitrary function of integrable square in a generalized Fourier series as in Eq. (A2-13). Thus it is useful to know when the $\phi_k(t)$ form a complete set. One sufficient condition for the characteristic functions of

$$\int_a^b R(s-t)\phi(t)\, dt = \lambda\phi(s) \qquad a \le s \le b$$

---

† See Courant and Hilbert (I, p. 50).
‡ Slepian (I, Appendix I).

to form a complete set is that $R(t)$ be the Fourier transform of a spectral density.† Thus, in the case discussed above, where $R(t)$ is the transform of a rational spectral density, the $\phi_k(t)$ are always complete.

Again, if the autocorrelation function which is used for the kernel satisfies Eq. (A2-18), the integral equation (A2-2) can be treated directly by elementary methods.‡ We do this here, restricting ourselves to the case of real $R(t)$. There is some similarity between this discussion and Slepian's treatment of the related characteristic-value problem.

Exactly as in the characteristic-value problem, it can be shown that a solution $x(t)$ of the integral equation must satisfy a linear differential relation. In this case it is

$$ N\left(\frac{d^2}{dt^2}\right) x(t) = D\left(\frac{d^2}{dt^2}\right) y(t) \qquad a < t < b \tag{A2-21} $$

We consider first the case $N(p) \equiv 1$. Then, from Eq. (A2-21), if a solution exists, it has to be

$$ x(t) = D\left(\frac{d^2}{dt^2}\right) y(t) \qquad a < t < b \tag{A2-22} $$

The problem now is to show under what conditions on $y(t)$ $x(t)$ as given by Eq. (A2-22) satisfies the integral equation. We find these conditions by substituting $x(t)$ from Eq. (A2-22) back in the integral equation and performing repeated integrations by parts.

First, we need to establish some facts about the behavior of $R(t)$. From Eq. (A2-18), for the case we are considering,

$$ R(t) = \int_{-\infty}^{\infty} \frac{e^{pt}}{D(p^2)} df \tag{A2-23} $$

and $$ R^{(k)}(t) = \int_{-\infty}^{\infty} \frac{p^k e^{pt}}{D(p^2)} df \qquad k = 0, 1, 2, \ldots, 2d - 2 \tag{A2-24} $$

These integrals converge absolutely and $R^{(k)}(t)$ exists for all t for $k \leq 2d - 2$. Also, by a conventional residue argument, if Γ is a closed contour containing all the residues in the top half f plane,

$$ R(t) = \int_{\Gamma} \frac{e^{pt}}{D(p^2)} df \qquad t > 0 \tag{A2-25} $$

From (A2-25) it follows that $R(t)$ has derivatives of all orders for $t > 0$, and

$$ R^{(k)}(t) = \int_{\Gamma} \frac{p^k e^{pt}}{D(p^2)} df \qquad t > 0, k = 0, 1, 2, \ldots \tag{A2-26a} $$

† See Youla (I, Appendix A) or Root and Pitcher (II, Appendix I).
‡ See Zadeh and Ragazzini (I).

Similarly, if Γ' is a closed contour containing all the residues in the bottom half f plane,

$$R^{(k)}(t) = \int_{\Gamma'} \frac{p^k e^{pt}}{D(p^2)} \, df \qquad t < 0, \, k = 0, 1, 2, \ldots \qquad \text{(A2-26b)}$$

Let us calculate $R^{(2d-1)}(0+) - R^{(2d-1)}(0-)$. Let the coefficient of p^{2d} in $D(p^2)$ be a_{2d}, then

$$R^{(2d-1)}(t) = \frac{(j2\pi)^{2d-1}}{a_{2d}(j2\pi)^{2d}} \int_{\Gamma} \frac{f^{2d-1} e^{j2\pi ft}}{(f - z_1) \cdots (f - z_{2d})} \, df \qquad t > 0 \qquad \text{(A2-27)}$$

$$= \frac{1}{a_{2d}} \sum \{\text{residues in top half plane}\} \qquad t > 0$$

where the residues in the summation are those of the integrand of the integral in (A2-27). Similarly

$$R^{(2d-1)}(-t) = -\frac{1}{a_{2d}} \sum \{\text{residues in bottom half plane}\} \qquad t > 0$$

where the residues in the summation are those of the same function as above. Hence

$$R^{(2d-1)}(t) - R^{(2d-1)}(-t) = \frac{1}{a_{2d}} \sum \text{residues} \qquad t > 0$$

and

$$R^{(2d-1)}(0+) - R^{(2d-1)}(0-) = \frac{1}{a_{2d}} \lim_{t \to 0} \sum \text{residues} \qquad t > 0 \qquad \text{(A2-28)}$$

$$= \frac{1}{a_{2d}}$$

From Art. 11-4 we know that $D(p^2)$ can be factored into $D_1(j2\pi f) \, D_2(j2\pi f)$ where $D_1(j2\pi f)$ has all its zeros in the bottom half f plane and $D_2(j2\pi f)$ has all its zeros in the top half f plane. Hence, from Eq. (A2-25)

$$D_2\left(\frac{d}{dt}\right) R(t) = \int_{\Gamma} \frac{e^{pt}}{D_1(p)} \, df = 0 \qquad t > 0 \qquad \text{(A2-29)}$$

$$D_1\left(\frac{d}{dt}\right) R(t) = \int_{\Gamma'} \frac{e^{pt}}{D_2(p)} \, df = 0 \qquad t < 0 \qquad \text{(A2-30)}$$

and

$$D\left(\frac{d^2}{dt^2}\right) R(t) = 0 \qquad t \neq 0 \qquad \text{(A2-31)}$$

We now write

$$D(p^2) = \sum_{k=0}^{d} a_{2k} p^{2k} \qquad \text{(A2-32)}$$

so
$$x(t) = \sum_{k=0}^{d} a_{2k} \left(\frac{d}{dt}\right)^{2k} y(t) \qquad a < t < b \qquad (A2-33)$$

Then,

$$\int_a^b x(\tau) R(t - \tau)\, d\tau = \sum_{k=0}^{d} a_{2k} \int_a^t y^{(2k)}(\tau) R(t - \tau)\, d\tau$$

$$+ \sum_{k=0}^{d} a_{2k} \int_t^b y^{(2k)}(\tau) R(t - \tau)\, d\tau$$

Integrating once by parts yields

$$\int_a^b x(\tau) R(t - \tau)\, d\tau = \sum_{k=1}^{d} a_{2k} y^{(2k-1)}(\tau) R(t - \tau) \Big|_a^t$$

$$+ \sum_{k=1}^{d} a_{2k} y^{(2k-1)}(\tau) R(t - \tau) \Big|_t^b + \sum_{k=1}^{d} a_{2k} \int_a^t y^{(2k-1)}(\tau) R'(t - \tau)\, d\tau$$

$$+ \sum_{k=1}^{d} a_{2k} \int_t^b y^{(2k-1)}(\tau) R'(t - \tau)\, d\tau + a_0 \int_a^b y(\tau) R(t - \tau)\, d\tau \quad (A2-34)$$

The terms evaluated at $\tau = t$ cancel in the two sets of terms which are integrated out. The other terms which are integrated out are

$$- \left[\sum_{k=1}^{d} a_{2k} y^{(2k-1)}(a) \right] R(t - a) + \left[\sum_{k=1}^{d} a_{2k} y^{(2k-1)}(b) \right] R(t - b) \quad (A2-35)$$

Integrating by parts once more, the third and fourth sets of terms in Eq. (A2-34) yield

$$\sum_{k=1}^{d} a_{2k} y^{(2k-2)}(\tau) R'(t - \tau) \Big|_a^t + \sum_{k=1}^{d} a_{2k} y^{(2k-2)}(\tau) R'(t - \tau) \Big|_t^b$$

$$+ \sum_{k=1}^{d} a_{2k} \int_a^t y^{(2k-2)}(\tau) R''(t - \tau) + \sum_{k=1}^{d} a_{2k} \int_t^b y^{(2k-2)}(\tau) R''(t - \tau)\, d\tau$$

$$(A2-36)$$

The terms evaluated at $\tau = t$ in the two sets of terms which are integrated again cancel. The other terms which are integrated out are

$$- \left[\sum_{k=1}^{d} a_{2k} y^{(2k-2)}(a) \right] R'(t - a) + \left[\sum_{k=1}^{d} a_{2k} y^{(2k-2)}(b) \right] R'(t - b) \quad (A2-37)$$

The process of integrating by parts is repeated $2d$ times, leaving the integrals of the form

$$\int_a^b y(\tau) R^{(n)}(t - \tau)\, d\tau$$

unintegrated as they occur at every other step. At each step except the last one, the middle terms in the integrated-out terms cancel algebraically. Linear combinations of derivatives of y at the two boundary points multiplying a derivative of $R(t)$, such as Eqs. (A2-35) and (A2-37), are left. Thus, after the $2d$ integrations by parts, we have

$$\int_a^b x(\tau) R(t - \tau)\, d\tau = Y_0 R(t - a) + Z_0 R(t - b) + Y_1 R'(t - a)$$
$$+ Z_1 R'(t - b) + \cdots + Y_{2d-2} R^{(2d-2)}(t - a) + Z_{2d-2} R^{(2d-2)}(t - b)$$
$$+ a_{2d} y(t) R^{(2d-1)}(0+) - a_{2d} y(t) R^{(2d-1)}(0-)$$
$$+ \int_a^b y(\tau) \sum_{k=0}^{2d} a_{2k} R^{(2k)}(t - \tau)\, d\tau \quad \text{(A2-38)}$$

where the Y_k's and Z_k's are linear combinations of derivatives of $y(t)$ at a and b respectively. But by Eq. (A2-31), the last integral in Eq. (A2-38) vanishes; and using Eq. (A2-28), we have

$$\int_a^b x(\tau) R(t - \tau)\, d\tau = y(t)$$
$$+ \sum_{k=0}^{2d-2} Y_k R^{(k)}(t - a) + \sum_{k=0}^{2d-2} Z_k R^{(k)}(b - t) \qquad a \le t \le b \quad \text{(A2-39)}$$

Thus $x(t)$ as given by Eq. (A2-18) satisfies the integral equation if and only if the linear homogeneous boundary condition on $y(t)$ given by

$$\sum_{k=0}^{2d-2} Y_k R^{(k)}(t - a) + \sum_{k=0}^{2d-2} Z_k R^{(k)}(b - t) = 0 \qquad \text{(A2-40)}$$

is satisfied identically in t, $a \le t \le b$. It is sufficient for Eq. (A2-40) that all the Y_k and Z_k be zero, but it is not necessary. In fact, since $D_2(p)$ is of degree d, and

$$D_2\!\left(\frac{d}{dt}\right) R(t) = 0 \qquad t > 0$$

it follows that each $R^{(k)}(t - a)$, $k = d, d + 1, \ldots, 2d - 2$, can be expressed as a linear combination of the first $d - 1$ derivatives of $R(t - a)$, $t > a$. Since the first $2d - 2$ derivatives exist everywhere and are continuous, this property also holds at $t - a = 0$, or $t = a$. Thus, the first sum in Eq. (A2-40) can be replaced by a sum of $d - 1$ linearly independent terms. An analogous argument allows the second sum in Eq. (A2-40) to be replaced by a sum of $d - 1$ linearly independent terms. Thus there are left $2d - 2$ linearly independent conditions to be satisfied by $y(t)$ and its derivatives at the end points a and b.

It may be observed that, since Eq. (A2-2) certainly has a solution if $y(s)$ is a characteristic function of the associated equation (A2-1) with the same kernel, all these characteristic functions satisfy the linear boundary conditions, Eq. (A2-40).

If Eq. (A2-2) with $R(t)$ as given by Eq. (A2-23) does not have a solution, properly speaking, because $y(s)$ does not satisfy Eq. (A2-40), it still always has a singular solution, i.e., a solution including impulse functions and their derivatives. To see this, add to the $x(t)$ of Eq. (A2-23) a sum

$$\sum_{k=0}^{2d-2} [b_k \delta^{(k)}(t - a) + c_k \delta^{(k)}(t - b)] \tag{A2-41}$$

with as yet undetermined coefficients. When this sum is substituted into the left side of Eq. (A2-2), there results

$$\sum_{k=0}^{2d-2} [b_k R^{(k)}(t - a) + c_k R^{(k)}(b - t)] \tag{A2-42}$$

Hence, by choosing $b_k = -Y_k$ and $c_k = -Z_k$, the boundary conditions are eliminated and the integral equation can always be satisfied.

Now we remove the restriction that $N(p^2) \equiv 1$; that is, we only suppose that $R(t)$ is given by Eq. (A2-18). We have stated already that a solution $x(t)$ of the integral equation must be some solution of the differential equation (A2-21). We show now that Eq. (A2-2) with $R(t)$ as given by Eq. (A2-18) has a solution if and only if some one of a set of associated equations of the type just considered has a solution.

Let $z(t)$ be a solution, unspecified, of

$$N\left(\frac{d^2}{dt^2}\right) z(t) = y(t) \tag{A2-43}$$

and consider the integral equation

$$\int_a^b x(\tau)\hat{R}(t - \tau)\, d\tau = z(t) \qquad a \le t \le b \tag{A2-44}$$

where
$$\hat{R}(t) = \int_{-\infty}^{\infty} \frac{e^{pt}}{D(p^2)}\, df$$

Since
$$N\left(\frac{d^2}{dt^2}\right) \int_a^b x(\tau)\hat{R}(t - \tau)\, d\tau = \int_a^b x(\tau) R(t - \tau)\, d\tau = y(t)$$

if $x(\tau)$ is a solution of Eq. (A2-44) for any $z(t)$ satisfying Eq. (A2-43), it is a solution of the desired equation (A2-2). Conversely, if $x(\tau)$ is a solution of Eq. (A2-2) with $R(t)$ as given by Eq. (A2-18) then

$$N\left(\frac{d^2}{dt^2}\right) \int_a^b x(\tau)\hat{R}(t - \tau)\, d\tau = N\left(\frac{d^2}{dt^2}\right) z_1(t)$$

where $z_1(t)$ is chosen arbitrarily from among the solutions of Eq. (A2-43). Let

$$\epsilon(t) = \int_a^b x(\tau)\hat{R}(t - \tau)\, d\tau - z_1(t) \qquad a \le t \le b \qquad \text{(A2-45)}$$

then

$$N\left(\frac{d^2}{dt^2}\right)\epsilon(t) = 0$$

Let

$$\hat{z}(t) = z_1(t) + \epsilon(t) \qquad\qquad\qquad \text{(A2-46)}$$

then $\hat{z}(t)$ is a solution of Eq. (A2-43), and by Eq. (A2-45) $x(\tau)$ is a solution of Eq. (A2-44) with $z(t) = \hat{z}(t)$. Thus $x(\tau)$ is a solution of Eq. (A2-2) if and only if it is a solution of Eq. (A2-44) for some $z(t)$ satisfying Eq. (A2-43).

The practical procedure for solving Eq. (A2-2) is therefore to operate on the general solution of Eq. (A2-43) by $D(d^2/dt^2)$ and substitute the resulting expression back in the integral equation. If the integral equation has a proper solution, it is then guaranteed that, for some choice of the undetermined constants, this expression is it. Otherwise, impulse functions and their derivatives must be added, as explained above.

BIBLIOGRAPHY

Bartlett, M. S.:
   I. "An Introduction to Stochastic Processes," Cambridge University Press, New York, 1955.

Bennett, William R.:
   I. Response of a Linear Rectifier to Signal and Noise, *Jour. Acoustical Soc. America*, **15** (3): 165–172, January, 1944.
   II. Methods of Solving Noise Problems, *Proc. IRE*, **44**: 609–638, May, 1956.
——— and S. O. Rice:
   I. Note on Methods of Computing Modulation Products, *Philosophical Magazine*, series 7, **18**: 422–424, September, 1934.

Blanc-Lapierre, André, and Robert Fortet:
   I. "Theorie des functions aleatoires," Masson et Cie, Paris, 1953.

Bode, Hendrik W.:
   I. "Network Analysis and Feedback Amplifier Design," D. Van Nostrand, Princeton, N.J., 1945.
——— and Claude E. Shannon:
   I. A Simplified Derivation of Linear Least-square Smoothing and Prediction Theory, *Proc. IRE*, **38**: 417–426, April, 1950.

Booton, Richard C., Jr.:
   I. An Optimization Theory for Time-varying Linear Systems with Nonstationary Statistical Inputs, *Proc. IRE*, **40**: 977–981, August, 1952.

Bose, A. G., and S. D. Pezaris:
   I. A Theorem Concerning Noise Figures, *IRE Convention Record*, part 8, pp. 35–41, 1955.

Bunimovich, V. I.:
   I. "Fluctuation Processes in Radio Receivers," *Sovietskoe Radio*, 1951 (in Russian).

Burgess, R. E.:
   I. The Rectification and Observation of Signals in the Presence of Noise, *Philosophical Magazine*, series 7, **42** (328): 475–503, May, 1951.

Callen, Herbert B., and Theodore A. Welton:
   I. Irreversibility and Generalized Noise, *Phys. Rev.*, **83** (1): 34–40, July 1, 1951.

Carnap, Rudolf:
   I. "Logical Foundations of Probability," University of Chicago Press, Chicago, 1950.

Chessin, P. L.:
   I. A Bibliography on Noise, *IRE Trans. on Information Theory*, **IT-1** (2): 15–31, September, 1955.

Churchill, Ruel V.:
I. "Fourier-series and Boundary-value Problems," McGraw-Hill Book Co., New York, 1941.
II. "Modern Operational Mathematics in Engineering," McGraw-Hill Book Co., New York, 1944.

Courant, Richard:
I. "Differential and Integral Calculus," I, rev. ed., 1937; II, 1936, Interscience Publishers, New York.

——— and D. Hilbert:
I. "Methods of Mathematical Physics," I, Interscience Publishers, New York, 1953.

Cramér, Harald:
I. "Mathematical Methods of Statistics," Princeton University Press, Princeton, N.J., 1946.

Davenport, Wilbur B., Jr.:
I. Signal-to-Noise Ratios in Band-pass Limiters, *Jour. Applied Physics*, **24** (6): 720–727, June, 1953.

———, Richard A. Johnson, and David Middleton:
I. Statistical Errors in Measurements on Random Time Functions, *Jour. Applied Physics*, **23** (4): 377–388, April, 1952.

Davis, R. C.:
I. On the Theory of Prediction of Nonstationary Stochastic Processes, *Jour. Applied Physics*, **23**: 1047–1053, September, 1952.

Doob, John L.:
I. Time Series and Harmonic Analysis, *Proc. Berkeley Symposium on Math. Statistics and Probability*, pp. 303–343, Berkeley, Calif., 1949.
II. "Stochastic Processes," John Wiley, New York, 1953.

Dwight, Herbert B.:
I. "Tables of Integrals," rev. ed., Macmillan, New York, 1947.

Emerson, Richard C.:
I. First Probability Densities for Receivers with Square-law Detectors, *Jour. Applied Physics*, **24** (9): 1168–1176, September, 1953.

Feller, William:
I. "Probability Theory and Its Applications," I, John Wiley, New York, 1950.

Friis, Harald T.:
I. Noise Figures of Radio Receivers, *Proc. IRE*, **32**: 419–422, July, 1944.

Fry, Thornton C.:
I. "Probability and Its Engineering Uses," D. Van Nostrand, Princeton, N.J., 1928.

Gardner, Murray F., and John L. Barnes:
I. "Transients in Linear Systems," I, John Wiley, New York, 1942.

Gnedenko, B. V., and A. N. Kolmogorov:
I. "Limit Distributions for Sums of Independent Random Variables," Addison-Wesley, Cambridge, Mass., 1954.

Green, Paul E., Jr.:
I. A Bibliography of Soviet Literature on Noise, Correlation, and Information Theory, *IRE Trans. on Information Theory*, **IT-2** (2): 91–94, June, 1956.

Grenander, Ulf:
  I. "Stochastic Processes and Statistical Inference," *Arkiv fur Matematik*, **1** (17): 195–277, 1950.

Guillemin, Ernst A.:
  I. "Communication Networks," **I**, 1931; **II**, 1935, John Wiley, New York.
  II. "The Mathematics of Circuit Analysis," John Wiley, New York, 1949.

Hanse, H.:
  I. "The Optimization and Analysis of Systems for the Detection of Pulsed Signals in Random Noise," doctoral dissertation, Massachusetts Institute of Technology, Cambridge, Mass., January, 1951.

Harman, Willis W.:
  I. "Fundamentals of Electronic Motion," McGraw-Hill Book Co., New York, 1953.

Haus, Herman A., and Richard B. Adler:
  I. Invariants of Linear Noisy Networks, *IRE Convention Record*, part 2, pp. 53–67, 1956.

Hildebrand, F. B.:
  I. "Methods of Applied Mathematics," Prentice-Hall, Englewood Cliffs, N.J., 1952.

IRE Standards:
  I. Standards on Electron Devices: Methods of Measuring Noise, *Proc. IRE*, **41**: 890–896, July, 1953.

Jahnke, Eugen, and Fritz Emde:
  I. "Tables of Functions," 4th ed., Dover, New York, 1945.

James, Hubert M., Nathaniel B. Nichols, and Ralph S. Phillips:
  I. "Theory of Servomechanisms," MIT Rad. Lab. Series, **25**, McGraw-Hill Book Co., New York, 1947.

Jeffreys, Harold:
  I. "Theory of Probability," 2d ed., Oxford University Press, New York, 1948.

Kac, Mark, and A. J. F. Siegert:
  I. On the Theory of Noise in Radio Receivers with Square-law Detectors, *Jour. Applied Physics*, **18**: 383–397, April, 1947.

Khinchin, A. I.:
  I. "Mathematical Foundations of Statistical Mechanics," Dover, New York, 1949.

Kolmogorov, Andrei N.:
  I. "Foundations of the Theory of Probability," Chelsea, New York, 1950.

Langmuir, Irving:
  I. The Effect of Space-charge and Initial Velocities on the Potential Distribution and Thermionic Current between Parallel-plane Electrodes, *Phys. Rev.*, **21**: 419–436, April, 1923.

Laning, J. Halcombe, Jr., and Richard H. Battin:
  I. "Random Processes in Automatic Control," McGraw-Hill Book Co., New York, 1956.

Lawson, James L., and George E. Uhlenbeck:
  I. "Threshold Signals," MIT Rad. Lab. Series, **24**, McGraw-Hill Book Co., New York, 1950.

Lee, Yuk Wing, and Charles A. Stutt:
  I. Statistical Prediction of Noise, *Proc. National Electronics Conference*, **5**: 342–365, Chicago, 1949.

Lehman, E. L.:
  I. "Theory of Testing Hypotheses" (Notes recorded by Colin Blyth from lectures by E. L. Lehman), Associated Students Store, University of California, Berkeley, Calif.

Levinson, Norman:
  I. A Heuristic Exposition of Wiener's Mathematical Theory of Prediction and Filtering, *Jour. Math. and Physics*, **26**: 110–119, July, 1947. Also Appendix C of Wiener (III).

Levy, Paul:
  I. "Theorie de l'addition des variables aleatoires," Gauthier-Villars, Paris, 1937.

Lindsay, Robert B.:
  I. "Introduction to Physical Statistics," John Wiley, New York, 1941.

Loéve, Michel:
  I. "Probability Theory," D. Van Nostrand, Princeton, N.J., 1955.

MacDonald, D. K. C.:
  I. Some Statistical Properties of Random Noise, *Proc. Cambridge Philosophical Soc.*, **45**: 368, 1949.

Magnus, W., and F. Oberhettinger:
  I. "Special Functions of Mathematical Physics," Chelsea, New York, 1949.

Marcum, J. I.:
  I. A Statistical Theory of Target Detection by Pulsed Radar, *Rand Corp. Report*, **RM-754**, December 1, 1947. Mathematical Appendix, **R113**, July 1, 1948.

Meyer, M. A., and David Middleton:
  I. On the Distributions of Signals and Noise after Rectification and Filtering, *Jour. Applied Physics*, **25** (8): 1037–1052, August, 1954.

Middleton, David:
  I. The Response of Biased, Saturated Linear, and Quadratic Rectifiers to Random Noise, *Jour. Applied Physics*, **17**: 778–801, October, 1946.
  II. Some General Results in the Theory of Noise through Nonlinear Devices, *Quart. Applied Math.*, **5** (4): 445–498, January, 1948.
  III. The Distribution of Energy in Randomly Modulated Waves, *Philosophical Magazine*, series 7, **42**: 689–707, July, 1951.
  IV. Statistical Methods for the Detection of Pulsed Radar in Noise, in Willis Jackson (ed.), "Communication Theory," pp. 241–270, Academic Press, New York, and Butterworth's Scientific Publications, London, 1953.

———— and Virginia Johnson:
  I. "A Tabulation of Selected Confluent Hypergeometric Functions," Technical Report No. 140, Cruft Laboratory, Harvard University, Cambridge, Mass., January 5, 1952.

———— and David Van Meter:
  I. Detection and Extraction of Signals in Noise from the Point of View of Statistical Decision Theory, *Jour. Soc. Industrial and Applied Math.*, part I: **3**: 192–253, December, 1955; part II: **4**: 86–119, June, 1956.

Mood, Alexander McF.:
  I. "Introduction to the Theory of Statistics," McGraw-Hill Book Co., New York, 1950.

National Bureau of Standards:
  I. "Tables of Normal Probability Functions," Table 23, *NBS Applied Math. Series*, Washington, 1953.

Neyman, J., and E. S. Pearson:
  I. On the Use and Interpretation of Certain Test Criteria for Purposes of Statistical Inference, *Biometrica*, **20A**: 175, 263, 1928.
  II. On the Problem of the Most Efficient Tests of Statistical Hypotheses, *Phil. Trans. Royal Soc. London*, **A231**: 289–337, 1933.

Nyquist, Harry:
  I. Thermal Agitation of Electric Charge in Conductors, *Phys. Rev.*, **32**: 110–113, July, 1928.

Paley, Raymond E. A. C., and Norbert Wiener:
  I. "Fourier Transforms in the Complex Domain," *American Math. Soc. Colloquium Publ.*, **19**, New York, 1934.

Phillips, Ralph S.:
  I. RMS-error Criterion in Servomechanism Design, chap. 7 of James, Nichols, and Phillips (I).

Price, Robert:
  I. The Detection of Signals Perturbed by Scatter and Noise, *IRE Trans. on Information Theory*, **PGIT-4**: 163–170, September, 1954.
  II. A Note on the Envelope and Phase-modulated Components of Narrow-band Gaussian Noise, *IRE Trans. on Information Theory*, **IT-1** (2): 9–13, September, 1955.
  III. Optimum Detection of Random Signals in Noise with Application to Scatter Multipath Communications, I., *IRE Trans. on Information Theory*, **IT-2** (4): December, 1956.

Rack, A. J.:
  I. Effect of Space-charge and Transit Time on the Shot Noise in Diodes, *Bell Syst. Tech. Jour.*, **17**: 592–619, 1938.

Reich, Edgar, and Peter Swerling:
  I. The Detection of a Sine Wave in Gaussian Noise, *Jour. Applied Physics*, **24** (3): 289–296, March, 1953.

Rice, Steven O.:
  I. Mathematical Analysis of Random Noise, *Bell Syst. Tech. Jour.*, **23**: 282–332, 1944; **24**: 46–156, 1945. Also reprinted in pp. 133–294 of Wax (I).
  II. Statistical Properties of a Sine-wave Plus Random Noise, *Bell Syst. Tech. Jour.*, **27**: 109–157, January, 1948.

Riesz, F., and B. Sz.-Nagy:
  I. "Functional Analysis," trans. from 2d French ed. by Leo F. Boron, Ungar, New York, 1955.

Root, William L., and Tom S. Pitcher:
  I. On the Fourier-series Expansion of Random Functions, *Annals of Math. Statistics*, **26** (2): 313–318, June, 1955.
  II. Some Remarks on Statistical Detection, *IRE Trans. on Information Theory*, **IT-1** (3): 33–38, December, 1955.

Schwartz, M.:
  I. A Statistical Approach to the Automatic Search Problem, doctoral dissertation, Harvard University, Cambridge, Mass., June, 1951.

Shannon, Claude E.:
  I. A Mathematical Theory of Communications, *Bell Syst. Tech. Jour.*, **27**, part I, pp. 379–423, July, 1948; part II, pp. 623–656, October, 1948.
  II. Communications in the Presence of Noise, *Proc. IRE*, **37** (1): 10–21, January, 1949.

Siegert, A. J. F.:
  I. Passage of Stationary Processes through Linear and Nonlinear Devices, *IRE Trans. on Information Theory*, **PGIT-3**: 4–25, March, 1954.

Slepian, David:
  I. Estimation of Signal Parameters in the Presence of Noise, *IRE Trans. on Information Theory*, **PGIT-3**: 68–89, March, 1954.

Solodovnikov, V. V.:
  I. "Introduction to the Statistical Dynamics of Automatic Control Systems," Goztekhteorizdat, Moscow–Leningrad, 1952 (in Russian).

Spangenberg, Karl R.:
  I. "Vacuum Tubes," McGraw-Hill Book Co., New York, 1948.

Stumpers, F. L.:
  I. A Bibliography of Information Theory (Communication Theory—Cybernetics), *IRE Trans. on Information Theory*, **PGIT-2,** November, 1953.
  II. Supplement to a Bibliography of Information Theory (Communication Theory—Cybernetics), *IRE Trans. on Information Theory*, **IT-1** (2): 31–47, September, 1955.

Thompson, B. J., D. O. North, and W. A. Harris:
  I. Fluctuations in Space-charge-limited Currents at Moderately High Frequencies, *RCA Review*, January, 1940 et seq. Reprinted in "Electron Tubes," **I:** 58–190, 1935–1941, *RCA Review*, Princeton, N.J., 1949.

Titchmarsh, E. C.:
  I. "The Theory of Functions," 2d ed., Oxford University Press, New York, 1939.
  II. "Introduction to the Theory of Fourier Integrals," 2d ed., Oxford University Press, New York, 1948.

Turin, George L.:
  I. Communication through Noisy Random-multipath Channels, *IRE Convention Record*, part 4, 154–166, 1956.

Uspensky, J. V.:
  I. "Introduction to Mathematical Probability," McGraw-Hill Book Co., New York, 1937.

Valley, George E., Jr., and Henry Wallman:
  I. "Vacuum Tube Amplifiers," MIT Rad. Lab. Series, **18,** McGraw-Hill Book Co., New York, 1948.

Van der Pol, Balth, and H. Bremmer:
  I. "Operational Calculus Based on the Two-sided Laplace Integral," Cambridge University Press, New York, 1950.

Van der Ziel, Aldert:
  I. "Noise," Prentice-Hall, Englewood Cliffs, N.J., 1954.

Van Vleck, J. H., and David Middleton:
    I. A Theoretical Comparison of Visual, Aural, and Meter Reception of Pulsed Signals in the Presence of Noise, *Jour. Applied Physics*, **17** (11): 940–971, November, 1946.

Wallman, Henry:
    I. Realizability of Filters, Appendix A of Valley and Wallman (I).

Wang, Ming Chen, and George E. Uhlenbeck:
    I. On the Theory of the Brownian Motion II, *Rev. Modern Physics*, **17** (2 and 3): 323–342, April–July, 1945. Also reprinted in pp. 113–132 of Wax (I).

Watson, G. N.:
    I. "A Treatise on the Theory of Bessel Functions," 2d ed., Cambridge University Press, New York, 1944.

Wax, Nelson:
    I. "Selected Papers on Noise and Stochastic Processes," Dover, New York, 1954.

Whittaker, E. T., and G. N. Watson:
    I. "A Course of Modern Analysis," 4th ed., Cambridge University Press, New York, 1927.

Wiener, Norbert:
    I. Generalized Harmonic Analysis, *Acta Mathematica*, **55**: 117–258, 1930.
    II. "The Fourier Integral and Certain of Its Applications," Cambridge University Press, New York, 1933. (Also Dover, New York.)
    III. "Extrapolation, Interpolation, and Smoothing of Stationary Time Series," John Wiley, New York, 1949.

Williams, F. C.:
    I. Thermal Fluctuations in Complex Networks, *Jour. Inst. Electr. Eng. (London)*, **81**: 751–760, 1937.

Youla, Dante C.:
    I. The Use of the Method of Maximum Likelihood in Estimating Continuous-modulated Intelligence Which Has Been Corrupted by Noise, *Trans. IRE Prof. Group on Information Theory*, **PGIT-3**: 90–106, March, 1954.

Zadeh, Lotfi A., and John R. Ragazzini:
    I. An Extension of Wiener's Theory of Prediction, *Jour. Applied Physics*, **21**: 645–655, July, 1950.
    II. Optimum Filters for the Detection of Signals in Noise, *Proc. IRE*, **40** (10): 1123–1131, October, 1952.

INDEX